Lecture Notes in Mechanical Engineering

Series Editors

Fakher Chaari, National School of Engineers, University of Sfax, Sfax, Tunisia

Francesco Gherardini●, Dipartimento di Ingegneria "Enzo Ferrari", Università di Modena e Reggio Emilia, Modena, Italy

Vitalii Ivanov, Department of Manufacturing Engineering, Machines and Tools, Sumy State University, Sumy, Ukraine

Mohamed Haddar, National School of Engineers of Sfax (ENIS), Sfax, Tunisia

Editorial Board

Francisco Cavas-Martínez●, Departamento de Estructuras, Construcción y Expresión Gráfica Universidad Politécnica de Cartagena, Cartagena, Murcia, Spain

Francesca di Mare, Institute of Energy Technology, Ruhr-Universität Bochum, Bochum, Nordrhein-Westfalen, Germany

Young W. Kwon, Department of Manufacturing Engineering and Aerospace Engineering, Graduate School of Engineering and Applied Science, Monterey, CA, USA

Justyna Trojanowska, Poznan University of Technology, Poznan, Poland

Jinyang Xu, School of Mechanical Engineering, Shanghai Jiao Tong University, Shanghai, China

Lecture Notes in Mechanical Engineering (LNME) publishes the latest developments in Mechanical Engineering—quickly, informally and with high quality. Original research reported in proceedings and post-proceedings represents the core of LNME. Volumes published in LNME embrace all aspects, subfields and new challenges of mechanical engineering.

To submit a proposal or request further information, please contact the Springer Editor of your location:

Europe, USA, Africa: Leontina Di Cecco at Leontina.dicecco@springer.com
China: Ella Zhang at ella.zhang@springer.com
India: Priya Vyas at priya.vyas@springer.com
Rest of Asia, Australia, New Zealand: Swati Meherishi at swati.meherishi@springer.com

Topics in the series include:

- Engineering Design
- Machinery and Machine Elements
- Mechanical Structures and Stress Analysis
- Automotive Engineering
- Engine Technology
- Aerospace Technology and Astronautics
- Nanotechnology and Microengineering
- Control, Robotics, Mechatronics
- MEMS
- Theoretical and Applied Mechanics
- Dynamical Systems, Control
- Fluid Mechanics
- Engineering Thermodynamics, Heat and Mass Transfer
- Manufacturing
- Precision Engineering, Instrumentation, Measurement
- Materials Engineering
- Tribology and Surface Technology

Indexed by SCOPUS, EI Compendex, and INSPEC.

All books published in the series are evaluated by Web of Science for the Conference Proceedings Citation Index (CPCI).

To submit a proposal for a monograph, please check our Springer Tracts in Mechanical Engineering at https://link.springer.com/bookseries/11693.

Maurizio Barberio · Micaela Colella ·
Angelo Figliola · Alessandra Battisti
Editors

Architecture and Design for Industry 4.0

Theory and Practice

Volume 1

Editors
Maurizio Barberio
Dipartimento di Meccanica
Matematica e Management
Politecnico di Bari
Bari, Italy

Angelo Figliola
Dipartimento di Pianificazione
Design e Tecnologia dell'Architettura
Sapienza Università di Roma
Rome, Italy

Micaela Colella
Dipartimento di Architettura
Costruzione e Design
Politecnico di Bari
Bari, Italy

Alessandra Battisti
Dipartimento di Pianificazione
Design e Tecnologia dell'Architettura
Sapienza Università di Roma
Rome, Italy

ISSN 2195-4356   ISSN 2195-4364 (electronic)
Lecture Notes in Mechanical Engineering
ISBN 978-3-031-36921-6   ISBN 978-3-031-36922-3 (eBook)
https://doi.org/10.1007/978-3-031-36922-3

© The Editor(s) (if applicable) and The Author(s), under exclusive license to Springer Nature Switzerland AG 2024

This work is subject to copyright. All rights are solely and exclusively licensed by the Publisher, whether the whole or part of the material is concerned, specifically the rights of translation, reprinting, reuse of illustrations, recitation, broadcasting, reproduction on microfilms or in any other physical way, and transmission or information storage and retrieval, electronic adaptation, computer software, or by similar or dissimilar methodology now known or hereafter developed.
The use of general descriptive names, registered names, trademarks, service marks, etc. in this publication does not imply, even in the absence of a specific statement, that such names are exempt from the relevant protective laws and regulations and therefore free for general use.
The publisher, the authors, and the editors are safe to assume that the advice and information in this book are believed to be true and accurate at the date of publication. Neither the publisher nor the authors or the editors give a warranty, expressed or implied, with respect to the material contained herein or for any errors or omissions that may have been made. The publisher remains neutral with regard to jurisdictional claims in published maps and institutional affiliations.

This Springer imprint is published by the registered company Springer Nature Switzerland AG
The registered company address is: Gewerbestrasse 11, 6330 Cham, Switzerland

Preface

Ten years after the introduction of the concept of Industry 4.0 at the Hannover fair in 2011, the enabling technologies of the Fourth Industrial Revolution are gradually being implemented in various industrial sectors. Among these, the AEC sector has also begun to accept the challenges dictated by the 4.0 paradigm, continuing that path of hybridization with other disciplinary fields and industrial sectors (aerospace, naval, automotive, etc.) that characterized the first digital era. The 4.0 challenge leads industry operators to introduce a series of new paradigms that affect the entire supply chain, from first-level training courses to the creation of innovative companies, passing through the design and construction of digital and performative architectures.

The affirmation of computational design, the increasingly transversal and multi-disciplinary flow of knowledge, and the democratization of machines open new possibilities in a historical moment where the combination of technological innovation and design can play an essential role in environmental sustainability and reduced consumption of resources. The development of CAD/CAM and robotics digital manufacturing technologies has helped to reduce the gap produced by the increase in computational power in the generation of the form compared to the materialization of the same. Through this process, architecture regains its own tectonic identity, and the architect can regain a material sensitivity which risks being dissolved in virtual space. The democratization of digital manufacturing tools and the ubiquity of computational design require a new material dimension and a new figure capable of controlling the entire process. In the post-digital era, where the essence of design lies in the control and information of the process that holistically involves all the aspects mentioned above, rather than in formal research, it is necessary to understand technologies and analyze the advantages that they can bring in terms of environmental sustainability and product innovation.

This book intends to systematize from a theoretical and practical point of view, the best contributions, and the best experiences in the professional and entrepreneurial, academic, and research fields of architecture and design based on this new design paradigm. The main purpose of the proposed systematization is to create a widespread awareness necessary to initiate technology transfer processes involving the public sector, universities and research centres, and the private sector consisting of innovative companies. The issues addressed in the book are central to the development of a total 4.0 awareness for architects, engineers and designers, and digital entrepreneurs: advanced and computational digital design, virtualization of the project and production and construction processes, use of cyber-physical systems, advanced and customized prefabrication, additive manufacturing, automated manufacturing and construction, artificial intelligence, as well as the story of significant experiences of public and private self-entrepreneurship.

Bari, Italy	Maurizio Barberio
Bari, Italy	Micaela Colella
Rome, Italy	Angelo Figliola
Rome, Italy	Alessandra Battisti

Contents

Theory

The Big Vision: From Industry 4.0 to 5.0 for a New AEC Sector 3
Micaela Colella, Maurizio Barberio, and Angelo Figliola

Achieving SDGs in Industry 4.0. Between Performance-Oriented
Digital Design and Circular Economy 19
Alessandra Battisti and Livia Calcagni

Industry 4.0 for AEC Sector: Impacts on Productivity
and Sustainability 33
Ilaria Mancuso, Antonio Messeni Petruzzelli, and Umberto Panniello

Programming Design Environments to Foster Human-Machine
Experiences 51
Giovanni Betti, Saqib Aziz, and Christoph Gengnagel

Designing with the Chain 67
Stefano Converso and Lorenzo Pirone

Adauctus Architectus Novus on the Definition of a New Professional
Figure 89
Giuseppe Fallacara, Francesco Terlizzi, and Aurora Scattaglia

The Future of Architecture is Between Oxman and Terragni 101
Mario Coppola

Open-Source for a Sustainable Development of Architectural
Design in the Fourth Industrial Revolution 113
Giuseppe Gallo and Giovanni Francesco Tuzzolino

Educating the Reflective Digital Practitioner 133
Ioanna Symeonidou

Teaching Digital Design and Fabrication to AEC's Artisans 151
Maurizio Barberio

**The Corona Decade: The Transition to the Age
of Hyper-Connectivity and the Fourth Industrial Revolution** 169
Alexandros Kallegias, Ian Costabile, and Jessica C. Robins

**Quasi-Decentralized Cyber-Physical Fabrication
Systems—A Practical Overview** 185
Ilija Vukorep and Anatolii Kotov

**Latent Design Spaces: Interconnected Deep Learning Models
for Expanding the Architectural Search Space** 201
Daniel Bolojan, Shermeen Yousif, and Emmanouil Vermisso

**From Technology to Strategy: Robotic Fabrication and Human
Robot Collaboration for Increasing AEC Capacities** 225
Dagmar Reinhardt and M. Hank Haeusler

**Overview on Urban Climate and Microclimate Modeling Tools
and Their Role to Achieve the Sustainable Development Goals** 247
Matteo Trane, Matteo Giovanardi, Anja Pejovic, and Riccardo Pollo

**Industry 4.0 and Bioregional Development. Opportunities
for the Production of a Sustainable Built Environment** 269
Luciana Mastrolonardo and Matteo Clementi

**Towards Construction 4.0: Computational Circular Design
and Additive Manufacturing for Architecture Through Robotic
Fabrication with Sustainable Materials and Open-Source Tools** 291
Philipp Eversmann and Andrea Rossi

**RFId for Construction Sector. Technological Innovation
in Circular Economy Perspective** 315
Matteo Giovanardi

Digital Tools for Building with Challenging Resources 331
Christopher Robeller

**Digital Deconstruction and Data-Driven Design
from Post-Demolition Sites to Increase the Reliability
of Reclaimed Materials** 345
Matthew Gordon and Roberto Vargas Calvo

**Impact and Challenges of Design and Sustainability in the Industry
4.0 Era: Co-Designing the Next Generation of Urban Beekeeping** 359
Marina Ricci, Annalisa Di Roma, Alessandra Scarcelli,
and Michele Fiorentino

**Resolve Once—Output Many (ROOM): Digital Design
and Fabrication at the Service of Social Equity** 373
Blair Gardiner and Sofia Colabella

From Analogue to Digital: Evolution of Building Machines Towards Reforming Production and Customization of Housing 387
Carlo Carbone and Basem Eid Mohamed

Virtual, Augmented and Mixed Reality as Communication and Verification Tools in a Digitized Design and File-To-Factory Process for Temporary Housing in CFS 411
Monica Rossi-Schwarzenbeck and Giovangiuseppe Vannelli

Digital Processes for Wood Innovation Design 431
Fabio Bianconi, Marco Filippucci, and Giulia Pelliccia

Technologies

Visual Programming for Robot Control: Technology Transfer Between AEC and Industry 453
Johannes Braumann, Karl Singline, and Martin Schwab

Design, Robotic Fabrication and Augmented Construction of Low-Carbon Concrete Slabs Through Field-Based Reaction–Diffusion 471
Roberto Naboni, Alessandro Zomparelli, Anja Kunic, and Luca Breseghello

Digitally Designed Stone Sculpting for Robotic Fabrication 485
Shayani Fernando, Jose Luis García del Castillo y López, Matt Jezyk, and Michael Stradley

MycoCode: Development of an Extrudable Paste for 3D Printing Mycelium-Bound Composites 503
Fatima Ibrahim, Giorgio Castellano, Olga Beatrice Carcassi, and Ingrid Maria Paoletti

3D-Printing of Viscous Materials in Construction: New Design Paradigm, from Small Components to Entire Structures 521
Valentino Sangiorgio, Fabio Parisi, Angelo Vito Graziano, Giosmary Tina, and Nicola Parisi

A Study on Biochar-Cementitious Composites Toward Carbon–Neutral Architecture 539
Nikol Kirova and Areti Markopoulou

DigitalBamboo_Algorithmic Design with Bamboo and Other Vegetable Rods 579
Stefan Pollak and Rossella Siani

Virtual Reality Application for the 17th International Architecture Exhibition Organized by La Biennale di Venezia 593
Giuseppe Fallacara, Ilaria Cavaliere, and Dario Costantino

Towards a Digital Shift in Museum Visiting Experience. Drafting the Research Agenda Between Academic Research and Practice of Museum Management 609
Giuseppe Resta and Fabiana Dicuonzo

Practice

The Humanistic Basis of Digital Self-productions in Every-Day Architecture Practice 651
Marco Verde

Digital Twins: Accelerating Digital Transformation in the Real Estate Industry 673
Mattia Santi

The Right Algorithm for the Right Shape 699
Inês Caetano, António Leitão, and Francisco Bastos

Volatile Data: Strategies to Leverage Datasets into Design Applications 733
Edoardo Tibuzzi and Georgios Adamopoulos

Simulating Energy Renovation Towards Climate Neutrality. Digital Workflows and Tools for Life Cycle Assessment of Collective Housing in Portugal and Sweden 747
Rafael Campamà Pizarro, Adrian Krężlik, and Ricardo Bernardo

Configurator: A Platform for Multifamily Residential Design and Customisation 769
Henry David Louth, Cesar Fragachan, Vishu Bhooshan, and Shajay Bhooshan

From Debris to the Data Set (DEDA) *a Digital Application for the Upcycling of Waste Wood Material in Post Disaster Areas* 807
Roberto Ruggiero, Roberto Cognoli, and Pio Lorenzo Cocco

From DfMA to DfR: Exploring a Digital and Physical Technological Stack to Enable Digital Timber for SMEs 837
Alicia Nahmad Vazquez and Soroush Garivani

Spatial Curved Laminated Timber Structures 859
Vishu Bhooshan, Alicia Nahmad, Philip Singer, Taizhong Chen, Ling Mao, Henry David Louth, and Shajay Bhooshan

Unlocking Spaces for Everyone 887
Mattia Donato, Vincenzo Sessa, Steven Daniels, Paul Tarand, Mingzhe He, and Alessandro Margnelli

Lotus Aeroad—Pushing the Scale of Tensegrity Structures 925
Matthew Church and Stephen Melville

Data-Driven Performance-Based Generative Design and Digital Fabrication for Industry 4.0: Precedent Work, Current Progress, and Future Prospects 943
Ding Wen Bao and Xin Yan

Parameterization and Mechanical Behavior of Multi-block Columns 963
D. Foti, M. Diaferio, V. Vacca, M. F. Sabbà, and A. La Scala

Theory

The Big Vision: From Industry 4.0 to 5.0 for a New AEC Sector

Micaela Colella, Maurizio Barberio, and Angelo Figliola

Abstract The contribution offers an overview of the key concepts of Industry 4.0 and the application of enabling principles and technologies in the AEC sector. In this sense, the most promising possibilities offered by the fourth industrial revolution are analysed, hinging them inextricably around the theme of sustainability, in the broadest sense of the term. Furthermore, the chapter addresses the issue of the transition towards the emerging Industry 5.0, proposing a refocusing of technological advancement in a human-centred and planet-oriented key. Moreover, the foundations are laid for a broader discussion around the issue of training the professional figures who will be called upon to manage such complex, interrelated, and systemic processes. What is prefigured is a new professional figure capable of managing the entire design process with a systemic vision through which human sciences are merged with technological research, embedded into a holistic vision necessary for guaranteeing a future of prosperity and economic progress while ensuring a sustainable tomorrow for our planet.

Keywords Industry 4.0 · Architecture 4.0 · AEC 4.0 · Sustainable architecture · Industry 5.0

United Nations' Sustainable Development Goals 8. Promote sustained, inclusive, and sustainable economic growth, full and productive employment, and decent work for all · 9. Build resilient infrastructure, promote inclusive and sustainable industrialization, and foster innovation · 12. Ensure sustainable consumption and production patterns

---

M. Colella
Dipartimento di Architettura, Costruzione e Design, Politecnico di Bari, 70124 Bari, Italy
e-mail: micaela.colella@poliba.it

M. Barberio (✉)
Dipartimento di Meccanica, Matematica e Management, Politecnico di Bari, 70124 Bari, Italy
e-mail: maurizio.barberio@poliba.it

A. Figliola
Dipartimento di Pianificazione, Design e Tecnologia Dell'Architettura, Sapienza Università di Roma, 00185 Rome, Italy
e-mail: angelo.figliola@uniroma1.it

© The Author(s), under exclusive license to Springer Nature Switzerland AG 2024
M. Barberio et al. (eds.), *Architecture and Design for Industry 4.0*, Lecture Notes in Mechanical Engineering, https://doi.org/10.1007/978-3-031-36922-3_1

1 Industry 4.0 Principles

The term Industry 4.0 (I4.0) was first mentioned in 2011, when H. Kagermann, W. Lukas and W. Wahlster presented their strategic proposal to strengthen the competitiveness of the German manufacturing industry at the Hanover Fair, which was subsequently adopted by the German federal government and entitled Industrie 4.0 [1]. However, although there is still no precise definition of what is meant by Industry 4.0, it tends to be generally understood as the set of new technologies and new factors of production and work organisation, which, in addition to changing the way production is carried out, will also profoundly alter relations between economic actors, including consumers, with significant effects on the labour market and social organisation itself [2]. To establish a more technical and in-depth definition of the founding principles of this new technological era, the literature review on I4.0 publications by Hermann et al. in 2015 [3] is very thorough and extensive. In this publication, the following are identified as key founding components of I4.0: Cyber-Physical Systems, the Internet of Things, the Internet of Services, and the Smart Factory.

Cyber-physical systems (CPS) are a fusion of physical and virtual space. It is a continuous cycle of data exchange, between physical processes and computer calculations, which makes it possible for the former to influence the latter and vice versa. Physical processes produce data that are collected, stored, and analysed by sensor-equipped and network-compatible systems. Thus, a physical process is followed by a computing process with a real-time associated response in the physical world. Cyber-physical systems are the elements behind the definition of the Internet of Things and the Internet of Services. Things, the objects incorporating technological devices that create an interface between the physical and digital worlds, can be understood as cyber-physical systems, whereby the **Internet of Things** (IoT) can be defined as a network in which cyber-physical systems interact and cooperate with each other according to pre-defined patterns [4]. The Internet of Things, a concept introduced by Kevin Ashton in 1999, thus refers to a new method of using the virtual network within physical space, i.e., the possibility of making parts of the physical world and objects interact with each other via the computer network.

With the **Internet of Services** (IoS), there is a leap in scale, in which cyber-physical systems no longer consist of individual objects, but of the individual activities of the enterprise value chain. Thus, the development of this technology enables a new mode of business management, characterised by a dynamic distribution of activities [5]. Based on the definitions already provided for the CPS and the IoT, the **Smart Factory** (SF) can be defined as a factory in which cyber-physical systems communicate through the IoT, to assist people and machines in performing their tasks [6]. CPS create a virtual copy of the physical world and its processes, so in the continuous connection and interaction between the physical and virtual worlds, between machines among themselves and with humans, within SFs, it is possible to make decisions remotely and reorganise processes in real-time. This creates the preconditions, for example, for flexible and 'intelligent' production, i.e., based on the real demand for a given product at a given time, in real-time. It can be argued, therefore, that the pivotal point of the current revolution is the entry of the virtual world into the real one, through the IoT and the IoS. We are witnessing the progressive fusion of the physical and cyber worlds, in a new concept of a cyber-physical system. Ultimately, it is the combination of these technologies that define the concept of I4.0. Thanks to the introduction of these new technologies, it is possible to gain an insight into the principles that characterise manufacturing in the new industrial era. As mentioned above, the topic is not historicised and the interpretations are multiple, with boundaries that are still very blurred, sometimes contradicting each other. In any case, given the need to make choices, for a clear systematisation of the topic, it was decided to distinguish between:

- the new technologies (CPS, IoT, IoS, SF)
- the new principles introduced into the world of production (Real-Time Capability, virtualisation, decentralisation, servitisation, interoperability, mass customisation)
- the new IT services underpinning the above innovations (big data, cloud computing, cognitive computing, artificial intelligence, machine learning).

As it is easy to deduce from the description of new technologies, they are based on the collection and analysis of an enormous amount of data. Data is often considered the essential asset of the new era. Therefore, fundamental elements for the establishment of 4.0 are big data and cloud computing, services for storing, processing, and transmitting data, without which, the production and collection of data become essentially useless. Recently, with cognitive computing, we have reached a new level of complexity in data processing, being able to substantially reproduce the functioning of the human brain, to make decisions in the face of a very high quantity and heterogeneity of data and variables. Several principles elevate the factory to '4.0', opening new and innovative scenarios. CPSs monitor physical processes by means of sensors and are at the same time able to analyse and compare the collected data with virtual simulation models. As a result, there is a constant check on the correctness of processes, and any errors or inconsistencies are reported in real-time (**Real-Time Capability**).

Interconnected plants can recalibrate the production plan, restart production on other machines and optimise processes. In this way, humans are supported by the machines themselves in their complex management and, whereas in the past, each change in the production chain required weeks of testing by highly skilled personnel, thanks to virtualisation, i.e., the use of simulation models, pre-production tests take place in the virtual world and the time during which machines are inactive is greatly reduced, resulting in enormous savings in economic terms, making them more sustainable [7].

SPCs are, therefore, able to make decisions and perform their tasks autonomously. Therefore, constant planning and control on site are no longer necessary, making decentralisation of production and decisions possible, which can be managed and controlled remotely. This, of course, means less effort and less time spent by workers on the factory floor, but at the same time more effort in the design of machines and processes. This will lead to a change in the way work is done, increasingly focused on highly skilled intellectual work and less and less on physical work (compensated by machine work). In Factory 4.0, robots and humans may work side by side, thanks to the use of intelligent human-machine interfaces. The use of these new generation robots, called smart robots, will encompass countless functions, from production to logistics to office management [8]. The constant connection between manufactured goods and the manufacturing company (again thanks to the installation of sensors) is leading towards a deep integration between physical goods and services. This process, known as the servitisation of manufacturing (service orientation), will favour entrepreneurial formulas that, in addition to selling a product, will offer constant support services. Customer service will be completely rethought, with an inversion of roles: it will be the company, in fact, that will remotely control the functionality of the product and intervene in the event of anomalies, failures, obsolescence or exhaustion of part of the products. In this way, we will increasingly move towards a market in which the purchase of goods will be replaced by the purchase of services, whereby the manufacturing company will remain the owner of the good and will guarantee its efficiency, maintenance, and eventual replacement. This type of market can only have positive consequences from the point of view of environmental sustainability, since, for example, the disposal of obsolete or end-of-life products will no longer be left to consumers but will hopefully be managed efficiently by the companies themselves. A factor of fundamental importance for the development of the 4.0 vision is interoperability, i.e., ensuring, through compliance with common standards, that all SPCs can communicate with each other, even if they belong to different manufacturers, to create an open network in which everyone speaks the same language [9]. The aforementioned principles describe a highly flexible production model, capable of modifying and reorganising production in a short time, thanks also to the high degree of modularity that will characterise its components. This, together with the ability to collect and process an enormous amount of data, including consumer needs and desires, in real-time, will make it possible to implement customised industrial production. Thus, the specific desires and needs of the individual consumer will once again occupy the central role they played in artisanal production. This scenario is referred to as mass customisation.

To summarise, besides the change in the role that humans will play in the production cycle, one of the most relevant consequences of the application of IoT and IoS to the industrial world is the profound change in production methods and volumes. Being able to know in real-time the demands and needs of consumers and having at the same time great flexibility of the factory, capable of varying for each production cycle the goods produced, the canonical serial productions with storage of large quantities of products lose their meaning, opening the way to customised and on-demand production, returning to a dimension of production volumes and customisation of the goods produced closer to the artisan world than to that of industrial seriality. Consequently, even if one considers only the changes related to the use of these technologies in industrial production, the benefits derived from them are manifold and of considerable magnitude; there is the possibility of a reduction in waste related to overproduction, the optimisation of energy and materials used, and a reduction in the need for built space for storing products. Unlike previous industrial revolutions, whose main outcome was the improvement of living conditions through increased productivity, this latest industrial revolution has the potential to significantly improve human (and probably planet Earth's) living conditions not through increased production, but through the optimisation of available resources for more conscious, targeted, and customised production. As with all previous technological revolutions involving a radical change in the way we live and relate to each other, but above all in the way we work and produce, its arrival is viewed by many with a mixture of scepticism about its true potential and fear that technology will eventually overwhelm man and replace him in his work, making him useless for productivity. However, this evolutionary process is already underway, and the question should be: how can this enormous potential be harnessed to solve cogent problems and ensure a prosperous future and widespread prosperity for mankind?

2 Industry 4.0 and Architecture

The new scenarios outlined by the advent of the fourth industrial revolution are producing a change in the way of thinking about society and the world of labour, with the definition, for example, of new professional figures; however, the most affected field by the change is the industry, with a rethinking of its management, production methods and the products themselves. Even though, in recent decades, it has been a field of phlegmatic with less technological progress, compared, for example, to the rapid evolution of the automotive industry, we believe that the construction sector and its related professionals are destined to become one of the main fields of application of I4.0 and, later, I5.0. There is no doubt that the most important innovation that has taken place in the last thirty years in the design sector is the spread of CAD (Computer Aided Design) systems first, and CAM (Computer Aided Manufacturing) later. Although several decades have passed, the still predominant use of CAD tools is simply a computerised version (computerisation [10]) of traditional drawing techniques. However, since the early 2000s, a new (perhaps fully digital) digital revolution has begun to take hold: from CAD design, we move to computational design. In fact, as Kostas Terzidis states, while the computerisation that has taken place in recent decades can be traced back to the digitisation of established, defined, and predetermined processes to improve their efficiency, precision, and workflow, in contrast, computational design concerns the algorithmic exploration of indeterminate processes [11]. Why is it important to emphasise this step from our point of view? This shift is necessary because computational design is the only one capable of fruitful dialogue with the virtualisation processes inherent to I4.0. In fact, truly computational creation involves the performance of an exploratory materialisation process, guided by cyber-physical responses, which extends the properties of the design rather than simply realising it [12]. Machines, therefore, are made 'intelligent' and become part of the computational design process, not only performing tasks according to predefined patterns but devising alternative responses to improve the efficiency of processes, consequently reorganising production.

Taking the technologies characterising I4.0 and imagining an application of their principles to the field of architectural design [13], it is possible to outline feasible scenarios that could configure the architecture of the 4.0 era. Starting from the principle of virtualisation, it is, therefore, possible to identify at least six levels of virtualisation:

1. **Virtualisation of conceptual genesis**, i.e., concept generation processes based on the use of artificial intelligence (AI) technologies, through the input of a textual description (prompt) to which correspond a series of outputs (and subsequent variants) expressed in the form of raster images or three-dimensional models or textual scripts [14, 15].
2. **Virtualisation of the design process**, i.e., forecasting processes that can address several factors simultaneously, thanks to the increasingly sophisticated development of three-dimensional and computational modelling programmes. To cite just a few examples, it is possible to virtualise the climatic behaviour of a building, simulating the position of the sun at a given time on a given day in a precise location on the globe and the resulting irradiation (under clear skies); or the phases of the construction process and the entire life cycle of a project, thanks to Building Information Modelling (BIM) software; or able to analyse the structural behaviour of buildings, and so on [16].
3. **Virtualisation of the CAD/CAM and robotic manufacturing process**, i.e., processes of prediction, control, and prior verification of Computer Aided Manufacturing phases, referring both to single machining and to several consequential machining operations [17].
4. **Virtualisation of the production process**, i.e., processes made possible by the advent of I4.0, in which the entire production process of the factory is foreseen and verified in the virtual world [18].
5. **Virtualisation of the maintenance process**, i.e., processes of prediction, control and preventive verification of the global and local behaviour of the building throughout its life [19].
6. **Virtualisation of the demolition process**, i.e., processes of analysis and design for the selective demolition of the building and the recycling or reduction of the impact on the environment of the building components that are no longer suitable for the purpose for which they were designed [20].

The affirmation of **integrated virtualisation** may therefore represent a possible way forward, given that designers are already accustomed to a working approach with an important virtualisation component, which would be a key element in the interaction between the design process and the industrial manufacturing process. In other words, the designer would be allowed to take part in the management of the industrial process, through an integrated design that would be able to contemplate manufacturing and industrial production methods right from the design stages. Vice versa, the computational design becomes the subject of work and 'critical' analysis by the intelligent machines involved in the various phases of the design and construction process, which can elaborate, and signal possible modifications aimed at optimising a process, concerning, example, the optimal use of a material according to its performance, or to reduce waste, etc.

This is an aspect that brings with it a series of critical issues, mainly related to the interchange of data between one process and another, hence the need to be able to work on common software platforms or those that can interface with each other. In a state of interconnection between design tools and manufacturing and construction tools, in a continuous exchange of data and control of physical processes, the entire process would become 'informed'—and not simply computerised—enriched and guided by the 'cloud of data' processed in real time.

In this context, the role of the designer should not be seen as marginal. The 'data cloud' can be a tool of extraordinary potential for the architect who can be a good director of the design and construction process. Imagining the application of the technologies characterising I4.0 to the field of architectural design, the architect's work would naturally be supported by the potential of the new tools, but it would at the same time become more arduous and burdensome in terms of responsibility, as the project becomes increasingly integrated, the result of a holistic conception that could no longer be evaded. It is enough to think of the possibility of designing with the support of a software tool that is always connected, capable of informing and consequently conditioning the project with data relating to the project site, the processes involved, e.g., environmental data (climatic data, seismic risk, presence of electromagnetic fields, etc.), and having to ensure that each condition is part of the project with an appropriate architectural response. At the same time, the designer's work would not be limited to the production of drawings useful for the construction site, but could (and should) allow for the management of manufacturing and construction processes through simulation tools of the processes themselves, which would ultimately also allow for the optimisation of available resources, be they material, energy, economic, etc.

Beyond the design moment, the benefits of 4.0 technologies could be no less fundamental if applied to the actual construction. In fact, the construction resulting from a 4.0 production process can be equipped with sensors and technological devices that make it a smart product, connected to the web and capable of reacting, according to different configurations, to the processing of data received in real time. New technologies could become part of the new generation of buildings, not only through minor elements such as furniture and household appliances, but also through the installation of sensors in structural building components ensuring the monitoring of their performance, in plant components to monitor their integrity and energy efficiency, and likewise ensure the control of values related to health and living comfort. New constructions would thus take advantage of the servitisation principle, whereby there would be the possibility for the manufacturing company to remotely detect wear and tear, malfunctioning, or impending failure of a building component, allowing timely intervention with targeted and less invasive maintenance. The characteristics outlined so far, although probably not described exhaustively given the topicality of the subject, make us realise how fundamental the almost exclusive use of prefabricated dry-assembled components is. Indeed, among the various construction techniques currently available, it is best suited to dialogue with the enabling technologies of the 4.0 era. The use of prefabrication, although 'advanced' and 'augmented' by 4.0 technologies, would be indispensable for the following reasons:

- Adequacy with respect to digital and computational design processes.
- Total adherence to digital or robotic fabrication processes, especially with respect to wet or mixed systems, which are known to be inaccurate and uncontrollable fabrication/construction processes.
- Greater accuracy in the fabrication of building components, made in a controlled environment and under consistently optimal conditions.
- Greater precision in the construction/assembly phase of building components.
- Greater adherence between design and actual structural and energy performance, thanks to the precision of all execution phases and the use of certified performance elements.
- Reduced production of waste and scrap material during manufacture and construction.
- Possibility, if appropriately foreseen in the design phase, of being able to replace parts that are no longer suitable for their intended function over time.

3 Advanced Prefabrication as a Tool for Sustainable 4.0 Architecture

Prefabrication, by its very nature, implies the concept of prediction. Envisioning the entire construction process during the design phase, and not a posteriori, means having to conceive the project through a method that cannot be based solely on formal considerations, but must constantly relate the whole and the individual parts, according to a coherent and synergic relationship. Prefabrication intrinsically entails a design methodology that cannot disregard considerations on the efficiency of the construction process, since it entails precise planning of the project, which naturally leads to a rationalisation of all the phases of the construction process, from optimisation of the use of materials to optimisation of construction times and costs, up to making forecasts on the discharge of the same with a view to a circular economy and reduction of the environmental impact at the end of the building's useful life. Manufacturing construction components in a factory ensures greater efficiency in the use of materials, drastically reducing waste. The quality of these construction components is also improved, with certifiable technical characteristics and performance, because unlike traditional construction sites, they are produced in dedicated factories, by qualified personnel, in a controlled environment, under conditions always maintained optimal and with the appropriate technical instrumentation, like what happens in the industrial production of any technological product. Off-site production allows for energy efficiency in all stages of the construction process, as the production of components is carried out in the factory, according to the company's energy-saving strategies, while the energy used at the construction site is significantly reduced, as the traditionally designed construction site is transformed into a rapid dry assembly operation of the constituent parts. The dry assembly of parts, typical of prefabricated buildings, also makes it possible to conceive of reversible buildings, which at the end of their useful life allow for the selective recovery of materials and their recycling or reuse, reducing or solving the problems generally associated with the disposal of construction materials.

Prefabrication, in short, is a means of producing buildings in a planned, fast, precise, efficient, and safe manner, as is the case for any other goods produced through an advanced industrial process.

To pursue the goal of building by the principles of sustainability through prefabrication practices, designers should be prepared to manage a more complex design process that is no longer consequential, but oriented towards integrated design. Integrated design is a holistic conception of design, in which all participants in the design process contribute simultaneously so that the design is the result of holistic thinking in which all parts are interdependent and contribute synergistically to the functioning of the entire architectural organism. For example, by ensuring that the building is energy efficient and sustainable, not only because highly efficient systems have been employed, but because environmental well-being is primarily pursued by passive strategies integrated into the building design itself. In this way, the final construction will not be the result of an addition of independent contributions and successive stages of project adaptation that, in most cases, distort the designers' original, albeit valid, conception.

Naturally, this new modus operandi entails a great propulsive thrust in the evolution of the architectural conception against a greater complexity and responsibility on the part of the designer, who with his design choices must succeed in synthesising all the contributing disciplines involved in the project itself. Traditional construction practices, slow and uneconomical, sometimes carried out by unqualified personnel according to an approximate execution, can defeat the effectiveness of design choices. They should therefore be mostly abandoned in favour of a largely industrialised process, even if this means profoundly changing the economic organisation of the construction sector. The affirmation of 4.0 prefabrication, however, in contrast to the widespread vision that imagines it as a means for the mass production of buildings that are all the same, to be reproduced indifferently in any climatic zone and sociocultural context, should be accompanied by important reflections on the design and technological solutions to be adopted, so that these are integrated and appropriate concerning those of the local architectural tradition. This step is important, not only as a means of visual integration, using materials and forms that belong to the local architectural tradition, but above all as a means of extrapolating from tradition the principles of passive architecture that respond to the place and its climatic characteristics. Combining an industrialised construction practice with architectural solutions linked to the context is not to be considered a forced objective, nor an unrealisable vision. New developments in technology are indeed moving towards total customisation of industrially manufactured products. In light of the considerations outlined so far, it is worth reflecting on one of the most relevant aspects that I4.0 could bring about and which, it is worth pointing out, could lead to overcoming the very limits of prefabrication processes as we have known them until now: mass customisation.

The application of I4.0 principles to architecture will necessarily pass through computational design and subsequent digitally controlled fabrication and construction. About the principles underlying these two technologies, it is possible to identify two possible construction processes that, not surprisingly, are becoming the focus of research and experimentation by academic and non-academic research centres, and of public and private funding and investment. We are talking about a new concept of prefabrication, customised digital prefabrication, and technology that is spreading relatively recently in additive manufacturing. These are the two scenarios within which, in our opinion, the construction of the near future will develop.

Additive manufacturing could play an important role in the transition to a more sustainable construction industry. Through additive manufacturing, it is possible to create elements with an optimised shape obtained through computational strategies, eliminating the waste of material, time and money required for subtractive manufacturing or the creation of necessary formworks and counter-moulds. It even becomes feasible and accessible to make complex shapes that would not even be conceivable using traditional manufacturing methods. Digital fabrication makes it possible of working extensively with non-standard products, conceived and create about a specific project or adapted to it, with the use of the materials most suited to the nature of the project or local availability. In addition, the possibility of manufacturing the building elements on the same site as the construction, transporting only the printers as a sort of mobile factory, or entrusting the manufacture of the components related to a given project to one of the digital manufacturing centres spread throughout the territory, would considerably reduce the energy consumption and pollution produced by the transport of all building components from the respective factories to the construction site. However, the centrality of the manufacturing companies in the transition from the general design to the executive project aimed at construction effectively excludes the designer from having a proactive role in the production part and from entering a relationship with the industry, except in more limited cases. Therefore, it is of fundamental importance that, in the affirmation and diffusion of customised digital prefabrication, as a means of moving towards a more sustainable construction process and life cycle of buildings, the irreplaceable work of the architect, who should assume a pivotal role in the transition to an architecture produced with mass production means but with the individuality of a handcrafted product, is not bypassed. The designer, through computational design, should be at the centre of the new design-production process enabled by the principles of I4.0 and digital fabrication. Thus, the construction of buildings with prefabricated I4.0 components lends itself perfectly to becoming a fully digitised process. The designer develops his or her own (computational) design, which is passed on to the digital fabrication machines, optimising the process for making the components to reduce waste of any kind. All components are manufactured and finally assembled to a precision only possible in an industrialised process, thus reflecting the high-quality standards of the design in the finished construction.

4 From 4.0 to 5.0: Towards a More Sustainable, Resilient, and Human-Centred AEC Industry

The 4.0 revolution introduced several cornerstones for the transformation of several leading industries in Western development models, among which we certainly find AEC. The pillars against which this transformation developed were centred on disruptive technological innovations and interconnected cyber-physical systems in which AI was the driving force for increasing the productivity and efficiency of the industrial system. The focus of this transformation centred on the optimisation of a business model is not entirely consistent with the European Union's Agenda 2030 for Sustainable Development or rather does not appear to be the correct framework for addressing emerging and complex challenges such as climate change, resource scarcity and increasingly acute social tensions. The evolution of such a development model must necessarily contemplate some emerging issues that require actions aimed at:

- Introduce a regenerative dimension, with the circular economy as a key element of the entire production cycle.
- Introducing a social and human-centred dimension, promoting technologies designed to assist the workers and not developed to replace them.
- Introducing an environmental and ecosystem dimension that goes from the exploitation of renewable energies and the restoration of biodiversity, as well as overcoming globalisation and the cultural flattening based on it, towards a new model capable of guaranteeing the preservation and evolution of identities and cultures.

Based on the above, it can be said that there is no need for a new industrial revolution, but rather for proper management of the transition from 4.0 to 5.0, which share the same operational tools, and some methodologies, but certainly not the same aims. From our point of view, it is necessary to distinguish between opportunities and goals. Opportunities should be understood as possible enablers of social and economic development and are essential to ensure prosperity and progress. However, it is not opportunities in themselves that generate progress and prosperity: they must necessarily be driven by a set of goals, which can only relate to the most pressing agendas facing humanity such as:

- Overcoming dependence on economic growth as the sole enabler for development and accelerating the development of effective policies and best practices aimed at the sustainable use of material resources.
- The great acceleration of biodiversity loss, climate change, pollution and loss of natural capital is closely linked to economic activities and economic growth.
- The increase in social and economic inequalities, both between industrialised and developing countries and within industrialised countries themselves, with clear differentiation between metropolitan areas and less urbanised territories.
- The fight against the scarcity of potable water by promoting better exploitation of water resources through the recovery and recycling of used rainwater and potable

water and through the development of technologically advanced and precision agriculture.
- Countering the phenomena of depopulation of entire territories and the consequent concentration of an increasing number of people in urban territories that are increasingly large, polluted, and close to collapse.
- The preservation of the cultural and historical identity of territories, and the activation of virtuous dynamics for their evolution thanks to technology; overcoming cultural globalisation.
- Managing migratory flows and eradicating the political, economic, and social causes that incentivise them, to enable the sustainable development of all territories, even those that currently appear most disadvantaged.

It is precisely the aims that represent the key turning point concerning I4.0: an industrial revolution that is not based solely on economic and technological aspects but rather on a multiplicity of factors that configure a vision of restorative and regenerative sustainable development with an inevitable shift of focus from the 'Internet of Things' to 'Digital for people-planet-prosperity'. This paradigm shift inevitably leads to a transition phase linked to the difficulty of clearly and systematically measuring the impact of this transformation on the social and environmental aspects as opposed to the economic and productivity aspects of the industrial system. Underlying this is the enabling technologies that affect virtually as well as physical space closely interrelated through human-machine interaction and the digital twins.

5 A Paradigm Shift in Education to Manage the Transition

Time after time, the act of designing and making objects has remained almost unchanged: the human mind comes up with a design, the hand sketches it out and, finally, the hand works to manually transform the concept into reality. Adding digital know-how to the process sets the basis for realising any mass-produced object. In this scenario, AI becomes a disruptive tool because it can assist the designer in the creative phase up to the materialisation of the digital form. While it is true that tools proliferate with disarming speed, what is lacking is an operational methodology capable of systematising the above and opening up new scenarios through which to respond effectively to the concrete problems of the contemporary era. The use of technology for 'educated' and 'creative' entertainment can unfortunately be configured yet another element of mass distraction concerning the concrete problems that humanity is called upon to face. This risk must be concretely stemmed from using solid technical education, which must now start as early as primary school and continue throughout people's lives. A key element of this process is the inter- and trans-disciplinary approach that must be reflected in the development of an operational methodology capable of providing the right know-how to meet the challenges of the current era governed by the exponential growth of 'disruptive technologies'. In this respect, the role of second-level training is crucial as it should foster the development of a holistic vision supported using tools and technologies that learners should at least be familiar with thanks to first-level training courses. Looking at Italy, for example, the Next Generation EU plan has paved the way for the adaptation of technological infrastructures in Italian secondary schools and the introduction of emerging technologies in education. The *'Scuola Digitale'* (Digital School) plan represents a driving force for this, even if what is currently lacking is a teaching methodology capable of integrating tools and technologies into the teaching programmes of the various disciplines, both humanistic and scientific. It is precisely this last aspect that represents a further transition from I4.0 to 5.0: the interdisciplinary approach, which is necessary to govern complex processes and transformations, must favour the inclusion of the humanities in technological research from the earliest stages of education. However, from the point of view of education, the continuous, rapid and relentless race for technological innovation is making it increasingly complex to update and bring up-to-date school and university curricula. Teachers and researchers are called upon to chase innovation and master it at once to be able to disseminate it seamlessly in the community of reference. This process, in any case, can be very difficult to put into practice since the acquisition of knowledge even by the teaching staff requires time and a necessary degree of theoretical and practical depth: it is not possible to teach something that one does not master effectively and that one does not govern morally and philosophically. Therefore, this perspective calls on universities and research centres to change their model and to open up more and more to the corporations that are the protagonists in the development of these technologies, to foster paths of open innovation and technology transfer. In particular, the theme of lifelong learning and the perpetual structuring of courses dedicated to the acquisition of new skills in the field of technological innovation will lead universities to become management hubs of knowledge transfer. No longer, therefore, almost exclusive holders of the highest peaks of knowledge, but actors capable of organising and facilitating the horizontal distribution of knowledge in partnership with technology companies and with institutions and governments. On the other hand, technological innovation can open up many as-yet-unexplored avenues that can foster self-entrepreneurship dynamics, especially among university students. In this sense, universities must play a fundamental role, systematically equipping themselves with business incubators and accelerators, which can provide fundamental support for all those students who are interested in developing their entrepreneurial idea through the launch of a start-up. Nevertheless, this dynamic should also be encouraged and facilitated among researchers, through incentives and career advancement for those who decide to open and run a spin-off company in parallel with their academic activities, perhaps creating new job opportunities for young graduates or PhDs. In this scenario, the university is called upon to become increasingly 'entrepreneurial', i.e., capable of incorporating into ordinary study courses one or more courses aimed at starting up and running a company or developing a business idea. Such a prospect has the potential to detonate the possibility, widely perceived, of an uncontrolled explosion of unemployment resulting from the endemic spread of technological innovations such as robotics and artificial intelligence. In fact, humanity has two roads ahead of it: the first is to stop the incessant development of such technologies because it is potentially too impactful from a social point of view, especially from the employment perspective; the second, making future workers ready, to be able to seize the opportunities that technological innovation can offer, in terms of creating new (albeit different) job profiles and positions or new business ideas. The key to harnessing the opportunities offered by I4.0 from a social point of view lies in how we manage to channel innovation from a socio-occupational point of view. Indeed, being able to utilise technology to significantly reduce repetitive and time-consuming jobs, while significantly increasing the number of creative and knowledge-intensive jobs, could be a hugely valuable achievement for all of humanity.

6 Conclusions

This contribution offered an overview of the key concepts of I4.0 and the application of enabling principles and technologies in the AEC sector. In this sense, the most promising possibilities offered by the fourth industrial revolution have been analysed, hinging them inextricably around the theme of sustainability, in the broadest sense of the term. Furthermore, the chapter addresses the issue of the transition towards the emerging I5.0, proposing a refocusing of technological advancement in a human-centred and planet-oriented key. Moreover, the foundations are laid for a broader discussion around the issue of training the professional figures who will be called upon to manage such complex, interrelated and systemic processes. What is prefigured is a new professional figure capable of managing the entire design process with a systemic vision through which human sciences are merged with technological research. The training process must therefore not only provide adequate knowledge of the enabling technologies but also deal with structuring an operational methodology to be developed concerning a holistic vision necessary for guaranteeing a future of prosperity and economic progress while ensuring a sustainable tomorrow for our planet.

Acknowledgements This chapter is the result of the combined work of three authors. The authors have revised all the paragraphs, and the paper structure and the research topics have been conceived together. However, the first part of the contribution was primarily written by Micaela Colella and reviewed and integrated by Maurizio Barberio and Angelo Figliola, while the second part (Sects. 4–6) was written by Maurizio Barberio and Angelo Figliola and reviewed by Micaela Colella.

References

1. Kagermann, H., Lukas, W.D., Wahlster, W.: Industrie 4.0: Mit dem Internet der Dinge auf dem Weg zur 4. industriellen Revolution. VDI Nachrichten **13**(1), 2–3 (2011)
2. Magone, A., Mazali, T. (Ed.): Industria 4.0. Uomini e macchine nella fabbrica digitale. Edizioni Guerini e Associati SpA, Milano (2016)

3. Hermann, M., Pentek, T., Otto, B.: Design principles for Industrie 4.0 scenarios: a literature review. Technische Universität Dortmund, Dortmund 45 (2015)
4. Giusto, D., Iera, A., Morabito, G., Atzori, L.: The Internet of Things: 20th Tyrrhenian Workshop on Digital Communications. Springer Science & Business Media, Berlin (2010)
5. Plattform Industrie 4.0., Industrie 4.0—White paper FuEThemen (2015). https://www.din.de/blob/67744/de1c706b159a6f1baceb95a6677ba497/whitepaper-fue-themen-data.pdf. Accessed 12 Apr 2023
6. Hermann, M., Pentek, T., Otto, B.: Design principles for Industrie 4.0 scenarios: a literature review. Technische Universität Dortmund, Dortmund 45, p. 10 (2015)
7. Blanchet, M., Rinn, T., Von Thaden, G., De Thieulloy, G.: Industry 4.0: The new industrial revolution-How Europe will succeed. Hg. v. Roland Berger Strategy Consultants GmbH. München. Abgerufen am 11 (2014)
8. Blanchet, M., Rinn, T., Von Thaden, G., De Thieulloy, G.: Industry 4.0: The new industrial revolution-How Europe will succeed. Hg. v. Roland Berger Strategy Consultants GmbH. München. Abgerufen am, 11, p. 8 (2014)
9. DIN e. V., DKE Deutsche Kommission Elektrotechnik Elektronik Informationstechnik in DIN und VDE (edited by): German Standardization Roadmap Industrie 4.0 (version 4), Berlin-Frankfurt, DIN e. V., DKE, (2020). https://www.din.de/blob/65354/57218767bd6da1927b181b9f2a0d5b39/roadmap-i4-0-e-data.pdf. Accessed 12 Apr 2023
10. Terzidis, K.: Algorithmic Architecture. Architectural Press, Oxford (2006)
11. Kostas, T.: Algoritmic Architecture. Elsevier, Oxford, p. XI (2006)
12. Menges, A.: The new cyber-physical making in architecture: Computational construction. Archit. Des. **85**(5), 28–33, 32 (2015)
13. Barberio, M., Colella, M.: Architettura 4.0. Fondamenti ed esperienze di ricerca progettuale. Maggioli Editore, Santarcangelo di Romagna (2020)
14. Del Campo, M.: Neural Architecture: Design and Artificial Intelligence. Applied Research & Design, New York, NY (2022)
15. Chaillou, S.: Artificial Intelligence and Architecture: From Research to Practice. Birkhäuser, Basel (2022)
16. Deutsch, R.: Data-Driven Design and Construction: 25 Strategies for Capturing, Analyzing and Applying Building Data. John Wiley & Sons, New York (2015)
17. Figliola, A., Battisti, A.: Post-Industrial Robotics: Exploring Informed Architecture. Springer Nature, Berlin (2020)
18. Rahimian, F.P., Goulding, J.S., Abrishami, S., Seyedzadeh, S., Elghaish, F.: Industry 4.0 solutions for building design and construction: a paradigm of new opportunities. Routledge, Milton Park (2021)
19. Maskuriy, R., Selamat, A., Maresova, P., Krejcar, O., David, O.O.: Industry 4.0 for the construction industry: review of management perspective. Economies **7**(3) (2019)
20. Elghaish, F., Matarneh, S.T., Edwards, D.J., Rahimian, F.P., El-Gohary, H., Ejohwomu, O.: Applications of Industry 4.0 digital technologies towards a construction circular economy: gap analysis and conceptual framework. Constr. Innov. **22**(3), 647–670 (2022)

Achieving SDGs in Industry 4.0. Between Performance-Oriented Digital Design and Circular Economy

Alessandra Battisti and Livia Calcagni

Abstract Design sits prominently at the heart of the circular economy and requires us to rethink everything: from products, to business models and cities. Since everything that surrounds us has been designed by someone—the clothes we wear, the buildings we live in, even the way we get our food—and mostly according to the linear model, almost everything needs to be redesigned in accordance with the principles of the circular economy. Circular design process comprises human-centred and performance-oriented approaches. Extending the life of a product allows it to remain in use for as long as possible, and may involve designing products to be physically durable or require innovative approaches that allow the product to adapt to a user's changing needs as time passes. Digital Design plays a crucial role in achieving quickly and efficiently, quality architectural projects both from the users' point of view and from a global perspective as it closes the loop of material flows.

Keywords Performance e oriented · Digital design · Circular economy · Architecture 4.0

United Nations' Sustainable Development Goals 11. Make cities inclusive, safe, resilient and sustainable · 12. Ensure sustainable consumption and production patterns · 13. Take urgent action to combat climate change and its impacts

1 Introduction

The digital revolution has set many challenges and opportunities before the architecture, energy and construction sector. Indeed the adoption of Industry 4.0 (I4.0) technology in sustainable production and circular economy has deeply transformed

---

A. Battisti (✉) · L. Calcagni
Dipartimento di Pianificazione, Design e Tecnologia Dell'Architettura, Sapienza Università di Roma, 00196 Rome, RM, Italy
e-mail: alessandra.battisti@uniroma1.it

L. Calcagni
e-mail: livia.calcagni@uniroma1.it

© The Author(s), under exclusive license to Springer Nature Switzerland AG 2024
M. Barberio et al. (eds.), *Architecture and Design for Industry 4.0*, Lecture Notes in Mechanical Engineering, https://doi.org/10.1007/978-3-031-36922-3_2

the system of project design, construction, building management and maintenance, and last but not least, the sharing of information. I4.0, digitization, smart building, augmented reality have become part of the evolved common vocabulary of design in the construction sector and in particular in all its different declinations related to circular economy. The very concept of circular economy is amplified by I4.0, that provides a strong support to the different project phases, thanks to the integrated approach that allows to evaluate in a virtual way—and in advance—all the information related to the entire life cycle of a project: from design, to construction, demolition and disposal. The concept of circular economy from the perspective of performance-oriented digital design has gained momentum among businesses, policy-makers and researchers due to its potential to contribute to sustainable development [1, 2] through a series of efficiency and productivity improvements, commonly known as circular strategies. The high value of digitally oriented design is also emphasized in some EU technical-analytical reports [3] that highlight how standardized information management can contribute to the prediction of environmental performances and thus improve decision-making concerning the future impact of a project. This is achieved by detailing the various activities and respective indicators that characterize the project from construction in terms of climate-altering emissions, resource consumption and waste production, to construction in terms of transport, construction site, realization, and eventually to operation in terms of operating energy consumption, heating/cooling, shading, ventilation, water and waste treatment, building life cycle, use and maintenance. Innovative technologies and paradigm shifts applied to digitized design contribute to extend the useful life of building artifacts, to reduce waste, to discretize performance and improve environmental impact, to identify opportunities to reuse building materials and/or components. The knowledge of physical-technical-performance characteristics, the availability of materials, and the configuration of optimized combinations of materials and components, as well as the use of a standardized language that ensures and/or increases interoperability between different technical formats, also contribute to better planning and organization of resources. In this sense, digital technologies play a crucial role in improving circular economy strategies and practices by overcoming environmental problems and fostering paradigmatic approaches towards the achievement of the SDGs.

2 Methodology

The authors carried out an analysis based on performance-oriented digital design and on 7 strategies of the circular economy—recover, recycle, reuse, regenerate, repurpose, reduce, rethink—to assess how and to what extent the achievement of the SDGs can be found in some projects and in particular goals 12 (Responsible Consumption and Production: Reversing current consumption trends and promoting a more sustainable future) and 13 (Climate Action: Regulating and reducing emissions and promoting renewable energy). At the very moment, recovery and recycling actions carried out in construction processes do not always necessarily promote a circular economy, even though they are among the most common applied strategies in the construction field to date. It is precisely in the medium- and long-term planning on recovery and recycling that I4.0 could enhance the circular economy and lead to the pursuit, albeit partial, of these two SDGs. In relation to these, the essay reviews experimental digital tools and practices, showing how they can reduce waste, increase energy and ecological efficiency, effectively and efficiently employ renewable energies, close production cycles and maximize the preservation of the economic value of materials and products. The proposed methodology follows that used by Bocken [4], which identifies three iterative steps for a practice and literature review: (1) identification of topics and categorizations by literature review, (2) synthesis through the development of an integrative framework, and (3) identification and mapping of practical examples to validate and further develop the framework. The logical framework is essential to extend the knowledge base by providing a detailed review of Industry 4.0 between sustainable production and circular economy by integrating three contemporary concepts in the context of supply chain management: Agile approach; IIoT stack and Technology ecosystems.

Agile approach. An approach that employs rapid iterations, fast failures and continuous learning. An approach whereby research teams work together extremely effectively by facilitating collaboration between different functions and turning use cases into self-learning examples that enable rapid innovation and renewal.

IIoT stack. Smart manufacturing system that enables seamless integration of legacy and new Industrial IoT infrastructures to build a stable and flexible technological backbone through intelligent use of existing systems with efficient integration within a new technological smart manufacturing process while limiting costs, consumption and waste.

Technology ecosystems. Technology ecosystems with access to vast datasets and co-innovation opportunities that enable collaboration between technology vendors, suppliers, customers and related industries to implement cutting-edge solutions and best practices.

3 Limitations and Implications of the Research

One must keep in mind that digital transformations are revolutionizing all aspects of production, affecting not only processes and productivity but also people. The right applications of performance-oriented digital design technologies can lead to more effective decision-making, new opportunities for upgrading, retraining and cross-functional collaboration.

In the construction field, the impacts can be identified especially in the reduction of production time, optimized process management with win-win benefits associated with reduced environmental impact, made possible by lower emissions and reduced waste, and more efficient consumption of energy, water and raw materials. Yet, evident risks remain: by pursuing digital transformation as a theoretical exercise, many research centers unwittingly create isolated and local operations which have little to do with future cycles of manufacturing excellence. They fail to access a broader network of production as they are more technology-driven rather than value-driven This results in a technology-first rollout where proposed solutions are deployed without a clear link to real value opportunities, business challenges or market capacity to absorb them. In fact, a large majority of the experiments deployed remain at the pilot project level, as they don't develop the full potential of transformation through performance-oriented digital design also in terms of return on investment.

With so much at stake, manufacturers are investing a lot of time and money in digital transformation. These investments are paying off for some, but most are unable to scale successful pilot programs or take full advantage of new tools and technology to see significant returns.

4 Case Studies

Given the emerging and widespread amount of applied and realized projects, our study investigated not only academic sources but also practice case study examples and grey literature [5]. As mentioned previously, the identified framework embraces circular economy (CE) and industry 4.0 (I4.0) paradigms, digital fabrication (DF) processes and digital technologies (DTs), and performance oriented design (POD), all with a view to sustainable environmental design. The integrative synthesis was followed by the identification and mapping of practical examples to validate and further develop the framework. Indeed, there are some paradigmatic European and international examples that provide evidence of how the integration of performance-oriented digital design and circular economy principles and strategies is increasingly implemented in the construction industry for the achievement of the SDGs, in particular goal 12 and 13. In analyzing these virtuous case studies we asked ourselves why they work, in which terms they contribute to achieving the 7 strategies of circular economy: recover, recycle, repurpose, re-use, regenerate, reduce, rethink; and eventually to what extent the three concepts of supply chain management occur in terms of agile approach, IoT stack and technology ecosystems.

More precisely the case study research method involved a cross-case analysis following the presentation of separate single-case studies [6]. The case studies have been selected according to their relevance to the criteria and topics mentioned above and have been analyzed on the base of the same indicators.

4.1 Maison Fibre

Maison Fibre (Fig. 1), exhibited at La Biennale di Venezia 2021, is the result of research on robotically manufactured fibre composite structures carried out by the Institute for Computational Design and Construction and the Institute of Building

Structures and Structural Design at the University of Stuttgart. It is a full-scale inhabitable hybrid structure combining laminated veneer lumber with fibre-polymer composites (FPC). It is the first multi-storey building system fabricated with this novel technique [7]. In terms of tectonics the entire structure consists exclusively of so-called fibre rovings: bundles of endless, unidirectional fibres made of glass and carbon.

IIoT stack. The manufacturing process involves the use of a coreless robotic winding process, which allows for locally load-adapted design and alignment of the fibres. Coreless filament winding is a robotic fabrication technique in which conventional filament winding is modified to reduce the core material to its minimum, thus enabling an extraordinary lightweight construction [7]. This technological smart manufacturing process allows to limit costs, consumption and waste.

Technology ecosystems. This structure marks a turning point in the transition from pre-digital, material-intensive construction that makes use of heavy, isotropic building materials such as concrete, stone, and steel—which are often extracted in distant places, processed into building components, and carried over long distances—to genuinely DFTs with locally differentiated and locally manufactured structures made of highly anisotropic materials [8].

Agile approach. The extremely low material consumption combined with the very compact, robotic production unit allow to run the entire production on-site without a significant amount of noise or waste, both during the initial construction process, and during expansion or conversions.

Furthermore, this novel material culture in architecture brings about its entailed ecological (material and energy), economic (value chains and knowledge production), technical (digital technologies and robotics), and sociocultural matters.

4.2 Aguahoja I

Oxman's study focused on water-based robotic fabrication as a design approach and on enabling technology for additive manufacturing (AM) of biodegradable hydrogel composites meant for manufacturing architectural-scale biodegradable systems [9].

Technology ecosystems. The research group aimed at applying material ecology (ME) research to the study and design of a new biocompatible material for architecture, characterized by a programmed life, therefore bound to be gradually reabsorbed into nature [10].

IIoT stack. The structure (Fig. 2) is digitally designed and robotically fabricated: 3D printed from biodegradable polymers. The construction process involves a robotically controlled arm and multi-chamber extrusion system designed to mix, process and deposit biodegradable-composite objects combining natural hydrogels (e.g. chitosan, sodium alginate) with other organic aggregates. More specifically, the architectural skin-and-shell is made of 5,740 fallen leaves, 6,500 apple skins and 3,135 shrimp shells, 3D printed by a robot, modelled by water and coloured with

Fig. 1 Maison fiber case study resume form

natural pigments. The result is an organism-matter that captures carbon dioxide, enhances pollination, increases soil microorganisms and provides nutrients.

Agile approach. These water shaped skin-like structures are designed and manufactured as if they were grown, therefore no assembly is required as well as no disposal issues occur.

Despite the urgent need for alternatives to fuel-based products and in spite of the exceptional mechanical properties, availability, and biodegradability associated with water-based natural polymers, AM of regenerated biomaterials is still in its early stage [9]. Moreover, these structures react to their environment, adapting their geometry, mechanical behaviour and colour in response to fluctuations in heat, humidity, and sunlight (time-based 'temporal' behaviour). This structure shows how ME presents new opportunities for design and construction that are inspired, informed, and engineered by, for, and with nature. An architecture capable of programmatically decomposing could introduce a new type of disposal in the construction sector that does not alter the ecosystem, and is perfectly in line with life cycle assessment principles oriented to the cradle to cradle design.

4.3 Harvard Science and Engineering Complex

The new Harvard University complex (Fig. 3) is designed to inspire learning and scientific discovery while showcasing sustainability. Thus, special emphasis was set on effective façade design, efficient energy performance as well as occupant comfort. These are mainly addressed through the stainless-steel screen envelope which is designed according to parametric simulations and using a novel manufacturing method, hydroforming [11]. Hydroforming is an industrial cold forming process in which a metal blank is driven into a single mold with hydraulic pressure to form extremely thin parts with exceptional structural stiffness. It has been developed in the automotive and aerospace industries where weight to strength ratio has a compound effect on production cost, safety, performance, and energy consumption, but it has not been widely used for architectural applications yet [12].

IIoT stack. In this case hydroforming was applied in a sun shading system that leverages the advantages of this technology: lower tooling costs, precise geometric definition, and superior structural properties. Calibrated to the extreme seasonal variations of the local climate, the system is precisely designed and dimensioned to temper solar heat gain in the summer while maximizing daylight and solar energy in the winter, reducing cooling and heating loads [12]. The screen also reflects daylight towards the interior while maintaining large view apertures.

Agile approach. The use of parametric performance studies and simulations, including rapid prototypes, full-scale visual and performance mock-ups, and advanced industrial design and simulation software (like CATIA), allows to maximise structural, material and manufacturing efficiency and at the same time enables rapid iterations, fast failures and continuous learning.

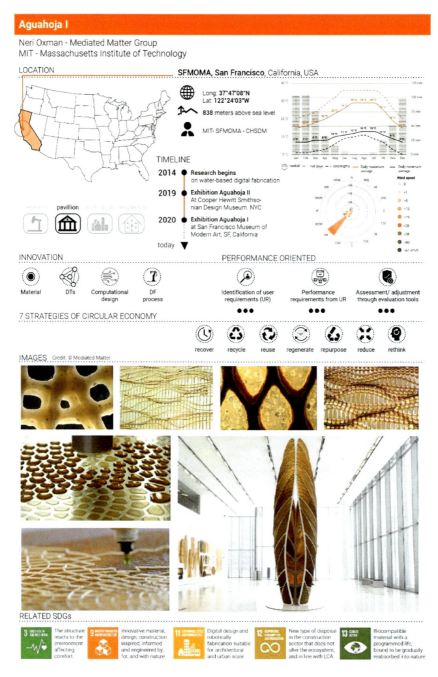

Fig. 2 Aguahoja I case study resume form

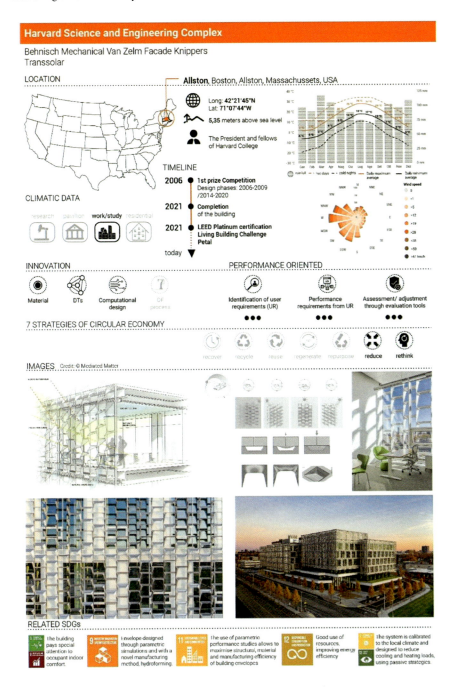

Fig. 3 Harvard science and engineering complex case study resume form

Technology ecosystems. This project proves how a coherent and appropriate combination of environment, technology and architecture can achieve excellent aesthetic quality, a high-performance building envelope and energy efficiency.

4.4 J-Office

The project (Fig. 4) consists in the conversion of a dilapidated building from a warehouse into an architectural design studio located in an old industrial park in Shanghai, China [13].

The concept of the Silk Wall, the external wall that surrounds the warehouse, was developed starting from the manipulation of simple materials using up to date DF processes. In particular the wall consists of cement blocks, angled to create an interesting texture that varies the amounts of light into the building. These cinder blocks are used throughout China since they are so inexpensive, but they are extremely rigid in form and dimension. Thus, the exploration of the material limits led to the use of the blocks with a different bricklaying method by creating stacking algorithms.

IIoT stack. Parametric processes have been used to superimpose the patterns, the contours and definition of silk forms while allowing the wind to enter. To develop the design concept, an algorithm was designed to force a rotation of each cement block.

Agile approach. After an issue appeared during the construction phase, thanks to the advantages of parametric design, a series of alternative results were soon produced by adjusting the parameters, and, after a short calculation, a range of options was identified. This project shows the advantages of parametric design not only in the initial design phase but especially in the construction and management phase: by simply adjusting some parameters, a short calculation is able to offer a range of options and display a series of alternative results.

Technology ecosystems. Computational optimization in the design phase helps to adjust fabrication layouts according to known computer numerically controlled (CNC) technologies [14].

4.5 Living Places

The project (Fig. 5) suggests a new way of thinking focused on building a better living environment that benefits both people and the planet.

Technology ecosystems. Assuming that all phases of the project development must be taken into consideration, the project is meant as an open-source development model that takes into account its entire lifecycle, enabling a new holistic approach to sustainable construction. In fact, by using low impact materials and by considering all stages of the building's life cycle and understanding the implication of each design choice, it is possible to reduce emissions by up to 75% while meeting the demand for

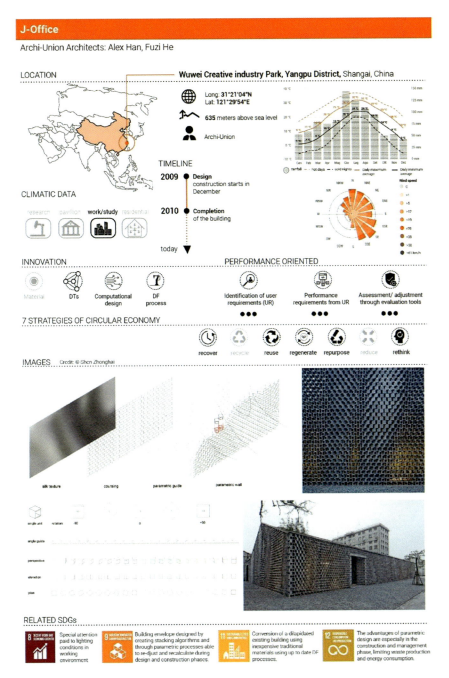

Fig. 4 J-office case study resume form

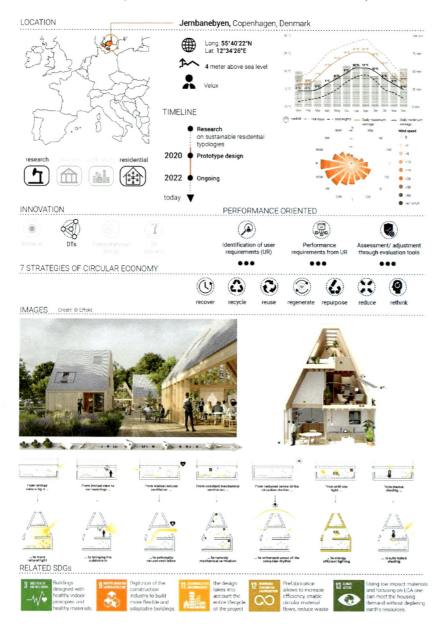

Fig. 5 Living places residences case study resume form

increased housing without depleting the earth's resources. By separating technical systems and building systems the prototype uses circular economy as a means to extend the building's lifetime and reduce cost, labor and waste [15]. The Build for Life approach comprises seven strategic drivers (flexibility, quality, environment, health, community, local, affordability) making up the Compass Model, which is meant to guide the design and building process, providing stakeholders with a framework for reaching an outcome that is sustainable on multiple levels [16]. This solution offers a simple modular building system that requires little to no maintenance and can easily be disassembled and thus repaired or retrofitted during its lifespan.

Agile approach. Moreover, Living places is conceived as a toolbox of different housing typologies that are context-responsive and designed to constantly adapt as occupants' needs change over the day, the year, and the lifetime. The project stands out for its people oriented approach in the definition of the demand-performance framework aimed at acquiring an holistic and integrated understanding of the occupants' needs. In fact, special emphasis is paid to daylight, thermal comfort, air quality, acoustics and outdoor connection [17].

IIoT stack. Overall, digitizing the construction industry and through prefabrication one can increase efficiency and enable more sustainable development, reducing waste, and enable circular material flows.

5 Conclusions

These projects all reveal the strong need—within the construction practice—for innovative processes and technologies that recognize the importance of sustainable design and overcome the inefficiency and lack of interoperability present in the sector [18]. The role of technological culture, together with a performance-driven approach, interdisciplinary dialogue and the creation of a growing information content are essential to manage innovation in the context of digital eras [19]. At the same time, DTs, such as the Internet of Things (IoT), big data, and data analytics, are considered essential enablers of the circular economy (CE) [20]. In fact, the combined methods of computational design and robotic fabrication have demonstrated potential to expand architectural design. Above all, factors such as material use, energy demands, durability, GHG emissions and waste production must be recognized as the priorities over the entire life of any architectural project [18]. To this end, the potential of DTs, DF, and performance oriented design is remarkable for achieving sustainability according to the well-known broad definition (Brundtland Report 1987) of SD as 'development that meets the needs of the present without compromising the ability of future generations to meet their own needs' [21].

References

1. Geissdoerfer, M., Savaget, P., Bocken, N.M., Hultink, E.J.: The circular economy–a new sustainability paradigm? J. Clean. Prod. **143**, 757–768 (2017)
2. Ghisellini, P., Cialani, C., Ulgiati, S.: A review on circular economy: the expected transition to a balanced interplay of environmental and economic systems. J. Clean. Prod. **114**, 11–32 (2016)
3. Poljanšek, M.: Building information modelling (BIM) standardization. European Commission (2017)
4. Bocken, N.M., Short, S.W., Rana, P., Evans, S.: A literature and practice review to develop sustainable business model archetypes. J. Clean. Prod. **65**, 42–56 (2014)
5. Adam, R.J., Smart, P., Huff, A.S.: Shades of grey: guidelines for working with the grey literature in systematic reviews for management and organizational studies. Int. J. Manage. Rev. **19**(4), 432–454 (2017)
6. Yin, R.K.: Case study research: Design and methods, 3rd edn, Vol. 5. Sage, Thousand Oaks, CA (2009)
7. Pérez, M.G., Früh, N., La Magna, R., Knippers, J.: Integrative structural design of a timber-fibre hybrid building system fabricated through coreless filament winding: maison Fibre. J. Build. Eng. 104114
8. University of Stuttgart Maison Fibre—Towards a New Material Culture. https://www.itke.uni-stuttgart.de/research/built-projects/maison-fibre-2021/. Accessed 02 May 2022
9. Mogas-Soldevila, L., Duro-Royo, J., Oxman, N.: Water-based robotic fabrication: large-scale additive manufacturing of functionally graded hydrogel composites via multichamber extrusion. 3D Print. Addit. Manuf. **1**(3), 141–151 (2014)
10. Milasi, M.: Aguahoja: il composto organico che ci salverà dalla plastic. https://www.domusweb.it/it/design/2019/03/25/aguahoja-il-composto-organico-che-ci-salver-dalla-plastica.html. Accessed 05 May 2022
11. Behnisch, S.: Architecture for research: humane and sensible building design should address scientists' needs and wishes. EMBO Rep. **23**(3), e54693 (2022)
12. Siu-Ching, M., Matthew R., Schieber, R.: Hydroformed Shading: A Calibrated Approach to Solar Control (2020)
13. Divisare Archi-Union Architects J-Office. https://divisare.com/projects/371379-archi-union-architects-zhonghai-shen-j-office. Accessed 17 May 2022
14. Wang, S., Crolla, K.: Disciplinary isolation-opportunities for collaboration between digital practices and manufacture in china's PRD
15. Effekt-Living Places. https://www.effekt.dk/buildforlife. Accessed 20 May 2022
16. Cutieru, A.: VELUX and EFFEKT Develop Strategic Framework for Designing Healthier and More Sustainable Build Environment. https://www.archdaily.com/971907/velux-and-effekt-develop-strategic-framework-for-designing-healthier-and-more-sustainable-build-environment. Accessed 20 May 2022
17. Build for Life, Velux. What if our buildings could be healthy for both people and planet? https://buildforlife.velux.com/livingplaces/. Accessed 20 May 2022
18. Agustí-Juan, I., Habert, G.: An environmental perspective on digital fabrication in architecture and construction. In: Proceedings of the 21st International Conference on Computer-aided Architectural Design Research in Asia (Caadria 2016) Caadria (2016)
19. Chiesa, G.: La prassi progettuale esplicito-digitale e l'approccio prestazionale. Techne **13**, 236 (2017)
20. Kristoffersen, E., et al.: The smart circular economy: a digital-enabled circular strategies framework for manufacturing companies. J. Bus. Res. **120**, 241–261 (2020)
21. Diesendorf, M.: Sustainability and sustainable development. Sustain. Corp. Challenge 21st Cent. **2**, 19–37 (2000)

Industry 4.0 for AEC Sector: Impacts on Productivity and Sustainability

Ilaria Mancuso, Antonio Messeni Petruzzelli, and Umberto Panniello

Abstract The Architecture, Engineering, and Construction (AEC) sector is experiencing an intense internal transformation, triggered by two topics that must be addressed for assuring a prosperous future of the industry. Specifically, on the one hand, the actors of the AEC sector need to commit themselves towards sustainability, in order to reduce the significant impacts that the industry causes on the environment in terms of pollutant emissions. On the other hand, work practices must be reviewed for improving their actual scant efficiency, which results for AEC sector in a very low productivity rate with respect to other industrial sectors. To ensure long-term business continuity based on the two pillars of productivity and sustainability, AEC actors are considering a deeper use of Industry 4.0 into their businesses. In fact, Industry 4.0 technologies are recognized to have positive impacts on both operational and sustainability performance in the AEC field, so much to determine the birth of the new Construction 4.0 paradigm, which is fascinating although scant adopted. In this context, the following chapter aims to analyze the changes that some of the most cutting edge and revolutionary technologies of Industry 4.0 bring in three phases of the construction life cycle. Specifically, the chapter explores how such technologies allow to address productivity and sustainability issues during the phases of architectural and design planning, works' execution, and support processes management, acting as levers for building more efficient and sustainable constructions.

Keywords Architecture · engineering and construction · Industry 4.0 paradigm · Technologies for sustainability · Technologies for productivity

---

I. Mancuso (✉) · A. M. Petruzzelli · U. Panniello
Department of Mechanics, Mathematics, and Management, Politecnico di Bari, Via Orabona 4, 70125 Bari, Italy
e-mail: ilaria.mancuso@poliba.it

A. M. Petruzzelli
e-mail: antonio.messenipetruzzelli@poliba.it

U. Panniello
e-mail: umberto.panniello@poliba.it

© The Author(s), under exclusive license to Springer Nature Switzerland AG 2024
M. Barberio et al. (eds.), *Architecture and Design for Industry 4.0*, Lecture Notes in Mechanical Engineering, https://doi.org/10.1007/978-3-031-36922-3_3

United Nations' Sustainable Development Goals Goal 9: Industry, Innovation and Infrastructure. Build resilient infrastructure, promote inclusive and sustainable industrialization, and foster innovation · Goal 11: Sustainable Cities and Communities. Make cities and human settlements inclusive, safe, resilient, and sustainable · Goal 12: Responsible Consumption and Production. Ensure sustainable consumption and production patterns

Suggested Topics IoT and Cyber-Physical Systems in Architecture · Digital Twin · BIM and Interoperability · Virtual Reality and Augmented Reality · Digital and Robotic Fabrication · Additive Manufacturing and 3D Printing

1 AEC Sector and Industry 4.0

The Architecture, Engineering and Construction (AEC) sector is undergoing an intense internal transformation, which starts from two issues that must be unravel for the future of this industry. On the one hand, there is the challenge of sustainability. In fact, AEC sector is responsible for 36% of global final energy use and 39% of all carbon emissions in the world [18], using large amounts of materials and energy to develop, operate, and demolish buildings. On the other hand, the sector is currently characterized by a very low productivity rate. Indeed, the construction sector presents one of the highest productivity gaps among the other industrial businesses, with a yearly growth of 1% versus the average of 2.8% [10].

To respond to these issues and ensure both sustainable and profitable business continuity in the near future, AEC sector can seize large opportunities deriving from Industry 4.0 paradigm. In fact, innovation driven by new technologies in AEC businesses can lead to an increase in productivity, simultaneously reducing environmental impact [6]. Actually, Industry 4.0 has ample maneuver margin for augmenting sustainability and operational performances. In particular, on the operational front, Industry 4.0 in AEC sector reduces inefficiencies and waste with better, faster, and safer collaborations, communications, and procedures, while on the sustainability front it ensures prudent use of resources with significant reduction in energy usage and emissions [14].

The potential of Industry 4.0 to relaunch the AEC market is evident so much so that it led to the concept of Construction 4.0 [14]. The Construction 4.0 framework borrows typical Industry 4.0 technologies to be used in construction sites, also employing new digital technologies specifically designed for the AEC field. Examples of Construction 4.0 applications are the adoption of IoT sensors and platforms to manage workers and buildings, the use of robots on construction sites, the spread of 3D printing of homes, as well as the integrated management of all the aspects related to a project within the Building Information Modeling software. The benefits of these Industry 4.0 technologies applied in the AEC field are remarkable, including increases in performances and transparency throughout the entire built environment value chain, as well as reductions in resource consumption and CO_2 emissions, with the result of more efficient and sustainable construction [17].

Despite Industry 4.0 technologies are available and mature [12], their application to the AEC sector is not straightforward [17]. This is because the construction industry is one of the least digitalized in the world [2]. In fact, the sector, due to attitudinal, industrial, and institutional barriers, remains strongly anchored to traditional business models, proving to be averse to risk and change [16, 17] and missing in the exploitation of new technologies to innovate processes and services [12, 20].

In this contest of great fascination with the uses and challenges related to the implementation of Industry 4.0 in the AEC sector for increasing the sustainability and profitability of businesses, the following chapter offers a review of the technologies that AEC companies can adopt throughout the entire life cycle of construction to simultaneously solve sustainability and productivity issues (see Fig. 1). Specifically, the chapter covers areas ranging from architectural and design planning to works' execution on construction sites, up to the management of support processes. For each phase of the life cycle experienced by construction (i.e., planning, execution, and management), the chapter investigates how Industry 4.0 technologies positively influence both the improvement of operational performance (through increase in quality and work safety and reduction of project times and costs) and the achievement of sustainability goals (through reduction of the energy consumption).

With this result, the work aims to increase both scientific and practical knowledge on how the Industry 4.0 paradigm enables productivity and sustainability in the AEC sector, paving the way for an innovation strategy where these two pillars are equally embedded into all areas of business.

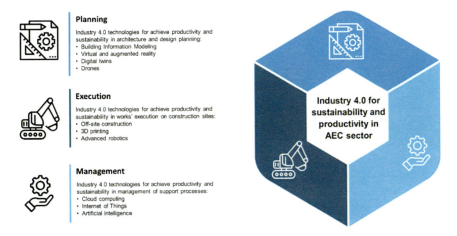

Fig. 1 Industry 4.0 technologies for sustainability and productivity in construction life cycle

2 Industry 4.0 for Architectural and Design Planning

Architectural and design planning represents the first phase in constructions lifecycle, as well as is among the most expensive task for professionals and companies. In fact, planning activities require studying design choices, assessing the effectiveness and feasibility of possible architectural solutions, and coordinating the work of multiple intra-company stakeholders (e.g., designers, architects, administrators, accountants) for satisfying the needs of clients.

The management of this complicated but indispensable phase of constructions' projects life cycle can be facilitated by the use of various Industry 4.0 technologies. In this section, the potential brought by Building Information Modeling (i.e., information system containing all the information relating to a project), virtual and augmented reality (i.e., apps and viewers that overlay digital information on the real world and simulate different scenarios), digital twins (i.e., copies of structures that allow virtual management of physical buildings), and drones (i.e., controlled aircrafts that map and control work sites remotely) are presented.

2.1 Building Information Modeling (BIM)

Building Information Modeling (BIM) is a software whose adoption has grown significantly in recent years in AEC industry [1]. BIM allows architects to design buildings by enclosing and linking, in addition to the drawings, all the information relating to the project that can be useful in different life cycle phases. For examples, as regards the phase of temporal and economic supervision of the project, thanks to the 4D and 5D dimensions of BIM, the model can be fully integrated with the Gantt diagram developed by the Microsoft Project software, used by almost all construction firms and professionals. Personnel benefits from this linkage because it can verify activities and times involved during a certain project phase, develop analysis scenarios for specific activities, and obtain a constantly updated overview of the work progresses and cost estimates. The 4D and 5D dimensions of BIM therefore optimize both the definition and updating phases of project releases (thanks to the possibility of quickly evaluating planned changes according to the clients' needs) as well as control ones (thanks to the possibility of analyzing and comparing the parameters of the work in real time to spot any deviation). In addition, the ability to use BIM models even on mobile applications allows supervisors to access the entire project, filtering the data of interest for coordination on the work site and viewing information that is always updated and updatable. The great interoperability offered by BIM is guaranteed by the open format IFC (i.e., Industry Foundation Classes), which assures the exchange of data between the various subjects involved in the life cycle of the commission (e.g., architects, engineers, designers, maintenance technicians) and between different software platforms. Thanks to this feature, the use of BIM generates a reduction in costs due to the zeroing of errors related to data fault or

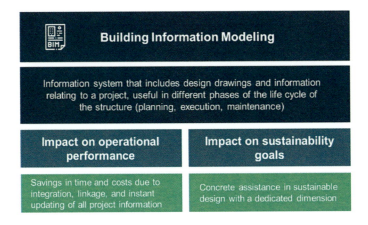

Fig. 2 Building information modeling for sustainability and productivity in planning

redundancy. In fact, the maximum degree of interoperability offered by technology translates into the possibility for each of the members of the project team to work using the best software solutions for his/her specific discipline, without any risk of data incompatibility or loss. This leads to significant reduction of errors generated by a lack of coordination and/or updating, and determines a contraction of global construction times from the start to the completion of the works. Furthermore, the BIM technology effectively reduces energy consumption and greenhouse gas emissions. In fact, the 6D dimension of the software, which allows to carry out energy and sustainability analyzes for projects, assists the realization of Near Zero Energy Buildings (i.e., building with high energy performances thanks to the construction, typological, and plant characteristics), saving energy and reducing CO_2 emissions. Figure 2 resumes the main characteristics of BIM, as well as the impacts of this technology on the productivity and sustainability of the architectural and design planning phase.

2.2 Virtual and Augmented Reality

Virtual and augmented reality (VR and AR) tools such as applications on mobile devices, glasses, and helmets designed for the AEC sector facilitate design activities as regards stakeholder engagement, design and construction support, design review, operations, management assistance, and training [5]. In particular, the uses of VR and AR for the improvement in the collaborations between offices and work sites are extremely interesting for the impact on productivity and sustainability. In fact, the integration of VR and AR devices with the 3D models of construction sites, elaborated with BIM or with AutoCAD, allows designers and foremen to view the same plans in real time and verify the validity of the design choices, simulating their presence in the virtual models. The result is an instant identification and communication of design errors, as well as the receipt of immediate feedback on the design changes, to be attached in the form of annotations, links, and videos to the elements of the project. Indeed, VR and AR devices speed up instant changes and scenario analysis, by allowing to update models and immediately share them with everyone involved. Furthermore, VR and AR tools can help contract managers in the updating of budget releases, as they enable the modification, in a single environment, of the project Gantts and the 3D models. In this way, the evolution of construction sites can be simulated, facilitating the identification of works' interferences (e.g., space and/or time overlaps), thus highlighting the risks generated by the contingent situation. The easy access to schedules and operational details, combined with the devices' feature of on-site measurement and direct comparison with work plans, automates processes. Specifically, AR applications allow to digitally carry out the measurements and checks against the prescriptions, confirming the status of the activities (i.e., in execution, completed, or delayed) and specifying the errors found on the site. Also correction actions are accelerated, since workers can suggest with interactive post-it all the information necessary for the correct execution (e.g., regulatory standards, digital instructions) as well as collaborate with remote experts for the implementation of particularly complex procedures. As regards virtual reality devices, they can optimize operators training phases, by transferring them from classrooms to interactive scenarios developed directly from BIM models or from the real site situations. The training in virtual reality, as well as the opportunity to guide staff on sites by means of AR, helps to improve safety standards on site, reducing the margin of error and speeding up operations, thus resulting in times and costs reduction. In reference to sustainability aspects, the collaboration and fulfilment of work functions with different remote teams translates into a reduction in the number of trips to and from the sites, as control activities are entrusted to operators already present on the sites and in contact with the staff in the offices. This factor, added to the ability to virtually view models without the need for continuous printing, reduces energy consumption. In Fig. 3 a synthesis of benefits released by the adoption of VR and AR for planning activities in AEC sector is presented.

2.3 Digital Twin

Digital twins consist in virtual representations of buildings, infrastructures, and plants, obtained through incorporation of data from digital scans, CAD drawings, BIM models, and Geographic Information System (GIS) software, as well as from Internet of Things sensors. Digital twins contain all the information of real artifacts, from the parameters relating to operating conditions and health state of structures to any anomaly and potential risk situation. Therefore, since they serve as the real-time digital counterpart of systems, they empower virtualization and optimization of design, construction, and maintenance of buildings and infrastructures [13]. In fact, digital twins allow to design every aspect of assets and test the actual performances

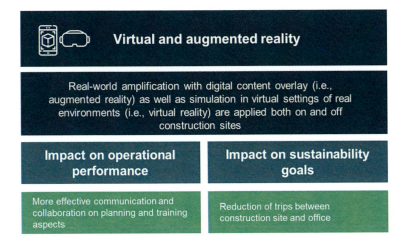

Fig. 3 Virtual and augmented reality for sustainability and productivity in planning

that would be obtained in the real world through virtual simulations. The virtual verification of the effectiveness of different scenarios on digital twins increases the quality of work, while errors related to design or modifications are reduced. This aspect also translates into a reduction in the time and energy required for the development of cost-benefit analyzes related to system changes. It should also be noted that buildings simulations can be used to assess energy needs, quality of indoor environments, and CO_2 emissions throughout the entire life cycle of the assets. These analyzes allow to increase constructions and machine performances in design phases, leading to savings in energy costs by means of reduction of energy wastes. Furthermore, digital twins' dashboards are available in real time on PC, tablet, or mobile device, illustrating the most varied metrics, from costs to energy efficiency values, up to planning and inspections. In this way, facility management activities after design phases can be optimized. In particular, real-time data from buildings and machines can be used to perform predictive analyzes, in combination with series of historical data. The possibility to intervene with preventive actions with respect to problems that could cause accidents improves the quality of work and reduces management and maintenance costs. The potential of digital twins for planning phase in the AEC sector is shown in Fig. 4.

2.4 Drones

The use of drones in construction sites is capable of effectively tracking and communicating the status of projects [15]. In particular, drones' adoption influences two critical AEC processes, that are place mapping and work packages monitoring. In

Fig. 4 Digital twin for sustainability and productivity in planning

fact, drones controlled by mobile applications and empowered with remote measurement techniques (e.g., laser scanners, laser imaging detection and ranging) reduce data acquisition times, corresponding to a reduction in mapping costs. Moreover, data collections are not only faster and cheaper, but also more precise than manual measurements, an aspect that contributes to the increase in the quality of work. As a result, drones can be used to create highly accurate digital models, reproducing the reality of indoor and outdoor environments in a very fast and reliable way. Moreover, images collected from drones can simply be integrated within the main design tools such as AutoCAD and BIM. This feature empowers design activities of architectural renovations with the access to reliable real-world scans and simplifies the comparison between the project models and the scanned reality, promptly assessing any design error. Drones positively influence work quality also during on-site checking phases, keeping track of the times and methods of works. In fact, foremen can instantly and remotely monitor (on smartphone, tablet, or PC) all the operational phases of the work, through video-inspections in the sites carried out with the aid of drones equipped with Ultra HD cameras. This work monitoring method eliminates the need to send personnel to control construction sites and generates a rapid identification of inefficiencies, thus accelerating the implementation of corrective actions. Therefore, the remote monitoring of work progresses by means of drones increases on-site safety by limiting the risks associated with falls (the most common cause of an accident on site) due to the reaching of particularly dangerous areas for inspections or checks. In addition, the use of drones assumes particular importance in complex sites (e.g., sensitive sites from an environmental point of view or an historical interest), where the invasiveness of the intervention must be minimized. The operational advantages connected to the use of drones in work sites are combined with a high degree of energy consumption optimization, as the electrical power supply of aircrafts limits the environmental impact of the devices and avoids noise pollution, ensuring silence

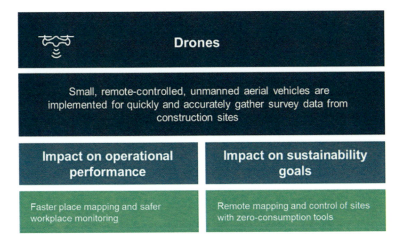

Fig. 5 Drones for sustainability and productivity in planning

in flight operations. Figure 5 exemplifies the considerations on the role of drones in increasing the productivity and sustainability of the AEC sector.

3 Industry 4.0 for Works' Execution on Construction Sites

The execution phase of construction works represents a highly traditionalist activity, guided by well-established and poorly efficient construction techniques based on an intense use of manpower. Actually, novel methods of works' execution arise in AEC sector, borrowing different Industry 4.0 technologies from purely industrial and manufacturing scenarios to improve productivity and business sustainability.

The Industry 4.0 technologies featured in this section are off-site construction (i.e., transformation of the construction process into the assembly of prefabricated modules), 3D printing (i.e., use of mega-devices for the construction of buildings or parts of them) and advanced robotics (i.e., intelligent machines that support staff in carpentry operations).

3.1 Off-Site Construction

Off-site construction involves the hybridization of construction with manufacturing. This method of works' execution consists in the production of modular components within an external manufacturing site and the erection of building structures throughout the modules' assembly on construction site [8]. The mechanism is particularly useful for carrying out building renovation activities, through components that are made in the factory and mounted directly on the old structure. Off-site construction requires an upstream 3D scanning activity of the building, followed by a design phase, then prefabrication, and finally assembly. In this way, complex structures are created safely and efficiently off-site, while the construction phase is limited to the installation of the components. As a result, off-site construction leads to paramount operational benefits, since it involves "just-in-time" philosophy, i.e., designing and building modular components only when necessary. This working method, combined with suppliers and operators of local origin, ensures shorter processing and delivery times, which in fact become independent from highly variable factors such as climate and weather conditions (i.e., one of the main causes of delivery delays in traditional projects). In addition, off-site construction, by transferring the performances of industrial processes to the construction sector, generates minimum stock levels and controlled waste. The result is a significantly improved quality of work as well as an overall cost decrease compared to traditional construction methods. The increase in the quality of work linked to off-site construction regards also not operational and procedural terms. Indeed, this technique affects the safety of workers who operate mainly in a closed and controlled environment, carrying out only the assembly of the modules on the working site. Concluding, the social impacts brought by off-site construction are added to the environmental ones. In fact, the production method optimizes energy consumption, guaranteeing energy savings thanks to a straightforward restoration and construction of structure without need of most heavy vehicles used in traditional construction sites. The influence of off-site construction on the productivity and sustainability of AEC businesses is summarized in Fig. 6.

Fig. 6 Off-site construction for sustainability and productivity in works' execution

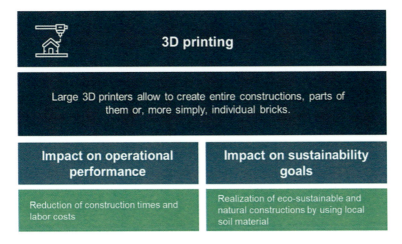

Fig. 7 3D printing for sustainability and productivity in works' execution

3.2 3D Printing

Large 3D printers designed for AEC sector can create entire constructions, parts of them or, more simply, individual bricks. The main advantage of 3D printing is the ability to reduce the execution times of the works. Indeed, the production of buildings with mega-printers directly on site shortens construction times from weeks to hours. The impacts on construction costs are also interesting, especially for the creation of customized geometries and components [19]. The economic savings also derive in part from the reduced use of labor, which in some cases is almost entirely replaced by the printer. This aspect increases safety on site and allows to direct resources towards less harmful tasks, with improvements in the quality of work. In addition, 3D printing can reduce the carbon footprint of the construction sector, creating eco-sustainable and natural constructions with almost zero impact. This is unlocked by the use of natural materials from the surrounding area as a raw input for printers. Therefore, 3D printing can become the basis for a circular housing model, entirely created with poor, waste or recycled materials collected from local soil. Figure 7 reports how 3D printing can be applied to the AEC sector for a more productive and sustainable execution of works.

3.3 Advanced Robotics

Advanced collaborative robots are used in the AEC sector supporting men in some of the most demanding phases of construction, such as, as reported by [11], site preparation (e.g., leveling), substructure works (e.g., foundations laying), and superstructure activities (e.g., load-bearing elements establishment). As regards foundation laying, in particular, machines are able to manage the loading, identification, cutting, rotation, and laying of all the bricks needed to complete the structures thanks to telescopic arms and according to a precise programmed logic and texture. The activities carried out can be controlled via mobile applications, which track operational parameters (e.g., laying times) and physical variables (e.g., bricks humidity), hence requiring minimal human intervention throughout the construction phase. In fact, human involvement can be planned to remotely control robots' works and refine activities, such as the recovery of cement in excess or the improvement of the walls' leveling. Moreover, workers can re-skill and up-skill towards less repetitive tasks. The added value of advanced collaborative robots in the AEC sector lies in the increase in on-site productivity and efficiency, in the reduction of wastes due to low-precision human activities, and in the decrease of the times and costs for carrying out the work. Furthermore, the role of advanced robots, whose arms are equipped with dynamic stabilization, is of great interest in the context of laying foundations for constructions in remote and difficult to access areas, where the safety of operators can be put at risk in traditional working methods. Finally, it has to be considered the benefits in terms of energy consumption associated with the use of advanced robots, given the possibility of using machines totally self-sufficient and ecological by virtue of the solar panel power supply established by some construction robots. The use of advanced robotics for the improvement of sustainability and productivity in AEC sector is well described in Fig. 8.

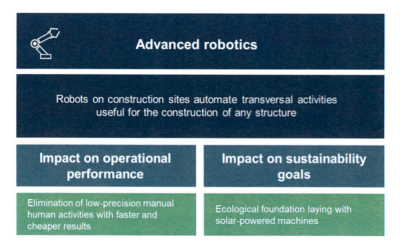

Fig. 8 Advanced robotics for sustainability and productivity in works' execution

4 Industry 4.0 for Management of Support Processes

Design and execution of the works are activities that substantially contribute to the functioning of AEC businesses. Along with these primary activities, there are numerous support processes (e.g., drafting and updating of documents, work control, costing, procurement) that are equally indispensable for a successful business. On these processes, Industry 4.0 technologies have impressive possibilities of use, configuring as tools capable of transforming multiple points of potential inefficiency into value-added and competitive assets.

The technologies explored in this paragraph for the optimal management of support processes are could computing (i.e., services for managing databases and software on demand), Internet of Things (i.e., systems for monitoring structures and work sites in real time), and artificial intelligence (i.e., algorithms for automating various manual procedures).

4.1 Cloud Computing

Cloud computing is one of the most significant technologies for the promotion of industries' digital transformation. In AEC sector, it allows the creation of databases automatically synchronized in real time, but also the shared access to software, calculation services, and on-demand applications from PC, smartphones, tablets, and other devices. Therefore, the adoption of this technology in AEC sector support coordination between people involved in different aspects of the projects, since it allows to speed up the transfer, updating, and sharing of information in real time [3]. In particular, the use of cloud computing platforms business ranges from shared design to control of works, up to reviews with customers. One of the most interesting applications consists in simplifying the communication between construction sites and headquarters, equipping foremen with mobile applications, accessible via tablets, to carry out instant reporting from sites. Open access to information optimizes the entire work, since it favors operations as well as the analysis of problems and possible solutions directly on site, with significant savings in times and resources. Furthermore, the possibility of using cloud management systems can improve the updates of budget releases and the evaluation of project changes requested by clients, being able to use a single environment to generate estimates, manage contract costs, plan and schedule interventions, and support the construction supervision. In addition, cloud computing solutions allow automatic backups and updates, by accessing all documents online on any device. This leads to a net decrease in project errors and conflicts between data, which are reported in a time-register form. In this way, it is possible to remotely control different versions and optimize coordination within the project team. Therefore, cloud computing contributes to increase the sustainability of operations in the AEC sector, reducing the need for travel and transfers for efficient project management. The advantages on operational and sustainability front

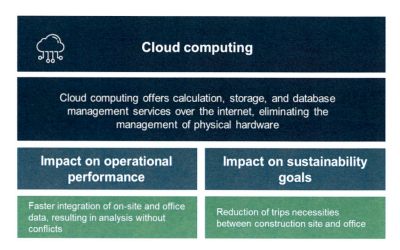

Fig. 9 Cloud computing for sustainability and productivity in management

resulting from the adoption of cloud computing in managing AEC support processes are presented in Fig. 9.

4.2 Internet of Things

Internet of Things in AEC industry simplifies both the analysis and the real-time monitoring of buildings and work sites, using sensors and a cloud platform for data storage and processing. Specifically, the technology finds ample space in the control of structural and environmental conditions of buildings and infrastructures [7], detecting any damage and foreseeing oscillations, subsidence, and structural variations. This is possible by integrating sensors directly inside the pillars, during the construction or restoration phases, to detect data regarding internal thermo-hygrometric characteristics, dynamic stresses, and construction variations. This information can be combined with environmental details gathered by sensors external to structures and relating to climatic data (e.g., temperature, quantity of rain, air quality, wind direction, intensity) but also machineries' performances (e.g., photo-voltaic systems). All data, collected by a cloud computing service, can be processed to obtain precious insights, generating a net reduction in maintenance costs and times. In addition, Internet of Things sensors influence one of the most expensive aspects of building management, namely energy consumption. Indeed, the connection of the sensors with the air conditioning systems and with the entire building management system allows to check various parameters in real time (e.g., temperature, humidity, space occupation) and to optimize the environments accordingly. In addition to this, Internet of Things enable to control operational processes, thanks to the instantly collection of information on the construction site and on the work in progress. In fact,

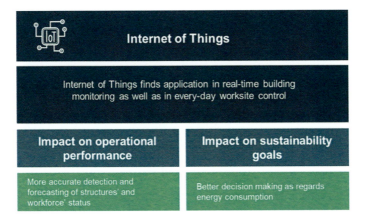

Fig. 10 Internet of things for sustainability and productivity in management

sensors can check accesses and monitor workers' safety, by tracking the processes involving machines, materials, and personnel. In fact, sensors incorporated into tools worn by operators (e.g., bracelets, helmets) can send alerts regarding lack of safety equipment in specific risky areas or interference possibility with dangerous equipment or materials. In this way, the technology clearly contributes to the reduction of the percentage of accidents at work, ensuring both greater safety for operators and lower costs for the company. Similarly, procurement management also improves, since constant monitoring on materials' location generates notifications for low stock conditions. It is therefore evident that thanks to the use of Internet of Things sensors, traditional construction sites are transformed into digital construction sites, where the use of resources, the control of the progresses, and the work scheduling can take place also remotely through a web interface. Figure 10 resumes the beneficial impacts of Internet of Things for AEC businesses.

4.3 Artificial Intelligence

Artificial intelligence algorithms can be applied in different ways by AEC actors, since they effectively respond to the industry's need to process huge amounts of heterogeneous data for extracting useful insights for decision making and tasks improvement [4]. A non-exhaustive overview of the uses of artificial intelligence in AEC ranges from the definition of the economic offer, to the revision of budgets and time schedules, up to the control of the construction site. For example, thanks to artificial intelligence software construction companies, technicians, and designers can use BIM files to automatically calculate project costs. In fact, the graphic entities of BIM models are used to define the bill of quantities, based on rules settable on the type of entity, processing, and measurement. The advantage of this technology consists in the connection of the output (i.e., the bill of quantities) to the input (i.e., the entities of the BIM model), which allows to automatically update the cost calculation with each modification of the original model. In this field, artificial intelligence aims to provide an integrated approach between the architectural design and the definition of bill of quantities, automating the operations of estimating and calculating project costs and reducing the time and margins of error in estimating and updating budgets. Furthermore, many artificial intelligence tools for defining economic offer exploit the set of data relating to previous orders to evaluate the existence of correlations and speed up the definition of the budget and time schedule on the basis of the most varied conditions, such as dimensions, costs, materials, structures and also sustainability-related constraints. Based on this information, the algorithms create thousands of solutions to the problem, analyzing the feasibility of each and eliminating the worst. The same mechanism can be used by the contract manager for the refinement of planning, automating the analysis of scenarios according to different factors (e.g., unexpected weather conditions, shortages of key resources including materials and manpower). In this way, the software can quantify the impact of changes on costs and schedules, simplifying the identification of the best strategies for managing the operations and responding to customer requests. In addition to being an excellent decision support tool, artificial intelligence algorithms simplify the management and monitoring of works' execution in various ways. By acquiring images directly from the construction site with a specific cadence (e.g., through the operators' tablet, drones or robot), it is possible to analyze materials, objects and structures, i.e., compare them with the planning given by the BIM models. This activity is useful, on the one hand, to determine the quality and progress of the work, immediately highlighting any anomalies to be corrected, and, on the other hand, to help project managers in identifying optimal work sequences to increase productivity and safety on sites. Figure 11 summarizes the main contribution of artificial intelligence for applications in AEC sector.

5 Conclusion

The present chapter explores how Industry 4.0 technologies can be applied in AEC sector to generate significant gains in productivity and sustainability. In fact, such technologies are recognized as capable of change construction processes, paving the way towards a new concept of Construction 4.0 [14], with significant advantages both in terms of performance and sustainability [9]. Specifically, the chapter analyzes the impacts of some of the most innovative technologies on three phases of the construction life cycle (i.e., architectural and design planning, works' execution, and management of support processes), illustrating how to use them for obtaining greater profitability and sustainability in the AEC sector. In this way, the chapter broadens the scientific knowledge relating to the concept of Construction 4.0, still in its infancy and still unaware of issues that go beyond the purely technical aspects

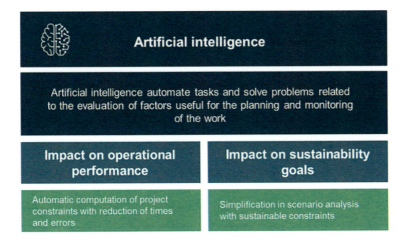

Fig. 11 Artificial intelligence for sustainability and productivity in management

of applying technologies [9], such as those relating to the modification of operations and sustainability of AEC sector.

References

1. Abdirad, H., Dossick, C.S.: BIM curriculum design in architecture, engineering, and construction education: a systematic review. J. Inf. Technol. Constr. **21**, 250–271 (2016)
2. Agarwal, R., Chandrasekaran, S., Sridhar, M.: Imagining Construction's Digital Future. McKinsey & Company (2016). https://www.mckinsey.com/business-functions/operations/our-insights/imagining-constructions-digital-future
3. Beach, T.H., Rana, O.F., Rezgui, Y., Parashar, M.: Cloud computing for the architecture, engineering & construction sector: requirements, prototype & experience. J. Cloud Comput. Adv. Syst. Appl. **2**, 1–16 (2013)
4. Darko, A., Chan, A.P., Adabre, M.A., Edwards, D.J., Hosseini, M.R., Ameyaw, E.E.: Artificial intelligence in the AEC industry: scientometric analysis and visualization of research activities. Autom. Constr. **112**, 103081 (2020)
5. Delgado, J.M.D., Oyedele, L., Demian, P., Beach, T.: A research agenda for augmented and virtual reality in architecture, engineering and construction. Adv. Eng. Inform. **45**, 101122 (2020)
6. European Commission: Second European Industry Day (22–23 February 2018). Summary of the discussions (2018). https://www.earto.eu/wp-content/uploads/EID_18_Summary_report.pdf
7. Gbadamosi, A.Q., Oyedele, L., Mahamadu, A.M., Kusimo, H., Olawale, O.: The role of internet of things in delivering smart construction. CIB World Building Congress, Hong Kong, China (2019)
8. Hosseini, M.R., Martek, I., Zavadskas, E.K., Aibinu, A.A., Arashpour, M., Chileshe, N.: Critical evaluation of off-site construction research: a scientometric analysis. Autom. Constr. **87**, 235–247 (2018)
9. Kozlovska, M., Klosova, D., Strukova, Z.: Impact of Industry 4.0 Platform on the Formation of Construction 4.0 Concept: A Literature Review. Sustainability **13**, 2683 (2021)

10. McKinsey Global Institute: Reinventing construction: A route to higher Productivity (2017). https://www.mckinsey.com/~/media/mckinsey/business%20functions/operations/our%20insights/reinventing%20construction%20through%20a%20productivity%20revolution/mgi-reinventing-construction-executive-summary.pdf
11. Melenbrink, N., Werfel, J., Menges, A.: On-site autonomous construction robots: towards unsupervised building. Autom. Constr. **119**, 103312 (2020)
12. Oesterreich, T.D., Teuteberg, F.: Understanding the implications of digitisation and automation in the context of Industry 4.0: a triangulation approach and elements of a research agenda for the construction industry. Comput. Ind. **83**, 121–139 (2016)
13. Rafsanjani, H.N., Nabizadeh, A.H.: Towards digital architecture, engineering, and construction (AEC) industry through virtual design and construction (VDC) and digital twin. Energy Built Environ. (in press)
14. Sawhney, A., Riley, M., Irizarry, J.: Construction 4.0: An Innovation Platform for the Built Environment. Routledge, Milton, United Kingdom. ISBN: 9780367027308 (2020)
15. Tal, D., Altschuld, J.: Drone Technology in Architecture, Engineering and Construction: A Strategic Guide to Unmanned Aerial Vehicle Operation and Implementation. John Wiley & Sons, Hoboken, New Jersey, United States. ISBN: 9781119545897 (2021)
16. Vennström, A., Eriksson, P.: Client perceived barriers to change of the construction process. Constr. Innov. **10**, 126–137 (2010)
17. World Business Council for Sustainable Development: Digitalization of the Built Environment Towards a more sustainable construction sector (2021). https://www.wbcsd.org/contentwbc/download/11292/166447/1
18. World Green Building Council: New report: the building and construction sector can reach net zero carbon emissions by 2050 (2019). https://www.worldgbc.org/news-media/WorldGBC-embodied-carbon-report-published#_ftn1
19. Wu, P., Wang, J., Wang, X.: A critical review of the use of 3-D printing in the construction industry. Autom. Constr. **68**, 21–31 (2016)
20. Zabidin, N.S., Belayutham, S., Ibrahim, K.I.: A Bibliometric and scientometric mapping of Industry 4.0 in construction. J. Inf. Technol. Constr. **25**, 287–307 (2020)

Programming Design Environments to Foster Human-Machine Experiences

Giovanni Betti, Saqib Aziz, and Christoph Gengnagel

Abstract The introduction of robotic construction methods in the building industry holds great promise to increase the stagnating productivity of the construction industry and reduce its current carbon footprint, in considerable part caused by waste and error in construction. A promising aspect is the implementation of integrated CAD/CAM processes where the design intent -expressed in CAD- is directly translated in machine instructions (CAM) without loss of information or need for intermediate translation and refactoring. As the introduction of robotics will hardly spell the disappearance of human workers, a key challenge will be orchestrating human-machine collaboration in and around the construction site or fabrication plant. Our contribution presents explorations in this field. Key to our approach is the investigation of the above-mentioned questions in conjunction with Mixed Reality (MR) interfaces to give access to both human and machine workers to the same dynamic CAD model and assembly instructions. This paper describes our work in this field through two artistic installations and ongoing research on additive manufacturing enabled construction. We probe principles for precise, customised, efficient, almost waste-free and just-in-time productions in the construction sector.

Keywords Human-machine-interface (HMI) · Human-machine-experience (HMX) · Computational design · Collaborative design · Mixed-reality · Robotic fabrication

United Nations' Sustainable Development Goals 8. Decent Work and Economic Growth · 9. Industry, Innovation and Infrastructure · 12. Responsible Consumption and Production

---

G. Betti (✉)
Department for Experimental and Digital Design, University of the Arts Berlin, Hardenbergstr. 33, 10623 Berlin, Germany
e-mail: Giovanni.Betti@henn.com

Henn Architekten GmbH, Alexanderstr 7, 10178 Berlin, Germany

S. Aziz · C. Gengnagel
Department for Structural Design and Engineering, University of the Arts Berlin, Hardenbergstr. 33, 10623 Berlin, Germany

© The Author(s), under exclusive license to Springer Nature Switzerland AG 2024
M. Barberio et al. (eds.), *Architecture and Design for Industry 4.0*, Lecture Notes in Mechanical Engineering, https://doi.org/10.1007/978-3-031-36922-3_4

1 Introduction

Design technology and culture has significantly evolved in recent years in the Architecture, Engineering and Construction (AEC) industry. Among other technologies, Building Information Modelling (BIM), parametric and computational design, real time immersive visualisation are among the technologies that have become commonplace in many practices. These advances in CAD don't seem to have translated yet into increased efficiencies in the construction sector, which has been lagging in productivity compared to several other industries [1].

We postulate that this is at least in part due to the information loss that happens between the design and the construction phase. The reasons for the persistence of this information loss are many and complex. Some relate to the economic and societal environment of a fragmented AEC industry, which lacks vertical integration. Some factors are technological and cultural. Despite the advances mentioned above, the main tool to communicate design intent to contractors and fabricators still is 2D construction drawings. Although often automatically created by BIM software, the 2D representations thus created are but a static snapshot of the design. The act of issuing 2D drawings of a design creates a break in the digital information systems generated in the design phases that impedes feedback loops with the construction phase i.e., subsequent changes to the design are not automatically reflected and there is no clear mechanism to register and reflecting differences between the design intent and the built reality. Furthermore, 2D drawings are extremely useful for communicating with people, but are of little to no use in transferring information to robotic arms or other CAM systems. With this in mind, we started experimenting with systems that would create direct links between design intent and robotic fabrication. An important focus in those explorations is the inclusion of a Non-Expert Robotic Operator (NERO) in the loop. This step is crucial to any experiment that aims to advance digital construction and assumes that construction workers will interact in some way or form with robotic entities in the dynamic environment of a prefabrication plant or a construction site. To facilitate the communication and cooperation between the NERO and the robot, we devised a series of ad-hoc Human machine Interfaces (HMIs) that free the user from cumbersome interactions via mouse and keyboard. In the first installation described in this paper we used a sound-based control system, while in the second installation and in the ongoing research on 3D Concrete Printed Slabs (3DCPS) we resorted to an AR interface. The NERO needs to be able to intervene in various phases of the construction process, from influencing the design (i.e., to react to changing requirements or differences between CAD model and the reality of a construction site), to controlling the sequencing of operations of their robotic collaborator, to access contextual assembly instruction for their own manual assembly operations.

In essence, our strategy is to give simultaneous access to robotic systems and NEROs to the same, constantly evolving, representation of the design intent and of the current state of fabrication. We split tasks between the robotic system and the NERO to leverage the specific skills and strengths of each. To this effect we allocate tasks requiring human capabilities such as tacit knowledge, high mobility, dexterity, and situational adaptability to the human actors; tasks requiring repeated operations and high accuracy are allocated to robotic agents. By allowing the NERO and the robots equal access to the design system and creating a bidirectional flow of information between the design system and the construction process, part of the design process is explicitly transferred downstream.

The shift from static 3D models toward smart, parametric, dynamic 3D models is becoming common in the AEC industry [2]. This shift means that designers author design systems rather than specific design solutions. In this perspective, the built outcome represents one negotiated instance of the design space enabled by a computational model. The specific instance emerges through a collective and collaborative navigation and exploration of the design space performed by various project stakeholders. Therefore, the notion of authorship and ownership of the final design becomes diffuse, with many stakeholders able to claim parts of it [3].

In our investigations we explored how this approach may affect the transfer of knowledge and information throughout the various design stages, replacing the current linear design process with a more iterative and dynamic system, minimising costs associated with design changes and enabling integrated decision-making processes. A key strategy in the workflows and experiments presented in this chapter is the use of highly visual and intuitive human-machine interfaces enabled by various Mixed Reality (MR) strategies. MR is leveraged as a way of providing to the human actors highly contextual (both in time and space) information and enabling them, through natural interfaces, to materialise and modify the computationally encoded design logic and to control the robotic actors [4] (Fig. 1).

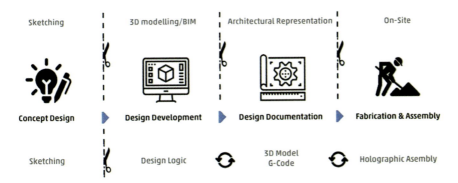

Fig. 1 Conceptual diagram of the current fragmented processes in AEC and consequent loss of information across the project phases (top) and an ideal bidirectionally connected process (below)

2 Description of the Installations

In the following chapter two art installations and one ongoing research project investigating additive manufacturing of prefabricated building components illustrate the application of MR and robotic fabrication driven interfaces. Originating from an artistic and intellectual exercise, those explorations challenge current collaborative design processes and foreshadow practical applications of such design environments in the construction sector.

2.1 Communication Landscapes

The participatory installation Communication Landscapes explores the epistemic potentials of emerging technologies related to Industry 4.0 and digital craft in the AEC industry and aims to foster an optimistic approach for future automation of mass-customizable production. The installation was conceptualised by the authors and others for the 2017 Seoul Biennale of Architecture and Urbanism, in the context of the exhibition Imminent Commons [5]. In the context of the exhibition theme, the installation explores how industry 4.0 can reintegrate manufacturing processes in urban contexts with advanced technologies and transition consumers to prosumers. The industrial revolution had far-reaching implications on society and urban planning. Production hubs started to grow and slowly migrate from the city centres to the suburbs. Consequently, craftsmanship and its peer-to-peer (P2P) model of production, trade, and consumption shifted into a business-to-consumer (B2C) relationship, restricting the artisan's access to new technological advancements in production and automation [6]. With the dawn of Industry 4.0 and the opportunities that the Internet of Things (IoT) brings, MR and robotic fabrication knowledge is being spread and democratised at an exponential rate. This can empower the artisan to explore and utilise new technologies that can also empower individual creativity in the production of mass-customised goods, distributed and co-designed throughout a wider population, promoting a new archetype: the prosumer—a producer and consumer of its own goods [7]. We devised an installation that could offer direct access to new modes of fabrication and digital crafting. To do so it is important to ensure that non-expert users can interact and use advanced equipment intuitively and with ease and playfulness. Natural and intuitive human-machine interfaces (HMI) are key to achieving this purpose. Research indicates that a well-designed interface enables users to intuitively navigate complex computer-aided design systems, bypassing the frustration and feeling of powerlessness of its users [8]. The disappearance of the classical computer model with its mouse-keyboard terminal might lead to more creative HMIs [9] derived by methods of gamification [10], which use new technologies, such as new mixed reality devices and applications [11]. Here, the entire body and/or sensory inputs such as voice-controlled interfaces can function as a bridge to control specific functions, such as the motion of robotic assemblies.

2.1.1 Installation Workflow

Based on these reflections, the installation, "Communication Landscapes", explores a future collaborative approach to distributed and participatory design. The installation consists of a microphone, a visual interface, and a robotic arm with an attached hot wire cutter. Over the course of five days, curious visitors were invited to engage in the installation by simply speaking into the microphone. This interaction triggers an algorithm to visualise the audio input and transform it in real time into a three-dimensional representation. By providing direct visual feedback of the created geometry, we were able to foster a fast and intuitive learning process, where users would learn to shape the virtual object by modulating their voice. Once the participants are satisfied with this design exploration, they can instruct the robotic arm to activate the production of the geometry. This automatically sets the robot arm in motion, and the hot wire cutter carves the design out of a block of foam, creating simultaneously a positive and negative instance of the designed object. Visitors are given the positive element as a gift, while the negative pieces are assembled into a sculptural wall. This sculptural wall represents a collective design featuring patterned surfaces, instantiated by the physical/digital translation of over 300 individual sound inputs collected over the duration of the installation. The position of each individual tile position is algorithmically determined based on their geometric similarities (Fig. 2).

In the design process of the installation, we made sure that we couldn't predict the exact form of the installation. Based on individual voices, each tile design is unique and unknown before it is created. As installation designers, we willingly relinquished control over the final product. Our role was to simply enable and only initialise the design system, with the specific intent of activating a collaborative process in which the "users" of the system could become co-creators of the final piece (Fig. 3).

We used the Grasshopper3D Firefly 3D plug-in to process the audio input in a suitable format for the design environment, allowing the pitch, volume, and frequency to be filtered in real time. The volume and frequency values of the input voices create two unique and opposing freeform curves that can be lofted to represent the recorded voice traces. The resulting ruled surface is allowed to rotate in the plane at a 90° angle, depending on the maximum peak frequency value of the sound input. The Grasshopper3D plug-ins Human and Human UI were used to create a custom user interface consisting of two main panels. The most prominent feature is the 3-D

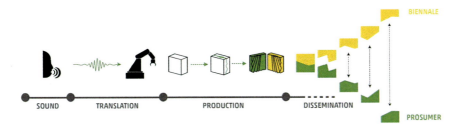

Fig. 2 Communication landscapes installation workflow

Fig. 3 From left to right: interface, interaction, and fabrication

representation of the foam block filling the entire left panel. In its rest state, the user can see a featureless box. As soon as a voice impulse is triggered, the shape starts to dynamically transform to represent the sound traces in real time, changing its colour and orientation to visualise the magnitude of volume and frequency. The right panel features three interactive infographics and two buttons: record and fabricate. The first infographic shows the custom placement map, indicating where the current design exploration would be placed in the sculptural wall, determined by its resemblance to previous design generations, allowing the participant to also play with the voice input to place the design at different locations. Beneath the placement map two scatterplots show the current magnitude of the frequency and volume. The record button allows recording the sound input, this act will determine the duration of the recorded sound and therefore the final traces that generate the shape. The fabricate button will translate the current design instance into g-code and trigger the robotic motion.

2.1.2 Results

The installation was collectively realised over the course of five days together with the participation of the visitors. It was satisfying to observe the ease of interaction for most participants regardless of age and nationality. Due to safety guidelines the fabrication process had to be supervised which did not hinder the experience but rather allowed to engage with the visitors. It was also rewarding to see the immediate response of joy and amazement at seeing their voice translated into geometry that could be fabricated within. Perhaps the most notable interactions took place between children and the robotic installation (Fig. 4), providing anecdotal evidence of the large user group that can be reached with careful design of the HMI. The installation focused mainly on the creation of a unique artefact by "talking" to a robot. The less developed part of the workflow involved the manual aggregation of the pieces on the wall, for which only a simple placement map was developed. Even if relatively simple, this installation contained many of the elements referred to in the introduction. We successfully established a direct pipeline between a mutable design intent and a robotic fabrication process. We included hundreds of exhibition visitors in a collective design process, enabling people of various ages and backgrounds, with little to no instruction, to interact both with a complex parametric model and an industrial robot.

2.1.3 Pop-Up Factory

With the installation Pop-Up Factory, realised for the festival Make. City Berlin in 2018 [12], we aimed at creating a richer design and assembly experience. To allow for more complex and nuanced interactions between NEROs and robots, we decided to explore the use of Augmented Reality (AR) devices such as the Microsoft HoloLens. This wearable device can offer highly contextual and complex information to the worker enabling not only to illustrate 3d representations or instructions but also allowing access to a wide variety of technical and non-technical information [13]. Such devices offer hand-free access to planning information and enable the workers to execute tasks and have a dynamic communication bridge to the design teams, thus enabling feedback loops between design intent and built reality.

Based on the previous installation described in Sect. 2.1 the above-mentioned concepts were integrated into the research scope and resulted in a prototypical AR interface supporting construction processes. It is a first modest case-study to validate the potentials, limitations, and epistemic findings towards a fully AR supported design and construction environment. The methodology here allows non-professional participants to collaboratively create a human-scale structure that can be geometrically changed while being constructed, enabling it to conceptually include last minute adjustments on the simulated construction site. To this end an intuitive holographic interface was designed to empower non-experts to playfully access more complex 3D modelling, on-demand robotic fabrication and AR-guided assembly processes in real-time.

2.1.4 Installation Workflow

The technical hardware features a Microsoft HoloLens headset, a robotic arm (ABB IRB 1200) with a custom hot-wire assembly, a microphone, and computing desktops. The installation design was created using the CAD modelling software Rhinoceros 3D alongside Grasshopper and custom Python scripts. Commercial add-ins to Grasshopper, like Fologram and HAL Robotics were used to create an AR interface and to control the robotic equipment. A design system was programmed to iteratively and collaboratively design a 2 m high tower-like sculpture, giving the participants control of the design outcome. This design narrative was aimed to firstly initiate a discussion challenging the authorship and/or ownership of design in the architectural practice on an intellectual level and to secondly probe the practicality of such interactive and reciprocal AR construction environments for the deployment in the building sector. The installation would exist simultaneously in physical and virtual space. Without the headset the visitors would see the hardware components and the partially built structure. The AR headset would reveal a rich holographic scenery offering a variety of functionalities and information. The holographic interface (HI) was structured around two workflows: a design workflow and a fabrication and assembly workflow. The user was let free to switch between the two workflows iteratively and continue to change the design of the unbuilt portion of the installation.

All the computing units were hidden from the visitors making the AR head-googles the only required interface (Fig. 5).

2.1.5 AR Aided Design Modes

Over the course of three days the visitors would build the sculpture as seen in Fig. 5. In the custom AR aided design environment all actions/selections had to be controlled using "air tap" and "eye gaze navigation" [14] which represent an unfamiliar interaction paradigm for the general public. In comparison to the simple voice interface of the previous installation, introducing the visitors to a new interaction paradigm was a more challenging task, one that could be nonetheless completed in 5–10 min. While this might be a high time frame in the context of a public and participatory installation, it seems a relatively short time frame to train a non-expert in the use of a new system.

Inside the AR design workflow, the visitors could edit control points that would modify the overall shape of the target branching tower structure. The anchors determined the end points of 5 vertical free-form curves and could be adjusted to shape a resulting complex geometry, creating a sort of "branching tower". This geometry would then be dynamically discretized in individual custom brick elements (See

Fig. 4 Installation interaction and results

Fig. 5 Pop-up factory, participatory mixed reality, and robotic fabrication installation

Fig. 6). When satisfied with the current design envelope the user could switch to the fabrication workflow. In this mode the user would then be able to access the fabrication and assembly instructions for the next modules. The system would constantly track the progress of the construction to offer the right sequence. The design workflow was conceived such that the built part would become no longer modifiable by the user, to ensure consistency in the conjoined design and fabrication process. Hence freedom to design what is not built would be preserved throughout the fabrication process.

The fabrication process would involve actions performed both by the human and the robot actors. Inspired by the first installation (see Sect. 2.1) a 3d surface texturing was enabled again using auditory impulses. The user could virtually tap a holographic 3d microphone to record the ambient acoustical noise in the room to texture the current set of brick-foams which would be integrated in the path generation of robotic fabrication sequence. Similarly, an air tap on a robot icon hologram would send the command to the robotic arm to start the fabrication process for the next set of custom blocks. The visitor would need to place a new stock of pre-cut foam material to be processed by the robotic hot-wire cutting. An algorithm would iteratively calculate the angles that are needed to assemble the foam bricks seamlessly to the new curvature "just in time". By virtue of the "just in time" nature of this process where only three to five custom bricks would be produced at any given time and then immediately assembly, labelling and inventory tracking could be kept to a minimum. After the robotic fabrication process each brick would need to be manually rotated and placed correctly relative to one another before this sub-assembly could be aggregated in the overall structure. Both the pre-assembly and the final aggregation steps would be augmented by holographic guides that would show the correct positioning and orientation of each element. In this process each actor -human and robotic- is used to the best of their skills: the hot wire cutting by the robot ensures submillimeter-scale accuracy and faithfulness to complex 3d information and the manual assembly process relies on the human dexterity in manipulating complex objects, mobility in an unstructured environment and higher positional tolerances.

Fig. 6 From left to right: AR-driven modulation, fabrication, and assembly

2.1.6 Results

The presented workflow uses AR to navigate through the design, fabrication, and assembly stage of a vicarious project in real-time, short-circuiting the digital information chain between the design phases aiming to minimise information loss. The installation is a first case-study evaluating AR-driven construction interfaces in controlled environments and would need to be developed further to be probed on a real construction site with e.g., safety requirements and construction regulations. There are also noteworthy technical shortcomings. For instance, the position accuracy and field of view of an overlaid holographic scene is still not as accurate as wished and spatially limited. Hologram visibility is dependent on the ambient brightness limiting the use in outdoor environments. The lack of a real inventory management system, even if reduced labelling labour for the installation, would be hard to easily scale to a more complex construction process, where various different materials and prefabricated elements would arrive at different times and located in less controlled settings. Furthermore, more work is required to explore more intuitive gesture-controlled metaphors and interface management to intuitively guide and structure the interactions with the AR environment for non-expert or skilled labour.

2.2 Large-Scale Additive Manufacturing and Mixed Reality Applications Towards a New Possible Construction Process

Alongside robotic fabrication large-scale additive manufacturing with mineral materials such as three-dimensional concrete printing (3DCP) has made significant leaps over the last decade and offers a promising alternative to produce highly optimised prefabricated building elements for the construction sector [15]. Having in mind that about 60% of the buildings and 25% of the infrastructure that will exist on earth in 2060 still need to be built, making the decarbonization of the building sector the most critical aspect to success in meeting the 1.5° target defined in the Paris Climate Agreement, and that Cement production alone accounts for 6–8% of all global emissions [16], 3DCP could unlock unexplored potential of mineral materials, fostering more resourceful and material aware solutions resulting in massive material and grey energy savings and an overall better life-cycle performance. For the above ground structure of commercial buildings in reinforced concrete, the slab construction contains a fair amount of the material mass and thus most of the embodied CO_2 emissions. Conventional reinforced concrete flat slab constructions are considered an economical and adaptable solution. However, this assessment neglects the enormous material consumption of this construction method, the poor ratio of dead weight and load-bearing capacity, and its poor performance in terms of room acoustics and building physics. We are currently developing a workflow enabling us to combine algorithmic geometrical modelling capacity with optimization processes based on

Finite-Element-Analysis (FEA) for the generation and fabrication of multifunctional reinforced mineral construction elements as ceilings and walls [17]. The process enables high adaptability and multi-performance optimization of the manufactured components, supplemented by a semi-automated pre-fabrication of formwork and reinforcement elements assisted by an AR-interface for assembly processes on-site. Since 47% of the construction work (cost and time) is due to erecting and dismantling formwork and 20% can be allocated to reinforcement work, there is a great potential for automation and/or process optimization [18].

In principle, the methodology can be applied to a wide range of structural elements. The case-study presented here shows an application for a biaxial load-bearing beam grillage made of reinforced concrete, which allows for a variety of applications through the combination of additive prefabricated and durable semi-finished parts and in-situ concrete supplementation. The case-study reimagines the design of the Maison «Dom-Ino» by LeCorbousier using 3DCP and applies structural optimization principles inspired by the work of Pier Luigi Nervi, implementing isostatic lines following the principal bending moments and hereby creating an optimised rib structure [19]. To ensure a scalable construction method for such a design, 3DCP elements form the basic framework of the supporting structure and at the same time integrate acoustic properties programmatically tuned via geometric articulation [17].

2.2.1 AR-Assisted Additive Manufacturing

Based on the bespoke workflow a physical mock-up was to be fabricated and assembled using additive manufacturing and a prototypical MR Interface. To this end a smaller portion of the slab construction derived from the design Fig. 7 was used. An AR-driven interface was designed to allow the interdisciplinary team of researcher to navigate through the geometry generation of the form pieces, allowing to make modifications to the geometry, to evaluate the programmed mechanical and acoustical optimizations, to validate the printability, check for any system bound collisions, visualise the current hydration/curing state, get feedback on the material usage/overall weight, and to bidirectionally simulate through the print-path evolvement to understand the current additive manufacturing process and foresee any system or logic failures by the means of spatial aware holographic aids. The current development of such an interface proved to be a necessary tool since most of the collaborating researchers were not familiar with the algorithmic design environment used to program the robotic system and generate the g-code. Hence a more intuitive and representative interface was needed to visualise the complex fabrication process. The zero-waste formworks can be optimised for manufacturing purposes validating the printability and structural integrity of the 33 individual bodies in fresh and hardened states [20] (Fig. 8).

By simulating the printing process, including cement hydration/curing and autogenous shrinkage [21], the usage of CAD and CAM can address many of these problems and offer more complex design features. Moreover, previous research pointed out that printing speed optimization is crucial to find a balance between buildability

Fig. 7 Case-study conceptual optimization

Fig. 8 Experimental AR—assisted fabrication interface

and interlayer adhesion, both of which are strongly dependent on the "waiting time" of deposition of a layer upon the previous one [21]. Therefore, the AR interface enables situational control allowing to alter the current speed parameter and material flow rate during the manufacturing process. This could massively help workers on site to first detect any small deviations in the ongoing construction work and use the accessibility of the design features to locally alter the individual and manageable geometry enabling the overall element to fit more precisely in the actual environment. Alongside the AR-interface a custom algorithm was developed to generate the G-code with high degree of controllability for the described process. Fologram was used to instantiate a first working and AR-driven interface allowing to integrate the above-mentioned design and analysis features.

2.2.2 AR-Assisted Assembly

Due to their low weight the printed modules can be easily managed by one or two workers. To further streamline the montage sequence all formwork bodies were tagged with Aruco markers containing information such as the individual position of the piece in the final construction and their weight. Regardless of their current position, each formwork can thus be identified via the associated marker and accurately positioned through holographic instructions (Fig. 9).

The first prototype AR interface was developed using Fologram and the HoloLens2 headset. Due to the imperfections of the projected scene compared to the accuracy in the physical environment, a custom MR interface is currently being

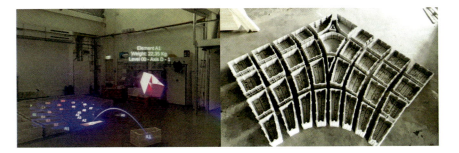

Fig. 9 Experimental AR—assisted assembly workflow

tested that uses a direct connection between Grasshopper and Unity 3d and the use of the Oculus 2 MR headset. The advantage of the e.g., oculus MR headset is that they can seamlessly blend between physical and digital scenes using "Passthrough API" [22]. Here the physical reality is recorded and digitally projected in the MR environment enabling to overlay entirely controllable, adjustable, and precise digital overlays. Research is currently investigating such MR implementation and with them ways to address safety concerns [23]. In future, such approaches could be deployed on construction sites and mobile robotic production hubs. The fabrication of 3DCP elements could happen on-site, reducing costs and efforts associated with transport. This availability and accessibility to on-site production hubs in combination with intelligent MR interfaces could also further streamline construction adjustments empowered by dynamic scanning of the construction site. Any local construction deviations could be identified and adjusted in the sequential manufacturing process immediately.

3 Conclusion and Discussion

Technological advances that have been shaping the AEC industry haven't translated yet into increased construction efficiencies. On the contrary, the lack of integration in the digital tools and workflows seems the main culprit. We identify current standard practices of transmission of design intents as one of the critical bottlenecks in the industry and we advocate for a new paradigm where information can freely and continuously flow in a bidirectional fashion between design software and construction site. We identify MR and related technologies as crucial elements to provide the worker on site with information that is highly contextual, both in space and time, up to date and easy to understand. MR can, furthermore, enable the worker to bidirectionally update the design in real time by registering differences between design intent and realisation and by directly integrating in the design required changes. Giving the worker direct visibility and agency over the design model would also enable it to more easily collaborate, communicate and control advanced robotics, whether those are robotic arm, industrial 3D printers or other.

In this paper we present a small number of investigations that point in this direction at increasing levels of complexity: from artistic installations to 1:1 prototypes of innovative construction methods. Through those we demonstrate different methods and levels of connecting the workers with the design model and increasingly sophisticated methods for interacting with construction robotics, assemblies, and inventory management. Many open challenges remain before those concepts can be demonstrated or implemented in a reliable manner on a real construction site. Among those there is the need to develop reliable and robust communication protocols to ensure that information integrity is maintained, and information is delivered in a contextually appropriate manner. New interaction metaphors and interfaces need to be developed that enable intuitive and robust interaction with information, construction materials, robotics, colleagues, and the surrounding environment.

References

1. Hasan, A., Baroudi, B., Elmualim, A., Rameezdeen, R.: Factors affecting construction productivity: a 30-year systematic review. Eng. Constr. Archit. Manage. **25**(7), 916–937 (2018)
2. Oxman, R.: Thinking difference: theories and models of parametric design thinking. Design Studies, vol. 52 (2017)
3. Schmitt, G.: A new collaborative design environment for engineers and architects. In: Smith, I. (Ed.), Artificial Intelligence in Structural Engineering. Lecture Notes in Computer Science, vol. 1454. Springer, Berlin, Heidelberg (1998)
4. Jahn, G., Newnham, C., Berg, N.: Augmented Reality FOR Construction From Steam Bent Timber (2022)
5. Betti, G., Aziz, S., Rossi, A., Tessmann, O.: Communication landscapes. In: Proceedings of the Conference: Rob|Arch 2018, Robotic Fabrication in Architecture, Art, and Design, Zurich, Switzerland, pp. 74–84 (2018)
6. Gershenfeld, N.: Fab. The coming revolution on your desktop. From Personal Computers to Personal Fabrication; Basic Books (2005)
7. Ritzer, G., Dean, P., Jurgenson, N: The coming of age of the prosumer. Am. Behav. Sci. **56**(4), 379–398 (2012)
8. Ben Shneiderman & Harry Hochheiser Universal usability as a stimulus to advanced interface design. Behav. Inf. Technol. **20**(5), 367–376 (2001)
9. Willis, K.D., Xu, C., Wu, K.J., Levin, G., Gross, M.D.: Interactive fabri-cation: new interfaces for digital fabrication. In ACM TEI (2011)
10. Savov, A., Tessmann, O., Nielsen, S.A.: Sensitive assembly: gamifying the design and assembly of façade wall prototypes. In IJAC (2016)
11. Francese, R., Passero, I., Tortora, G.: Wiimote and kinect: gestural user interfaces add a natural third dimension to HCI. In: Proceedings of the International Working Conference on Advanced Visual Interfaces, pp. 116–123. ACM (2012)
12. Betti, G., Aziz, S., Ron, G.: Pop-up factory: mixed reality installation for the MakeCity festival 2018 in Berlin. In: Proceedings of the Conference: eCAADe/SIGraDI 2019, Architecture in the age of the 4th Industrial Revolution, pp. 115–124 (2018)
13. Kivrak, S., Arslan, G.: Using augmented reality to facilitate construction site activities. In: Mutis, I., Hartmann, T. (Eds.) Advances in Informatics and Computing in Civil and Construction Engineering (2019)

14. Microsoft Mixed Reality Gestures. https://docs.microsoft.com/en-us/windows/mixed-reality/gestures. Accessed 13 May 2022
15. Hansemann, G., Schmid, R., Holzinger, C. Tapley, JP., Kim, HH., Sliskovic, V., Freytag, B., Trummer, A., Peters, S.,: Additive fabrication of concrete elements by robots: lightweight concrete ceiling. In: Fabricate 2020: Making Resilient Architecture, UCL PRESS, London, pp. 124–129 (2020)
16. Lehne, J., Preston F.: Making Concrete Change: Innovation In Low-Carbon Cement And Concrete. Chatham House Reports (2018)
17. Aziz, S., Alexander, B., Gengnagel, C., Weinzierl. S.: Generative Design of Acoustical Diffuser and Absorber Elements Using Large-Scale Additive Manufacturing, Architectural Acoustics and Sound, Rome (2022)
18. Weiss, M.: Kennzahlen für Stahlbetonarbeiten-Anwendung bei Hochbauprojekten, na (2010)
19. Halpern, A.-B., Billington D.-P., Adrianssens S.: The ribbed floor slab systems of Pier Luigi Nervi. In: Proceedings of the International Association for Shell and Spatial Structures (IASS) Symposium 2013 "BEYOND THE LIMITS OF MAN" 23–27 September, Wroclaw University Of Technology, Poland, pp. 127–136 (2013)
20. Aziz, S., Kim, J-SStephan. D., Gengnagel, C.: Generative structural design: a cross-platform design and optimization workflow for additive manufacturing. In: The 3rd RILEM International Conference on Concrete and Digital Fabrication, Loughborough University, UK (2022)
21. Buswell, R.A., Leal de Silva, W.R., Jones, S.Z., Dirrenberger, J.: 3D printing using concrete extrusion: A roadmap for research. Cem. Concr. Res. **112**, 37–49 (2018)
22. Oculus. Mixed reality with passthrough (2021). https://developer.oculus.com/blog/mixed-reality-with-passthrough/
23. Riva, G., Maria, R.E.F.: EXTEND: resolution revolution to extend reality. Cyberpsychol. Behav. Soc. Netw. **24**(1), 74–75 (2021)

Designing with the Chain

Stefano Converso and Lorenzo Pirone

Abstract This paper will describe the case studies of Design-Build projects to discuss the use of digital and computational technologies as facilitators of a new role in AEC process, devoted to establishing the "missing chain" in the AEC sector, too often affected by fragmentation in several, conflicting actors. The "Design Mentality", by nature oriented at synthesis, and at bringing topics together can be the most effective "productivity boost" for the sector, but it needs a change in relationships and a different framework of responsibility. The architect needs to enjoy the management role, that means "make things happen", designing the context in a wider sense, including the selection of partners, of companies, packages, and solutions, and establishing the right "rules for form" that can keep the project identity. The "wireframe method" will be discussed as a geometrical method to manage the deployment and aggregation of parts in an adaptive process.

Keywords Off-site construction · File to factory · Wireframe coordination · Design management · Procurement · Design packages

United Nations' Sustainable Development Goals 9. Industry, Innovation and Infrastructure · 8. Decent Work and Economic Growth · 12. Responsible Consumption and Production

1 Introducing a Mechanical Approach into Architecture

The notion of introducing a "mechanical" industry approach to Architecture is not new in the discourse around digital innovation in our domain. US office Kieran Timberlake, for example, stated clearly already in 2003 that a method based

---

S. Converso (✉)
Dipartimento di Architettura, Università degli Studi Roma Tre, 00153 Rome, Italy
e-mail: stefano.converso@uniroma3.it

L. Pirone
RIMOND Engineering Procurement and Construction Management, Milan, Italy
e-mail: lorenzo.pirone@rimond.com

on assembly, prefabricated components and a custom manufacturing chain had to be applied more consistently, showing it in their work titled "Re-Fabricating Architecture" [1].

BIM software itself, in special projects of that pioneering time, turned out to be mechanical, when projects had to really deal with fabrication. Parts of the model were not just "representative" of the object, but identical to the final object geometry. The notion of "part and assembly", typical of parametric modelling environments, turned out to be crucial in these projects. It was a time when BIM was, in fact, a sort of "younger brother" of these mature, complex environments. As opposed to the "representational" BIM models, these ones immediately suggested a sense of tectonics and materiality. While BIM insisted on classifications and taxonomy on top of objects, the mechanical environment pushed for geometry definition down to every little detail providing an almost tactile feeling. The only abstraction that can be found in these models lies suddenly in the set of relationships that is shown on the screen as a list on top of those "physical" objects. While 3D started in the world of animation and found in the simulation of dynamics and the relationship with movie industry its first start, and BIM somehow started from classifying, from trying to reduce objects to lists, here, in the mechanical models, the objects became somehow true to themselves, even independent, from their final assembly.

In other words, digital environments can provide different interpretations in relation to architecture and construction and therefore contrast the idea of a clear opposition between a physical and a purely digital environment and the related worry, well diffused in the 90s and never really abandoned, that Architecture would somehow dissolve in a virtual domain. Antoine Picon highlighted in his seminal book on the notion of Materiality in architecture more subtle possible interactions and influences that our current intense use of digital environments is providing with a material world that is being hybridized rather than becoming fully virtualized. So, in reply to a question if we are facing a "material turn" of architecture, after a "digital break" he argues that "(…) *This dimension was never forgotten, despite the concerns of theorists and historians like Kenneth Frampton, who assumed that digital tools would jeopardize the special relation between architecture and matter encapsulated in the notion of tectonics. It has simply become more present and, above all, is now interpreted in different terms than it was at other moments in the history of the discipline (…)*".

So mechanical models in that sense, to recall Frampton's studies, are not related to a textile, Semperian interpretation of tectonics but in general to the process of digitizing the material components, in every sense, being it formwork, steel structures, glazing, as well as more subtle still physical parts like material textures, real colours, and every kind of physical manifestation of reality into the design digital synthesis. The fact that these components, like in the manufacturing industries, are mainly prefabricated, is less relevant than it seems, in this context. It is more about the presence of digital objects that have a strict, often direct relationship with reality. A relationship that has changed as much as the tools themselves have changed, in a dynamic balance. As opposed to many approaches that looked for philosophical roots, here the investigation moves on to a more sudden, unspoken, almost instinctual, and direct inspiration.

2 Wireframe Aesthetics and Master Geometry

For many years, in the digital era of 3D modelling for Architecture, wireframe was the only way to go. Lynn argues very well how much this era should not be seen as an anticipation to something more complex to come, but as an era that produced projects that were inspired by the aesthetics, means and culture of that time. Lynn challenges through the exhibition the perception of digital instruments in architecture as something to be located sometime "in the future" and rather focuses on analysing projects. This design approach to software looks at how the features already present in programs at the time were translated by the architects into actual features in buildings and projects. He finally provided a key to link aspects like folds, surfaces, frames of these projects to forms found and explored through software, by breaking the "tool" approach and rather referring to the computer as an instrument, to something that enters the thinking process in the design as pure possible formal inspiration (Fig. 1).

The introduction of Solid modelling seems to have somehow erased that type of aesthetics. That clear, light, and dynamic structure made of "sticks and frames" became massive: objects attached on top of each other or even, more recently "pushed and pulled" out of a unique form. The notion of "bodies" came into the digital environment. Again, Lynn observed the evolution from simple to complex bodies,

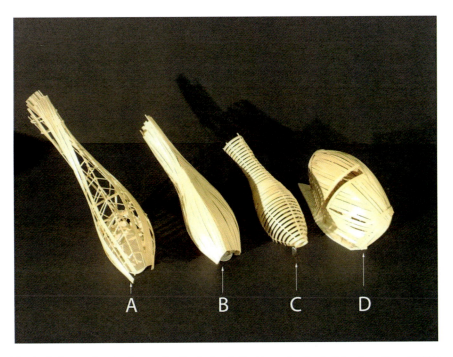

Fig. 1 An image showing four studies for the Lewis Residence, by Frank Gehry from the exhibition "Archaeology of the Digital", held at CCA Montreal. Courtesy of Frank Gehry and Associates; taken from: https://project6rosemary.weebly.com/blog/archives/04-2015

controlled by subdivided mesh or Nurbs geometry [2]. Despite some attempts to recall the idea of structure (skeleton animation, for example) the attention shifted, somehow suddenly, to the "unique body". In general, what used to be a system turned into a single, "one" model. Wireframe remained, though, in the world of mechanical model, as "embedded sketch", hidden in almost every solid, and as "reference geometry": reference planes, axes, "traces" tell the story of the position of objects, of their underlying structure, of the sequence of actions that brought them where they are in the model. The introduction of solid modelling never fully replaced such an approach, since as already stated many parallel modelling techniques survived together.

In this parallel between modelling history and its influence on architecture, emerging tools matter, and two of them somehow broke the existing schema at the end of 90s: Generative Components by Bentley Systems, and a bit later, Grasshopper. Two examples of a way to establish a rule-based modelling technique, based on a sequence of steps and operation: a shift that broke the sense and idea of uniqueness, even though such a procedure was anyway ending with one single model. It's important to remember how the very beginning of Grasshopper, for example, lies in a tool called "Explicit History": a way to save the structure of a file, in the sense of the sequence of commands that defined its geometry generation. A tool that was born as a personal initiative of a young architect and software developer, David Rutten. This self-generation came to define a style, called "generative design". Instead of defining the final form, I define the rules for it. An almost "verbal" approach to design. One of the immediate formal effects of this approach was the emergence of the word "populating": there was a geometrical structure, and then something supposed to fill it with objects. There was a pure and sincere fascination with these procedures: the repetition with variation of a single component became the first attempt produced in these environments. Curvy triangulated and tessellated structures became widespread. In fact, this evolution brought back the idea of structure, and opened the discussion on the so called "Architectural Geometry" [3] discipline. that was raised in publications and in a series of conferences, from the "Smartgeometry to the AAG (Advances in Architectural Geometry) series. Wireframe was back, with a renewed identity of "active structure".

A completed project that could somehow be representative of this period is the Great Court of the British Museum, designed by Foster and Partners, where a clear mathematical model was developed by prof. Chris Williams according to the architectural intention to cover the entire span with a single surface, that was optimized and further optimized using the "relaxation" method [4] (Fig. 2).

At this point another important feature was also back, that is the link between wireframe and set of actual building components, numerically fabricated but as "variable parts". Single beams of variable length, and single nodes, with variable angles in different directions. A practice that fully embraced this approach as its primary identity was one of the pioneering offices of the digital era: the Swiss firm called "DesignToProduction". They worked with many architectural firms to help and finalize design geometry streamed to fabrication. Wireframe for them was the starting point of the process of clarification, rationalization, and full exploitation of the design geometry.

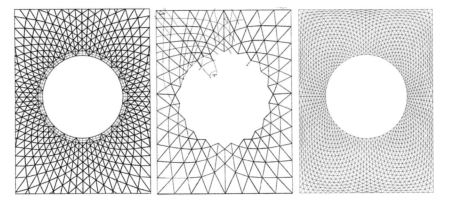

Fig. 2 Sequence of geometrical grid optimization according to structural behaviour, provided by Professor Chris Williams for the British Museum courtyard design by Foster and Partners. Courtesy of Professor Chris Williams taken from the scientific paper: http://www.math.chalmers.se/Math/Grundutb/CTH/mve275/1314/Chris2.pdf

An important step in their strategy towards the definition of a "master geometry" is the case of the fabrication process engineering of the Mercedes Museum, by UN Studio. In that case, it was interesting not even how much the process was technically efficient (even if it was), but how much the story behind it was made explicit, described, embodied and put to the surface by the presence of Arnold Walz, the project "specialist in geometry", and one of the founding partners of the DesignToProduction office. It must be highlighted, again, how much this identity started with the design approach to the project by Van Berkel and Bos, who put the geometrical intertwining of ramps at the core of the space generation. A mentally focused design. Following that experience, wireframe light models managing parts detailed development were applied by DesignToProduction to all other projects, working with all materials, including concrete: the experience of custom CNC manufactured formwork, already tested in the German Museum reached a high scale in the massive production organized for SANAA's EPFL centre, again conceived as a surface, fully walkable.

Their Swiss origin makes evident how much their main material of reference is wood, and its tradition renewed by the possibility to be worked out through numerical machinery. It is not by chance that the works that more relevantly represent an architectural approach to these possibilities come from Japanese architects: the Pompidou Metz, by Shigeru Ban interesting melted the experimentation on a set of wooden net with the form finding design procedure used to generate a roof surface embracing the entire museum space. The construction and material-oriented approach of Ban helps in the tectonic feeling, often missing in many "experimental surfaces", maybe still structurally working, but less dramatically expressed (Fig. 3).

The result is an impressive "wooden tent" that lights up interior space and features a feeling of continuity, despite being composed of many different "variable" parts perfectly joined together. Fabian Scheurer, the second founding member of the Swiss

Fig. 3 Interior of the "wooden tent" designed by Shigeru Ban for Pompidou Metz Museum. The whole structure was controlled and fabricated through a lightweight digital model prepared by designtoproduction. v1—Photo by Didier Boy de la Tour, taken from AREA Magazine online. v2—Courtesy of Jean De Gastines Architects

office, in several publications and public debates lamented later how much the rise of BIM took the lead towards manufacturing-oriented models [5], by somehow missing the full exploitation of the potential behind the detailed control over form definition of the parts. Even if perfectly defined, though, the parts they had to engineer for many projects were still "formally servant" to the whole. A further step in this formal exploration requires architects to explore a less strict relationship between parts and their "reference geometry": the project itself becomes expressed as a kit of parts.

3 Project as a Kit of Parts

So, a shift can be imagined if the "model structure" can move towards becoming an architectural and aesthetical guide for a project, a building, or a pavilion.

In this case, the project can be seen as "expressed by its parts": yes, the uniqueness and the quality of the whole is still there, but it comes from a tension with the individual parts. While some of the early exercises of the so called "parametric surfaces" always left the substantial aesthetical leadership to the whole (it's full of examples), the parts became increasingly important, with their shape and in some cases materiality in many different projects over time. Examples of this logic can be found in the pavilions designed and built for the P.S. 1 original YAP competition in New York. Projects, despite many differences, featured very often an interesting tension between an overall shape and parts that were original, customized, and present in their shape and materiality somehow independently of the whole. Or, to better state it, that fully participates and contributes to the project identity (Fig. 4).

Of course, the best projects showed a balanced relationship between the perception of the whole and the emergence of parts, while somehow blurring the difference and the separation between the two. Examples of the most successful projects range from the first SHoP's proposal to the Public Farm n.1, by Work. AC Ny based office. In both projects, but most prominently in the latter, the tension between a whole and

Fig. 4 Two snapshots of the space generated by the installation called "Public Farm n.1", winning proposal of YAP competition in New York's MoMA Queens branch. Courtesy of Work.AC Architects

the parts became interesting: while the approach to the courtyard led as well to an urban approach, down to different scales of space (more monumental, intimated and shaded, open, and finally the roof), approaching the structure made also possible to appreciate the quality of the objects: tubular cardboard structures, sectioned in different manners and assemblies, hosting plants, lights, or just empty shaders. The possibility of having objects emerge with their singularity is helped in this case a lot by the typology of pavilions and of open shelters, where there is no need to achieve certain performance issues, such as the thermal envelope, just to mention the most constraining one. An interesting approach came thanks to the research of Kengo Kuma, who paid since its first projects an attention towards the balanced composition of parts, and over time shifted the attention towards a formal prominence to the part, and its aggregation logic, in a balance with the whole that evolved and still changed from project to project over time. Kuma's move to wood helped this path and reached an interesting point for this text with the cake shop SunnyHills in Tokyo where the notion of envelope was challenged formally by the stacked wood mountain that apparently took to formal leadership on the building behind it (Fig. 5).

In this project interiors are quite rough; Kuma was still exploring the expressive potential of such a tectonic mechanism, but he avoided the effect of "surfacing" by bringing the tectonic wood assembly to participate in the structural behaviour of the building. So, a three-dimensional effect is perceived and adds a layer of complexity and mystery to the building's overall perception that remains urban. Kuma's provocative approach in the cake shop is part of a strategy of investigation of the aggregation logic, which in this case is rooted in a search for a return to Japanese tradition. Wood is taken seriously as construction material for a contemporary architecture closer to nature, and the research moves on from the construction logic, based on local

Fig. 5 Exterior view of Kengo Kuma's SunnyHills cake shop in Tokyo, Japan, 2013. Courtesy of Design Gallerist, Blog section https://www.designgallerist.com/blog/sunny-hills-kengo-kuma/

studies on component shapes and connections, that are investigated Stacked components provide a sense of freedom: they are free enough to allow for some formal uncertainty, even if they are precisely joined and prepared exactly like they were in traditional construction. The key is in the balance between the identity of the part and the whole, in its "controlled fracture", that comes from the rebellion to a strict established relationship between a building and its components.

At a different scale, the will to break up" the established relationship between parts and a whole in a context like housing has been the target of the approach of the first projects by the Danish office of the former OMA lead architect Bjarke Ingels. In that sense BIG's design approach can be seen under the logic of the tension between "prominent" shapes and cellular aggregation, where the whole takes the lead. But the starting move has been regaining to the part, in this case the residential cell, an autonomy, a sort of formal independence. The project approach shares the attempt

Designing with the Chain

Fig. 6 Exterior view of the dwelling complex called "The Mountain", designed by Bjarke Ingels Group (BIG) office, with JDS Architects, 2008. Courtesy of City of Copenhagen, photo by Ty Stange

to "play with the program" with other Dutch offices such as MVRDV or Neutelings, but BIG's approach reaches a level of geometrical purity and abstraction that will be later revealed in the Serpentine Pavilion that Ingels designed as a stacked series of empty metallic cells (Fig. 6).

The further, natural step of such a play between the poles of whole and parts happens when the relationship extends to larger structures and the form tries to reach and play with the idea of networks that are by definition "open".

4 From Parts to Networks

The notion of the tension between a whole and its parts, the design of a "structure" for a flexible set of components is a paradigm that can be easily found in contemporary design culture outside the specificity of Architecture. In many domains, networks and aggregations happen at different levels: from web portals to energy communities, the sense of a macro-structure hosting multiple contributions became a true means of expression. This is at the same time an organizational and a formal principle: it is practically allowed and pushed by digital networks, as well as by the growth of travel capacity worldwide.

How can architects and designers deal with these networks?

Can the project itself be seen as a structure that is related to its parts, without losing its architectural sense? In fact, the network of specialists that increases at many levels could subtract "design power" to the architectural work. They can be consultants or contractors, or even a mix between the two. In his constant reflection on modernity, and on the conditions of contemporary architectural practice, Koolhaas expressed such condition in the famous OMA/AMO diagram, expressing in two directions the extension of contemporary architectural practice as having to deal and manage a cloud of possible contributions and inputs (Fig. 7).

The type of interpretation of such a paradigm can be in two opposite ways: for some designers, the complexity drives towards a pure disciplinary approach, where architects become just a node of the network. On the opposite side, there is the chance, often recalled by several authors, of architects, designers embracing digital means to shorten and blur recurrent boundaries and bring back to themselves the power over the control of production, being its construction, in architecture.

Some design practices run over a "renewed crafting" approach. The work of Studio Mumbai is innovative despite being non-digital since it innovates the design-build

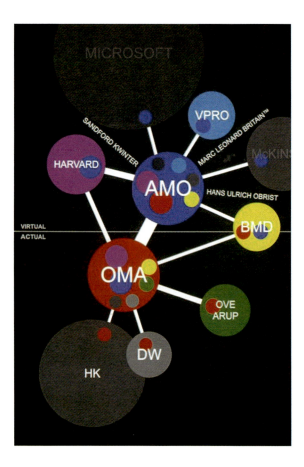

Fig. 7 The "OMA/AMO diagram, showing the network of relationships and interaction of Koolhaas lead architectural practice. Courtesy of AMO

Fig. 8 An image of the Palmyra House, the home built as a twin of pure boxes, animated by a variable series of components, that modulate light, opaqueness, enclosure and melting with the surrounding palm tree forest. Courtesy of Aga Khan Award for Architecture, photo by Rajesh Vora

chain through a day-to-day link to craftsmanship, in their case. The beautiful images of the office courtyard in the woods show a continuity between design, object, and part production, where somehow the distinction between the whole and the part is blurred in a diffused sense of materiality. This tactile feeling is so intimately connected to the office architectural expression that when asked to design an installation for the V&A Museum in London, Studio Mumbai built every single part in their own space, to deliver, then, everything to the UK (Fig. 8).

The relevance of the part in the work of Studio Mumbai, was shown by their first appearance at Venice Biennale, where they exhibited their office, made of tools, materials, and prototyped project parts. It also was evident in their Palmyra House, that melted with the surrounding forest of palm trees as an "open frame" populated of details, sunscreens and penetrated by wind and filtered natural light.

In these contexts, the management is no longer a cold activity, it can truly become a source of sense. If, on one hand, it is true that the amount of information to manage extends, and so does the responsibility, on the other hand it is also true that all these cases show that merging management and design, and eventually also formal exploration with construction prototyping can become a true space of expression.

This paper chooses as case study the work of RIMOND, a newborn company that was founded in 2015 with the somehow crazy intention to position itself in the gaps of the AEC process, where networks normally fail. This mainly happens in the biggest fracture, the one between Design and Construction, that is why the

Design-Build soon became its main field of interest and approach. The Gap, then is identified exactly in the same position that DesignToProduction occupied with its "bridging" coding and geometry activity. In this case, though, even if the company later evolved into a group, with a dedicated division for Construction services, the practice maintained a transversal collaboration, with people moving fluidly between design and field activity, developing a workflow based on negotiation and constant interaction with a network of partners.

This management role would normally be abstract, based on time, deliverable and cost approach. The choice of RIMOND was to strictly link such management with form development, with geometry, parts and design control. A pure manufacturing management model, brought into an AEC practice. This approach applied to constantly changing collaborations, where design can be either developed internally or in collaboration: what matters is to get inspiration from circulation of information and guarantee a steering approach. It turns out as an "Open" Design-Build approach.

The case of Al Wasl plaza became the framework to test and deploy a method to manage and generate design geometry in combination with physical prototyping, all under a "technical management" role.

5 The Al Wasl Plaza Dome: Managing a Network of Components

The competition design, by American office Adrian Smith and Gordon Gill, featured the central plaza of Dubai 2020 Expo as a gigantic dome, set up as a 360-projection surface, through a set of dedicated projector "pods" located all around the structure at a height of 20 m. The structure, being more a pavilion than a real "building", helped the process being it separable into few, clear packages: the concrete base, the prefabricated steel tubular structure, recalling the geometry of Expo 2020 logo, the textile projection surfaces, the two maintenance gantries and the huge projection and speaker pods. RIMOND approach was to establish since the very beginning a common ground of all parts: a wireframe geometry was created as "hanging centreline" to host and position exactly the mentioned packages. The first step of the process was the geometry adaptation to follow manufacturing optimization of steel components, in constant talks with fabrication specialists working directly from the company headquarters. The process led to apparently light changes that drove a huge simplification of welding profiles. During these trials the wireframe model allowed dynamic and efficient exchange with structural engineers of Thornton Tomasetti and physical visualization of structure with lead architects AS+GG. The ability to understand everyone's technical feelings was crucial to such a role. No exchange would have been possible without such a human design factor (Fig. 9).

The main model, then, once optimized in these first rounds, started to receive additional "hanging nodes", to host secondary structures, from inner MEP conduits to connections for outer components such as CCTV Cameras, LED lighting, shading

Designing with the Chain

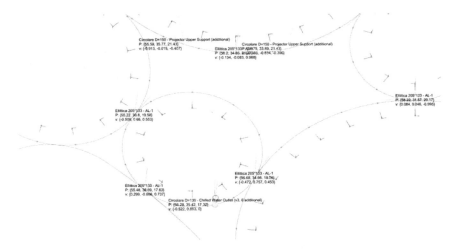

Fig. 9 A screenshot of the "coded centreline model" of the Al Wasl coordination model. Courtesy of RIMOND, created by Lorenzo Pirone

fabric panels' brackets, all custom-designed to parametrically adapt to the changing angles and orientation.

This work was conducted indifferently to components defined and manufactured internally, such as the brackets, and to components coming from the network of subcontractors, mainly using an interchange with Manufacturing Models. The interaction with Mexican Manufacturers of Kinetica for the Speaker and Projection Pods was a very good example in that sense. Two digitally oriented manufacturers were able to act in a "build before it's built" manner, simulating final configuration and negotiating the points of connections (Fig. 10).

This project was driven by constant interaction, on top of a dynamic integrated but light model, based on clear design principles. The experience can be interpreted as the seed of a possible creative method, in a design-build framework, where the population of components from partners can influence the design development process as much as the process itself could constantly send them updated information to take into account.

Most importantly the population this way gets the flavour of construction and the feeling of tectonics into the digital environment, one of the effects of the use of mechanical models into the digital environment, that gets more underrated due to the "BIM abstraction". As indicated by Fabian Scheurer in a recent article commenting on the lack of fabrication models in the BIM evolution, if you remove from the model connections, detailing, often you might lose the true sense of construction logic and therefore of tectonics, in the architectural simulation.

At the same time, this experience showed, once again, how much the digital model cannot guarantee any automation in exchange activity alone: breaking certain barriers means an organizational change and fully embracing manufacturing (and its engineering) as a dynamic link to design, it's a matter of responsibility.

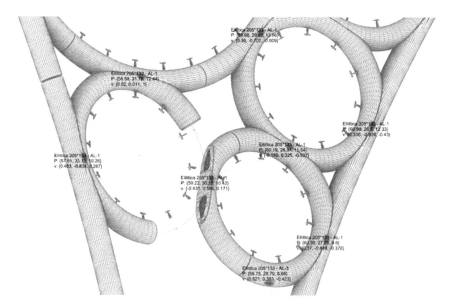

Fig. 10 A screenshot of the population of constructive components from external sources on top of the "coded centreline model" of the Al Wasl Plaza dome. Courtesy of RIMOND, created by Lorenzo Pirone

Even if there is a technical need behind the shared detail definition, the simple possibility to directly use fabrication components into architectural models and negotiate their position, when not their shape, opens a design space. How about understanding, since the very beginning, the feel of tectonics in the architectural definition?

6 Beyond Design Management, Towards Design Freedom

The mentioned seed of Al Wasl Process recalls the possibility to overcome the "Design Management" and the "value engineering" approaches, both rooted in a mentality that somehow imagines a freezed design and a rigid set of phases. A new Design Space can on the contrary emerge into a continuously evolving process where construction definition can become imaginative and get an active role in the architectural development. Italy can look at its post-WW2 design tradition, when the position of delay in modernization linked to industrial mass production produced a generation of brilliant architects dealing with the potential of craft and its use in defining custom architectural detailing and original tectonic solutions, thinking of people like Ponti, Albini, or Nervi and Morandi but an interesting moment came later, when the crisis of modernism became a source of avant-garde inspiration. The season of the so-called Radical Architecture, that challenged the disciplinary boundary and the relationship between architecture and the urban environment also made its main actors become designers of objects, parts, and independent components.

Archizoom office expressed its positions in their research on the No Stop City, where their drawings were showing intentionally unbounded fields in which objects, parts could freely develop inside a set of geometrical and spatial set of references. In this research, building or even abstracting the frame of the "reference geometry"—to recall a previous introduced definition—is aimed at generating a freedom of expression of the individuals, and focusing on their own immediate environment, hosted by a continuous space described and drawn in analogy to the artificial, almost infinite space of supermarkets and factories (Fig. 11).

In his reference to the issue of Bigness in Architecture, Koolhaas evidenced the typological built consequence of what Archizoom foresaw, linking the dissolution of the perception of the envelope to the rise of artificial lighting and air conditioning, and attempted to deal with those issues. An interesting set of projects to analyze in that sense are a series of pre-pandemic projects for their Headquarters done by the big players of the Digital Scene. These projects came when these big players had just functional and anonymous designed buildings and came to deal with representing a new model of work, but also with the aim to represent their values at large, first to their employees and then to the general public as well. Significantly, Facebook selected Gehry for its new Campus but imposed to him its vision of a "Warehouse" (Fig. 12).

The relationship with such a client forced somehow Gehry back to his L. A. origins [6] when working on a local, bottom-up language made out of composition of parts that become architectural episodes, recalling the urban idea in their random sequence. The same issue of scale was faced and proposed by Google when looking at firms to design their own new location. The company has been in talk for a while with ShOP Architects and later ended with a design by BIG and Thomas Heatherwick. Google campus moved by a similar attempt to generate a "bottom up" space, made of parts, and establishing an almost urban inner condition, that strongly recalls Archizoom's radical space images. While SHoP's attempts in the Charleston East R&D extension focused on a dynamic layout, keener on the overall logic of pedestrian fluid connection through ramps shaping the building shape itself, the final design represented a move towards space continuity in every direction.

In BIG's and Heaterwick's design the framework, prepared to allow such freedom, is neither invisible nor cold. A brilliant and dramatic, tent-shaped roof covers a continuous array of spaces spanned in just two levels and organized by courtyards. Each courtyard hosts a vertical support reaching one edge of the hanging petals of the roof. This interconnection between elements, though, seems not to reach its full potential. The overall shape given to the roof to makes it appear in the landscape as a "giant Bedouin tent", forces its relationship with its bottom component parts, the support, and the courts. The whole in that sense takes the lead and the vision of the roof interior appears randomly from the bottom landscape made of slightly different levels. Its majesty can almost never be appreciated from inside as it is from far, but it fails as well to become "infinite but intimate" (Fig. 13).

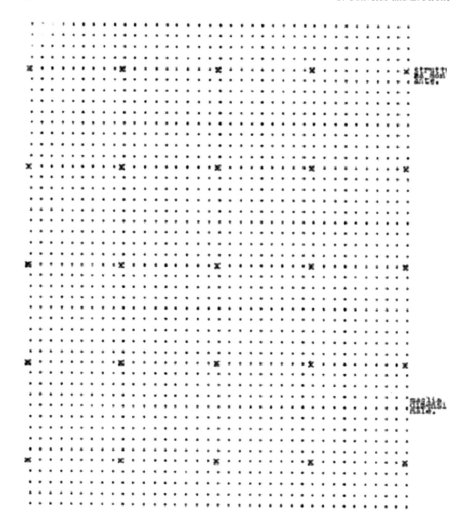

Fig. 11 An extract of one of the most abstract attempts of No-Stop City project by Archizoom Associati office, 1970–1971. Courtesy of FOG Associates

The space has an undoubtful attempt for an open form, they could somehow find a possible variation in the interaction between petals and freely aggregated space beneath them. The architecture uses a language of extreme lightness in the light columns supporting an almost tensile and textile surface. A formal language that could remind the relationship between the light columns and the ceramic light vaults of Labrouste's Bibliothéque Nationale in Paris if you look at the impressive images of the empty space. When filled up, though, most of the magic gets lost in the relationship between the hanging roof and the chaotic space beneath. Despite a very clear logic in the search for a relationship between courtyards and the columns, the

Designing with the Chain 83

Fig. 12 Plan of Facebook campus in Menlo Park, California, designed by Gehry Partners. Courtesy of Google

Fig. 13 Interior view of Google's New Headquarters, designed by BIG and Heatherwick Studio. The photo is taken in the main building. Courtesy of EU Mies Award

overall effect ends up in separating the roof and the bottom world. This is probably the result of the lack of a strategy to allow a full openness of the layout, since the project forces the part free development to be packed to achieve the powerful image of the tent, mostly appreciated from outside, and undoubtedly powerful. This causes, though a strange effect in the campus: since the space was not falling all into one building, two more are added, smaller, and the overall image is composed by three, separate objects of different sizes generating a residual space in between and missing the continuity provided, for example by a similar attempt like Frei Otto's Olympic campus in Munich in 1970s with free roofs covering indifferently open space in the park, pavilions, and the stadium itself. Somehow, the whole campus misses the opportunity to become fully urban, recalling the problem, that, again Archizoom's vision fully anticipated in the No Stop City: "(…) *the exterior image of these organisms does not exist: the facade does not constitute the linguistic structure of the building, therefore it does not make explicit the functions happening inside. With the No-Stop City, the city becomes a continuous residential structure, that has no voids, and therefore no architectural images: big furnished plans, theoretically infinite, interiors illuminated artificially and micro-conditioned. Inside such a space it is possible to organize new dwelling typologies, open and continuous, for new forms of communities. The No-Stop City is a project for an amoral city, with no quality. Inside its big furnished plans, the individual can finally realize his own habitat as a finally set free, fully creative activity.*" Google search engine introduced in a very structured market a new approach user-focused, and flexible and dynamic down to the extreme of complete hiding and putting in background its supporting structure, to just show a blank starting page focusing and taking shape around each single user and its tailored needs. It represented, in digital terms, the closest concept to an open environment, with no "predefined" paths or shapes in data structures. Its headquarter though has a stronger overall image than a full feeling of space openness.

This, probably, due to the private and bounded feature of these spaces. When there is no public space, public access, there can't be a full urban effect, even in spaces structured around an interior path that allows free wandering and almost urban episodes, like in Gehry's quite successful interaction with Facebook. A company, though, that in digital terms, in comparison with Google's gully open approach represented for many people a step back, towards a more visible graphical and formally defined "framework" that encapsulates its content: Facebook is "first" Facebook and then its users. The third pole, in this balance between digital identity and its architectural representation is the manifestation of closed, controlled environment and form, with no ambition to open itself, provide independence to its parts and neither search or simulate an urban simulation is the absolute finiteness, and controlled closed form, embodied by the circular form of the "fortress" that Foster and Partners designed for the Apple Headquarters, designed by Foster and Partners, where no space is left to emergence of individual components neither in constructive neither in the layout attempt to foster visually the idea of collaboration and informal meetings. In that case, it certainly matches physical and digital approach clearly on a top-down approach.

The web itself, that was the stage for these companies to grow as global players, seems to develop in its next evolution more distributed, bottom-up structures. Dynamics like the so called web3, or the blockchain, but also the previous examples of peer-to-peer systems like Napster, BitTorrent and their heirs, are back on revealing the fundamental network-oriented nature of the global technical interaction, that as we have discussed opens practical possibilities of new work methodologies but also generate new sensitivities, or new materialities, as Antoine Picon brilliant definition suggests.

In that sense another way, probably most promising, is to find a dynamic and lightweight approach to form evolution, that allows some freedom in materiality definition. Can the structure of a design allow some freedom in its material definition?

Can some parts receive changes over time, inspire materiality without losing a sense of overall definition of the architecture? The emergence of a "rough" approach can be felt as following this procedure and authors suggested a possible future in such a "rough and provisional, but precisely built architecture [7] (Fig. 14).

One of the contemporary researchers that, although probably zero digitally focused, mostly succeeded in looking for freedom of expression in constructive terms can be found in the Parisian office of Lacaton and Vassal. Their work started simultaneously with the search of architectural components that could in their case make them exit the already established building process and associated architectural language. Anne Lacaton reminds in an interview their discovery of the technology of greenhouses as a possible exit strategy to gain control over the link between architecture and its materiality, breaking up the established process, their prices and get new space for their clients. What makes them fully part of the attempt of most of the digital research of the origins, somehow lost in the evolution of the technology into a commodity, is the revolutionary character, of its possibility to challenge traditional habits and gain back power to architects. An attempt to use technology for a political target, meaning with politics a change in the power balance, a play with it, in architectural terms. In fact, they achieved in a consistent and coherent research path,

Fig. 14 Interior view of Palais de Tokyo, the operation of "reduction to frame" conducted by Lacaton and Vassal in their design for the new museum of contemporary arts in the former nineteenth century building. Courtesy of EU Mies Award

Fig. 15 Exterior view of Lacaton and Vassal "Manifesto City", in Mulhouse, France, 2014

many of the targets that can be shared with the digital pioneers. Another, significant one of them is performance, or to better state it a "performative approach" to design: they worked to achieve more square meters at the same cost, and to achieve a mostly passive energy behaviour in indoor climate control (Fig. 15).

This "reduction" to action can be seen much closer to a "Google approach" than many pioneers of the digital revolution tried to achieve. In Lacaton and Vassal architecture, there is a strong presence of what can be defined as a "built framework", that seems to encounter the request of people for spaces of freedom, of regained freedom, being it gaining more space, like in the PLUS projects or more freedom of expression, movement, like in their provocative but meaningful intervention at Palais de Tokyo in Paris. They often mention how much, for them, the possibility to manage a technology, to force the process was crucial to the success of their attempt. "It was not possible to achieve under constraints and costs of established architecture".

But significantly, they started their feeling and their reflection in Africa, with projects starting with simple moves. finding out that moving a chair in a shade or setting up a minimal meaningful environment with few simple moves interpreting local available things, somehow brings us back to the essence of being architects. An essence that is at the same time political, fully physical, and immensely rich, that keeps on existing even when complexity, number of parts, procurement chain and actors become richer, larger, and stronger.

Acknowledgements This article is one of the outcomes of the collaboration between Università degli Studi Roma Tre—Department of Architecture, and the R&D Division of RIMOND. The topics of this article are shared among the authors. Converso wrote paragraphs 1-2-3-6-7 and Pirone worked on 4 and 5.

References

1. Kieran, S., Timberlake, J.: Refabricating Architecture: How Manufacturing Methodologies are Poised to Transform Building Construction. McGraw Hill, New York (2003)
2. Lynn, G.: Folds, Bodies & Blobs: Collected Essays. Books-by-architects), La Lettre volée, Paris (1998)
3. Pottmann, H., Eigensatz, M., Vaxman, A., Wallner, J.: Architectural geometry. Comput. Graph. **47**, 145–164 (2015)
4. Williams, C.J.K.: The analytic and numerical definition of the geometry of the British Museum Great Court Roof. In: Burry, M., Datta, S., Dawson, A., Rollo, A.J. (eds.) Mathematics & Design 2001, pp. 434–440. Deakin University, Geelong, Victoria, Australia (2001)
5. Symposium "DigitalFUTURES Talks: Material agency in wood architecture & BIM to Fabrication?" with Fabian Scheurer (ETH Zurich), Martin Self (AA and Xylotek), moderated by Philip F. Yuan (Tonji University). https://www.youtube.com/watch?v=S66RBJIFWAg. Accessed 10 Jan 2023
6. Zardini, M.: (Ed.): Frank O. Gehry: America as Context: No. 20. (Lotus Q S.), Electa Publishers, Milan (1995)
7. Hopkins, O.: Post-digital architecture will be rough, provisional and crafted by robots, DEZEEN, digital edition, 12 December 2018. https://www.dezeen.com/2018/12/12/post-digital-architecture-owen-hopkins-opinion/. Accessed 10 Jan 2023

Adauctus Architectus Novus on the Definition of a New Professional Figure

Giuseppe Fallacara, Francesco Terlizzi, and Aurora Scattaglia

Abstract This contribution proposes a reflection on the evolution of the professional figure of the architect thanks to the implementation of technologies brought by the fourth industrial revolution. Technology has continually transformed the architect, expanding his skills and potential through digital fabrication tools: these allow the figure of the architect to be elevated from a simple designer to a complete realizer of the work. The contribution is divided into three parts: the first investigates, in a critical key, how technology has been a constant that has assisted human evolution in its various phases, modifying, changing, evolving and, perhaps in some ways, dis-evolving man; the second part proposes to present the definition of the new professional figure of the architect, fruit of the progress brought about by technological evolution through the application of digital fabrication tools in the management of the architectural process; furthermore, among the new tools that enhance the figure of the architect, we can now register the impact of artificial intelligence. This can represent a new starting point in the reflection of the design process, making an essential contribution to what comes before the project itself, that is, the creative phase.

Keywords Technological innovation · Digital Fabrication · Artificial Intelligence · Human–machine interaction · Generative design

United Nations' Sustainable Development Goals 4. Quality education · 9. Industry, innovation and infrastructure · 11. Sustainable cities and communities

---

G. Fallacara (✉) · A. Scattaglia
Department of Architecture, Construction and Design, Politecnico Di Bari, 70124 Bari, Italy
e-mail: giuseppe.fallacara@poliba.it

A. Scattaglia
e-mail: a.scattaglia1@studenti.poliba.it

F. Terlizzi
Higher Education Institute 'A. Badoni', Lecco, Italy
e-mail: francesco.terlizzi@istruzione.it

© The Author(s), under exclusive license to Springer Nature Switzerland AG 2024
M. Barberio et al. (eds.), *Architecture and Design for Industry 4.0*, Lecture Notes in Mechanical Engineering, https://doi.org/10.1007/978-3-031-36922-3_6

1 Introduction

We are experiencing a new evolutionary phase, perhaps revolutionary, among the many that have characterised the evolution of humanity: it is a process that stems from the fourth industrial revolution, and it's leading to fully automated and interconnected industrial production, in which products and means of production become networked and can "communicate," enabling new methods of production, value creation, and real-time optimization. The interaction between man and machine, which is diminished in the more authentic concept of augmented reality and into the transition from "digital" to "real" through additive manufacturing, 3D printing, robotics, communications, and machine-to-machine interactions, is certainly one of the directions of development on which Industry 4.0 products are having a profound impact. The link between humans and machines is marked by intricate reciprocal impacts. Following classical cause-and-effect reasoning, technology cannot be considered only a product of man. It feeds back on the man in an advanced form of circular causality. A circular interaction between man and technology requires one to look for the evolution of man in the complex interactions between biological evolution (Darwinian), socio-cultural evolution (Lamarckian), and development or technological evolution. For this reason, we can speak of human–machine co-evolution, identifying biological, socio-cultural and technological evolutions as interacting factors.

Technology has become an environment to live in, an extension of the human mind, a world that is intertwined with the real world, impacting and being influenced by how we design and build spaces. The introduction of digital technologies and media into architecture has sparked a design revolution that has resulted in substantial advancements in the discipline. Particularly, digital fabrication has enabled architects to overcome the gap between representation and construction, allowing for a seamless connection between design and production. The introduction of new technologies has forced a redefinition of the architect's professional identity. As a result, it is essential to evaluate the role of technology innovation in architectural design and implementation, as well as how it impacts how architects approach their profession.

The employment of digital fabrication and material techniques to achieve a balance between virtual models and actual artefacts is a topic of critical importance and urgency in this setting. As discussed by Robin Evans in "Translations from Drawing to Building" [1], the gap between drawing, the standard design medium, and building, the result of an architect's work, is the void in which significant innovation occurs. In architecture, digital approaches such as digital fabrication and generative processes can facilitate a seamless connection between design and manufacturing, allowing for better efficiency and innovation.

Against this backdrop, the reflection proposed in this paper is strongly related to the scope of this book, which seeks to systematize and analyze the greatest contributions and experiences in the professional, academic, and research sectors of architecture and design based on this new design paradigm. It has already been proven that the tools that have come with technological progress have changed the profession and made it necessary to work under a new definition of the professional figure, related to the potential and difficulties of the post-digital era.

2 From *Homo Sapiens* to *Homo Technologicus*. Reflections on the Cognitive, Emotional, and Social Consequences of the Pervasiveness of Technology.

"It's a turp of decadence!" [...] "Who cares about the Greeks and the Latins, good at most to furnish some roots to the words of modern science!"—exclaimed the old professor Richelot, teacher of classical languages, the protagonist of a posthumous narrative composition by J. Verne "Paris in the twentieth century". Written in 1863 and set in 1960, the novel describes the French capital as a hyper-technological metropolis where the invasion of machines has reduced human capital and the triumph of capitalism, of technical and utilitarian culture, has generated a grandiose, productive, and winning system, in which art, reduced to mere entertainment, no longer counts for anything, except as a ridiculous damsel in the service of science, with its lyrics to progress.

The paradox of the story lies in the fact that it is a futuristic fantasy that, with extraordinary foresight, in the mid-nineteenth century, predicts a future that is not the radiant one of the progresses promised by the sciences but is instead the gloomy one of the deaths of the humanistic dimension of human consciousness. Verne's distrust of machines and technology, which transpires with prophetic lucidity in this sort of visionary novel, made reflect on what the logician and the philosopher of language, Ermanno Bencivenga, considers to be the threat that is most undermining our time: the ability to reason which risks to disappears. For Bencivenga, this 'gentle catastrophe', as silent as it is devastating, is gripping the new generations, especially those who are more exposed to the frenzied proliferation of technological tools, as well as of information and communication media, which have now become too fast and powerful compared to the time that logical thinking requires. The disturbing result is that young people get increasingly used to the idea that someone else, or something else, will reason for them.

During our evolution, we have learned to speak, read, and write. According to anthropologist Arnold Gehlen and many others, technology has always been the means man uses to make up for his physical and mental deficiencies and therefore it can be considered as the extension of our senses [2]. In other words, technology constitutes for man an extension of his body, of his physical and psychic abilities; it is a prosthesis that allows the creation of augmented reality. For example, the hammer extends our hand for strength; the car extends our foot for speed; the mobile phone extends our ears and mouth to increase our communication skills. *Homo technologicus* has therefore grown up using digital technologies since birth. Among the latter, the remote control, the mouse, and the mobile phone: these tools, in fact, lead to different management of all information: today's children are skilled in controlling information flows, dealing with its over-availability, selecting it appropriately and according to their needs. Their non-linear behaviour, the control of information, the knowledge of how to navigate through information efficiently and effectively, how to communicate, and how to effectively build a network of peers lead to the development of crucial skills for a chaotic and creative society.

The paradigm of how technological science is changing us as people and how it is shaping our minds is represented by social media (Facebook, Pinterest, YouTube, Twitter, etc.): "social networks" that forge in their image and likeness an orthodox way of thinking and acting which, together with the automatic prompts of cell phone keypads, are drastically thinning our vocabulary [3]. Sigmund Freud argued that "it is impossible to know men without knowing the power of words" and "if words become thinner", as the psychiatrist Paolo Crepet points out, "they simplify, become ugly and horrible neologisms; if our emotions become small icons coloured on a mobile phone screen, even the bricks of our unconscious will inevitably become sand" [4].

Homo sapiens, with its rationality, which has always been its pride, has therefore transformed itself, since its appearance on planet Earth, into *Homo technologicus*, given the close, almost symbiotic relationship that has linked it to the various technologies from he discovered and built, and with which it can be said that he co-evolved.

However, it is necessary to "start thinking again" because we are all participating in the definition of a new evolutionary matrix that deprives the new generations of what they need more than anything else: the ability to think.

Nevertheless, what is meant by thinking? It derives from the Latin verb *"pèndere"*, which means: to weigh, to estimate; alluding to the act of evaluating things with the intellect; thinking is reasoning, it is arguing, or "giving an account of what one is saying". To think, after all, is to preserve one's dignity as a human being because, without thought, there is no conscience.

In today's digital, globalized and connected world, we must preserve this ability in order not to become software, we must regain the ability to process thought, open up to the slow spirit of our brain and induce the new generations by educating them to focus attention on something other than a display in order to avoid—as caustically noted by the anthropologist Niola—inadvertently making them pass from the age of the "*Cogito ergo sum*" to the age of the "*Digito ergo sum*".

3 *Novus Architectus Adaucto* and "Adjacent Possible" of New Stone Architecture.

In the age of digital transformation that is characterizing these decades, changing entire industries and classes of professionals, the figure of the architect is the one who, thanks to the digital revolution, has undergone and will continue to undergo significant transformations.

It is now unthinkable to design structures without the help of computers. They are used throughout the whole of the architectural process, from conceptual design to actual building. The digital processes used by architects and building consultants include three-dimensional modelling and visualisation, generative form discovery, programmed modulation systems, structural and thermal evaluations, project management and coordination, and file-to-factory manufacturing. Digital Fabrication deals not only with additive technologies, which are only one of many pieces that make up a much more complex picture. In particular, it includes all those technologies that allow moving from a simple idea to its realization. These also include subtractive technologies—milling machines, numerical control machines, mechanical cutting and laser cutting—and all the technical knowledge related to the digital world necessary to juggle such machinery properly.

The combination of past and current technologies knowledge directs architecture towards an informed approach to ensure maximum performance, whether energetic or structural, but also aesthetic and cultural, prefiguring new paradigms for buildings, especially those in stone. The stone, compared to the modern materials, is considered difficult to submit both for the product design and for contemporary architecture. In addition, it is considered to represent an old material, or rather strongly connected to the forms of the past and tradition. On the contrary, it is known that, for a designer, there are no ancient or modern materials, but it is the way the material is transformed, worked, and shaped that makes it new and innovative rather than old and decrepit.

Certainly, the renaissance of the using stone nowadays is related to the evolution of digital fabrication in the dual aspect of both technological evolution of the parametric and generative three-dimensional modelling software and thanks to the robotics and numerical control machine tools applied to stone manufacturing. In the project-product process, a new and unprecedented direct relationship between designers and final product is established thanks to robotic production of stone components. The contemporary designer is a "*Novus architect adaucto*" or an "expanded designer" in the sense that he possesses new robotic arms which allow him to cut and shape the stone according to his direct requirements without any external mediation. In this perspective, the famous "dress architect" (the role of the architecte—*habit de l'architecte*) changes both because of the new tools at its disposal and of the new forms that he can produce.

In this new perspective, the role of the "fabricator", the person who creates the work, disappears. His role interposes between the author and the final work, but the designer becomes the maker of his work thanks to his new robotic arms that allows him to cut and shape the stone and to mount and assembly the masterpiece. The direct control of the making of the work by the author is the first step to redefine a new profession, paradoxically similar to the architect—master of the past.

Although there is great excitement, shown by the growing interest on the part of professionals in understanding how it is possible to integrate these technologies within their cultural background, the limits due to the construction process have not yet been completely overcome.

In this new scenario, the new architect of the future is the author and creator of his own work or at least the one who ensures the final work the most suitable correspondence with the original idea of project without mediation and/or constructive interpretations by others. The new designer takes on new critical skills, goes back to the real construction and assumes new ethical and civil responsibilities of his job. Thanks to the development and dissemination of digital knowledge related to manufacturing (FabLab—Fabrication Laboratory), this process of redefining the role of the designer could occur at any scale, from architecture to design product, and at any level of business activity, from big industries to the small workshop of the province.

Now, it is an incessant cultural process, not only within the stone world, which is based exclusively on a new culture of doing and of the strong contact with the materials, of learning and supporting the process concerning the incubation and birth of ideas. The latter is the most delicate and important aspect which is a key point for the reformulation of university programs and teaching methods.

The lithic prototypes come to life under the influence of research and innovation aimed to expand the "adjacent possible" of stone, according to what Steven Johnson argues in "Where Good Ideas Come From: The Natural History of Innovation". The adjacent possible, according to the author's statements, is a "kind of shadow future, hovering on the edges of the present state of things, a map of all the ways in which the present can reinvent itself". That is identify for us new possibilities and potentiality of stone even with the risk to exceed in the morphological/structural boldness, absorbing the lessons of constructive-technological areas, including external contributions not strictly related to the logic of stone. As Steven notes, the adjacent possible "captures both the limits and the creative potential of change and innovation."

Johnson identifies seven creativity "models", through which it is possible to search for innovation. Some of them can be identified within the designer intellectual work:

Slow intuition, which is preserved and stored in mind for a long time before it is shaped by the lightning of the immediate intuition.

Serendipity, which is a neologism indicating the feeling you get when you discover something unsought and unexpected while you are looking for another.

Serendipity is not only a feeling, but it also indicates the typical element of scientific research when important discoveries have been made while looking for something else. The term **exaptation** was coined in 1971 by the evolutionary biologists Stephen Jay Gould and Elizabeth S. Vrba: an organism develops a trait optimized for a specific use, but then that trait is redirected to a completely different function. A classic example are the feathers of birds, which initially appeared in dinosaurs to regulate body temperature and later evolved into flying instruments: a tool created by the evolutionary needs for a given purpose reveals unexpected utility for another purpose. A pen or feather adapted to warm up is "exaptated" for the flight.

According to Johnson, intuitions arise slowly over the time, but they materialize rapidly thanks to the composition of the last tile to finish the mosaic, giving the complete vision of the scene. The last tile comes like a thunderbolt! The innovative environments are those that encourage their residents to explore the adjacent possible, making available a wider and more versatile sample of spare parts—mechanical or conceptual—and promoting new ways to recombine them. There are many novel solutions and brilliant ideas, at your fingertips. The matter is to use the available resources in a different manner, in order to create new combinations. Most of the times innovative ideas do not arise from the strokes of genius, but from a good *bricolage*.

This adjacent possible concept is interesting and very useful in developing innovative routes. Although we are accustomed to think about innovation as a leap ahead in time and space or sudden swerving due to the genius of the inventor, we must agree that the history of cultural, artistic, and scientific progress is comparable to the "story of a door leading to another door, exploring the palace one room at a time". Researching, designing, making, building you may encounter errors or serendipity phenomena, which are both fundamental for scientific research: from mistakes and "casual discoveries" are born the best innovative ideas of our society (see Fig. 1).

4 Architect with *"Extended Mind"*: What Happens When Artificial Intelligence and Human Creativity Meet?

Just as digital manufacturing technologies have made it possible to enhance the figure of the architect, metaphorically equipping him with robotic arms that exponentially elevate his architectural abilities, a further upgrade can be accomplished at an intellectual level by implementing artificial intelligence within the design process. Artificial intelligence (AI) has permeated every aspect of contemporary life. It is used in everything from online recommendation systems to pricing algorithms, from providing personalized news and ads to suggesting the perfect ending to a sentence written by you.

In the beginning, it was science fiction that introduced us to the concept of artificial intelligence and the so-called "technological singularity". The term "singularity" was introduced by the science fiction writer Vernor Vinge in 1983, and it was brought into wider circulation by Vinge's influential 1993 article "The Coming Technological Singularity" [5]: it refers to the point in time when the advancement of technology will accelerate beyond the capacity of human beings to comprehend or predict it. Artificial beings, created by humans who share with them the ability to process complex thoughts, come into conflict with their creators. On the other hand, this event has already taken place in Mary Shelley's Frankenstein, published for the first time in 1818, and considered to be one of the most important literary forerunners of science fiction. But it was in Samuel Butler's dystopian book "Erewhon", published in 1872, where intelligent robots were first mentioned, perhaps forecasting our own technological culture. "Compared to the machines of the future, those of today are like the first dinosaurs to man. The largest, in all probability, will shrink a lot" wrote Butler a century and a half ago. Since then, science fiction books and movies have looked at the topic from many different perspectives, including philosophical and ethical ones. In an effort to both amuse and raise issues about the future, film and

Fig. 1 a, **b** G. Fallacara 2016, Adauctus architectus novus, Interpretation from Habit de l'Architecte, engraving attributed to Nicolas II Larmessin, probably made in the seventeenth century.

literature have given us many complicated concepts in recent years. This translates into what is still the current debate between "technophobia" and "technofilia", where the optimistic hopes of "futurologists" clash with the visions of the most pessimistic "techno sceptics" who see artificial intelligence as the invention that would lead to the end of humanity.

Among the various sectors that have been permeated by artificial intelligence-based solutions, there are also the art and architectural ones. Art has always had a complex and evolving relationship with science and technology. Recent developments in machine learning have led to an acceleration in the exploration and discovery of the potential possibilities of AI applied to art through the adoption of technologies such as neural networks and deep learning. In recent decades, and with particular acceleration in the last five years, rendering algorithms designed to modify images in several ways have been developed. Furthermore, in recent months, social network boards have been flooded with images generated by artificial intelligence systems: algorithms capable of generating spectacular photorealistic images from text input.

The use of artificial intelligence in architecture design is not new. Attempts to replicate an architect's design abilities using a computer date back to the late 1960s [6]. However, AI research has gone through several phases of rapid decline, but it is currently on its way to a successful future [7]. As a result, particularly because to modern computers' enhanced capabilities [8], AI will soon massively empower architects in their day-to-day practice [9]. In any case, the employment of AI in the architectural profession will not be universally welcomed. Some may dismiss it, just as some did when computers first entered the mainstream architectural culture around thirty years ago [10].

In architecture, a typical project goes through conceptual or pre-design to the operations in order to manage the building itself [11]. Architectural design is a complex process that requires imagination and skills to generate new ideas [12]. Since the design criteria are not yet fully defined at the conceptual stage, the application of artificial intelligence to this process should not be directed toward finding a solution in a specified search area; this method should instead be seen as a study of the requirements and potential solutions to achieve those criteria during the conceptual design phase [13].

Today the designer has new tools at his disposal to create, and an "inspiring muse" who offers new opportunities to investigate his own emotions, and new languages capable of mixing reality with the product of the reinterpretations of neural networks. Artificial intelligence and machine learning play a fundamental role in today's lives, thanks to the development of increasingly advanced systems inspired by the human way of thinking, acting, solving, and creating. To talk about artificial intelligence means talking about what it means to be human in the age of AI. If there is one characteristic that belongs to humanity, it is creativity. Creativity is not limited to art, and creativity is very human. There is no precise definition of creativity, but one of them is that creativity could be defined as the production of new effective knowledge from existing knowledge, which is achieved through problem-solving [14]. Creativity is also a form of care and recognition of the different ways of being. Precisely for this reason, it is natural to ask ourselves whether it is legitimate to approach a term so strongly connoted by our sensitivity as living beings to something artificial such as a machine or software.

Creativity is one of the most mysterious and, simultaneously, most noteworthy qualities of all human existence. We can define it as the ability to create something that is innovative and recognized as such by the community. Indeed, human creativity is also the result of the stratification of acquired data because every artist draws, even unconsciously, from the fabulous creations of the past. Therefore, no work of human ingenuity is entirely self-sufficient and purely original. However, the human artist has free will. He chooses which field of art to dedicate himself to and the purpose of his work. On the contrary, an artificial intelligence system follows an algorithmic procedure that tells it where, how, and what to look for.

AI is about origination and novelty, doing things that nobody anticipated and that we did not expect. It has an emergent intelligence. We cannot predict quite the outcome that it is going to offer us. It is based on everything it was fed, but it not only giving back exactly what has been given. It is recombing it and giving back to us a in a new outcome. From this, it can be stated that there is something about a sort of autonomous creativity.

Nevertheless, what happens when machines, algorithms and artificial intelligence come into play? There is the need to imagine a new type of creator, a new breed of artisans who are at ease with these evolved instruments. Artificial neural networks are more than a tool; it has an intelligence of its own, and we must understand that when working with it. It can be seen as a medium, as something we work within. Indeed, there is a need to understand its properties. When a sculptor works with a piece of marble, he or she must understand the marble's feel. With AI, we must understand the way it sees the world and the way it thinks. It's something to work alongside.

It is essential to consider that an AI system is not self-sufficient in deciding what to do: at the base of its creation, there is information that man enters. Therefore, it is possible to believe that creativity can be improved by the collaboration and interrelation between man and machine.

AI can be seen as a kind of "extended mind": a prosthetic device that can enhance the natural intelligence of the human being [15]. It's another form of intelligence to work with. In this way, for the designer comes the possibility to work not only with his own mind, but with this broader sea of possibilities which can take him to a new place.

5 Conclusions

The impact of digital design and fabrication techniques on architecture is already far-reaching. The integration of digitally generated data to produce precise and complex geometry, to direct making and assembly process, and exploit material performance is returning architects to a position that had disappeared with the master-builders of medieval times. The tools provided by digital technological progress have allowed the designer to take control of and retool the entire design, fabrication, and manufacturing process, leading to the generation of the architect of the future: a professional figure based on a combination of the skills of the architect, augmented by computers and computer-driven machines. With these new powers, architects are now able to craft the digital tools and processes required to make architecture for the post-digital age.

The language of architecture, the figure of the architect as a professional, and the method of design have evolved, in tandem with the changing purposes that different societies have had for this profession and art, which is destined to undergo radical transformations in the coming decades. It is hoped that this contribution will not only spark the curiosity of readers, but also motivate them to engage in further research and participate in lively debates on the subjects at hand.

Acknowledgements This paper is the result of the combined work of three authors. The authors have revised all the paragraphs, and the paper structure and the research topics have been conceived together. However, paragraph 2 was written by Francesco Terlizzi; paragraph 3 by Giuseppe Fallacara; paragraph 4 by Aurora Scattaglia.

References

1. Evans, R.: Translations from Drawing to Building and Other Essays. MIT Press, Cambridge (1997)
2. Gehlen, A.: Man in the Age of Technology". Columbia University Press (1980)
3. Fusaro, D.: Pensare Altrimenti. Einaudi, Torino (2017)
4. Crepet, P.: Il Coraggio. Vivere, Amare, Educare, p. 69. Mondadori, Milano (2017)
5. Vinge, V.: The coming technological singularity: how to survive in the post-human era. In: VISION-21 Symposium Sponsored by NASA Lewis Research Centerand, Ohio Aerospace Institute, USA (1993)
6. Negroponte, N.: The Architecture Machine. The MIT Press, Cambridge, USA and London, United Kingdom (1970)
7. Lee, K.F.: AI Superpowers China, Silicon Valley, and the New World Order. Houghton Mifflin Harcourt, Boston and New York, USA (2019)
8. Trabucco, D.: Will Artificial Intelligence kill architects? An insight on the architect job in the future. J. Technol. Arch. Environ. (Special Issue No 2) (2020)
9. Chaillou, S.: ArchiGAN: artificial intelligence x architecture. In: Yuan, P.F., Xie, M., Leach, N., Yao, J., Wang, X. (eds.) Architectural Intelligence. Springer, Singapore (2020)
10. Leach, N.: Architecture in the Age of Artificial Intelligence: An Introduction to Ai for Architects. Bloomsbury, London (2021)
11. As, I., Pal, S., Basu, P.: Artificial intelligence in architecture: generating conceptual design via deep learning. Int. J. Archit. Comput. **16**(4), 306–327 (2018)
12. Dixon J.R., Simmons M.K., Cohen P.R.: An architecture for application of artificial intelligence to design. In: 21st Design Automation Conference Proceedings, pp. 634–640, Albuquerque, NM, USA (1984)
13. Dekker, A., Farrow, P.: Creativity, chaos and artificial intelligence. In: Dartnall, T. (ed.) Artificial Intelligence and Creativity. Studies in Cognitive Systems, vol. 17. Springer, Dordrecht (1994)
14. George, G.H.: Philosophical Foundations of Cybernetics. Abacus Press, Tunbridge Wells, Kent (1979)
15. Leach, N., Del Campo, M.: Machine Hallucinations: Architecture and Artificial Intelligence. Wiley, Oxford (2022)

The Future of Architecture is Between Oxman and Terragni

Mario Coppola

Abstract What will be the future of architecture? Architecture is among the major culprits of the CO_2 emissions that cause the current ambiental crisis. The discoveries of coding, digital design and digital fabrication—including topological optimization, reactive skins, and lightweight technologies such as 3D printing of natural and environmentally friendly materials—are essential to finding an alternative path to those used so far. A road that not only addresses the "technical" issues of the climate crisis but is capable of proposing a new vision of the world, a spatial system that is structurally, physiologically and symbolically based on the concept of coexistence, of symbiosis, between the world of man, the house of man, and the rest of the biosphere. The work on biomaterials from the digital fabrication world assumes crucial importance in this perspective. It is the first research capable of demonstrating that the inert envelopes of architecture can become organisms and no longer in a metaphorical key: buildings like trees programmed to become skyscrapers, houses like pods and fibres that transport people by capillarity along with water and nutrients. Energy from domesticated photosynthesis processes, eyelids and hairs that grow to shield excess light. Architecture that reconfigures itself as a living form similar to what already happens in nature for the calcareous secretions that we call shell. Yet before we can indulge ourselves in transforming architecture within these new ways, there are at least two issues that cannot be ignored so as not to repeat the mistakes made by Modernism a hundred years ago.

Keywords Posthuman · Anthropocene · Ecologic · Regenerative · Computational

United Nations' Sustainable Development Goals 7. Affordable and Clean Energy · 11. Sustainable City and Communities · 13. Climate Action

---

M. Coppola (✉)
Dipartimento di Ingegneria Civile, Edile e Ambientale, Federico II, 80125 Napoli, Italy
e-mail: mario.coppola@unina.it

© The Author(s), under exclusive license to Springer Nature Switzerland AG 2024
M. Barberio et al. (eds.), *Architecture and Design for Industry 4.0*, Lecture Notes in Mechanical Engineering, https://doi.org/10.1007/978-3-031-36922-3_7

1 Today as Yesterday. A Brief Introduction

The discoveries of coding, digital design and digital fabrication—Industry 4.0 foundations—are probably essential to finding a new road for architecture and design, a road that not only addresses the "technical" issues of the climate crisis—CO_2 production in the first place—but that is capable of proposing a new vision of the world, a spatial system that is structurally, physiologically and symbolically based on the concept of coexistence, of symbiosis, between the world of man, the house of man, and the rest of the biosphere. Yet before we can feel free to revolutionize architectural and design languages and tectonics, we must respond to some crucial questions so as not to repeat the mistakes made by Modernism a hundred years ago, when the discoveries of industry—for example, the mass reproducibility, that for the first time was free from size and cost limits—changed forever the face of the landscape and of the urban peripheries, generating immeasurable damage to human well-being and to terrestrial ecosystems.

Indeed, the profound changes that have affected Western society in recent years trigger an interesting comparison with the great transformations that took place in the first decades of the twentieth century. At that time, the Industrial Revolution changed not only the production method but the collective imagination itself: in the space of a few years, we have gone from a world of small and different architectures, made above all by hand, to a world populated by mammoth buildings children of the assembly line.

The architects of Modernism exploited these unprecedented production tools to give life to a new urban and architectural vision, formally based on the standard, of the repetition of simple and economic elements. But behind the desire for "a house for all"—a need that continues to animate a large part of the construction world, for example in China—and the figures and languages invented by the masters of the modern, there was not only the will to exploit new construction techniques to offer an architecture more suited to post-mechanization lifestyles nor just the intention to increase the size of buildings and reduce costs. Argan wrote in 1947:

> We can therefore consider the so-called architectural "rationalism" as a close analysis or critique of tradition, aimed at tracing its most authentic and original foundations, at restoring its essential values: therefore, it leads back, albeit against academic classicism, to an ideal classicism and against a customary naturalism at the very foundation of the idea of nature. When European culture wants to go beyond the rationalistic limit of scientific Cubism and the architecture that is connected to it, it has only one way: to overturn the problem, to oppose the value of consciousness to the value of the unconscious. It is the closed road of Surrealism. [...] Wright does not know enough about the history of art to be able to give a precise historical objective to his irritated aversion to classical art and in general to the great Western figurative tradition; but he is keen enough to identify the cause of the aversion in the principle of authority on which that tradition is founded. But faced with the basic argument of the European anti-traditionalist polemic he remains doubtful: even mechanical civilization has its myths and its principle of authority. Wright sees the symbol of the principle of the authority of classical and Catholic civilization in the dome of St. Peter; in the skyscraper he sees the symbol of mechanical civilization. Wright does not flatly condemn the mechanical character of modern civilization, but he wants the machine to serve man in his work and not the other way around. [1]

The desire to re-propose and restore the essential values of tradition, of the anthropocentric law inscribed in classical tectonics—composed of orthogonal geometries, not existing in nature, and detached from the ground through the *crepidine*—is the deep cultural movement that constitutes the substratum, the humus from which the "abstract", "minimalist" research of modern architecture takes shape. A law issued by a world two thousand five hundred years old, in which man's main need was to be emancipated from the rest of nature, a hostile habitat from which to get out in every possible way, starting with the perceptual, cultural, and symbolic-spatial one [2].

Therefore, if it is true that the current urgency is, on the contrary, to find a way to coexist, a symbiosis between human civilization and the rest of the biosphere—the biosphere of which humanity is an integral part, of which humanity needs to exist while the opposite is not true—it would seem evident that the discoveries of digital design and digital fabrication brought by the great technological revolution of the last few decades are the key to overturning this relationship, transforming submission and exploitation into a harmonious and balanced relationship based on the idea of interdependence and non-differentiation between man and nature. A relationship that should be structurally different from that set up by contemporary society, based on the capitalist consumer economy. But in history it frequently happens that great transformations produce tsunamis and violent settlements and also involuntary self-sabotage, internal reactions that are contrary to the intentions made explicit at the outset. Just think of the damage perpetrated in suburbs by the naive claim to build buildings that were hundreds of meters long, without differentiations, unable to establish empathy and produce psychosomatic well-being, focusing everything on the use of structural elements (pillars, trilithons, etc.) and typological archetypes from the great Western architectural tradition that, alienated from decorations, materials, proportions, should have been enough to make the places thus obtained habitable.

For all these reasons, before concentrating the creative energies on the possibilities offered by the new tools, and looking for the ecological revolution we desperately need, it is necessary to ask ourselves about two questions. The first: is it just designing slicing, parametric arrays, and topologically optimized structures enough to produce an architecture that is in harmony with nature? And, even before that, what should we mean today by "nature"? The second: can we produce, in the name of coexistence and symbiosis, figures that are radically, epistemologically different from those shaping every urban scenario, proposing once again the cultural mechanism of the *tabula rasa*?

2 Parametricism and the Global Market Society

On May 6th, 2010, Patrik Schumacher, Zaha Hadid's partner and founder of DRL master at the Architectural Association, published in the Architects' Journal a long text entitled "Let the style wars begin". The text contains the description of Parametricism the German architect considers the next hegemonic architectural style on a global scale. The article ends with a lot of details about the new language rules and a series of heuristic principles, divided into "dogmas" and "taboos".

More or less from the publication of this manifesto, many architectural practices belonging or tangent to the international movement (so far summarily defined as "digital" architecture) started to rally in the label coined by Schumacher.

However, after the appearance of Folding Architecture by Greg Lynn in 1993, all the subsequent attempts at defining Hadid's, Himmelb(l)au's and other post-deconstructivists' architectures had already changed their expressive register while abandoning the concept of the "fold" as an inter-individual continuity-complexity, a nature-culture continuum. From these attempts, several definitions of architecture came out: "blob", computational, algorithmic, generative, procedural or, simply, computer-aided. All of them lost sight of the profound reasons for a hybrid, multiple, open composition: they focused only on technique, therefore encouraging the mass distribution of mannerist-by-definition research. The result was a language accessible to anyone who, though in a lack of culture, talent, and sensitivity, can use morphogenetic software. With the coming of Parametricism—that in a very short time acquires global importance, transforming parametric architecture into a label known by most architects—focus definitely shifts onto the tools, and technology, in a perspective, that sees architecture as a technical affair devoid of cultural variables, in which the search for new architectural "phenotypes" can be carried out automatically through scripting and genetic algorithms. In this perspective a living architecture is being theorized but it does not represent the real change of a cultural paradigm, first of all in an ecological and therefore, as Edgar Morin clarifies, anti-capitalist key.

As a matter of fact, the term "parametric" refers explicitly to the information technologies related to post-deconstructive research [3] since setting those technologies as fundamental and characterizing reasons. So, this global movement is based on a design that comes from the selection and manipulation of numerous parameters that define the final status of the project. It is an architecture that becomes, according to Schumacher, an instrument serving customers, as if the only function of architecture should be meeting the needs of the dominant economic/productive model in the industrialized Western society. Schumacher says that parametric architecture is the most effective tool—he repeatedly underlines, in recent years, architecture is not art but a tool to organize space in the most "productive" way—to ensure better, faster operations of the contemporary production based on a network society.

So, architecture is just an instrument of "progress", which suggests a spatiality made to support, optimize, and maximize an economically intended "productivity", accepting that the architect's role is not to propose an innovative political vision or even criticize the socio-economic political existing model, but just improve the users' performances according to the needs of today's business production. From Schumacher's perspective, that appears to be very different from Zaha Hadid's vision (i.e. the MAXXI, the way it connects both with the city tissues both with the river), in other words, parametric architecture aims to "reinforce" and improve the processes of a society that Zygmunt Bauman defines, with a very different connotation, liquid [4]; i.e. it deals with those processes of post-industrial production in their hyper-capitalist evolution, which produce such phenomena of relocations as needed by a globalized exploitation of human and natural resources, governed by elusive, constantly moving super-managers.

The target of Parametricism seems therefore that to give a physical body to a society in which competition and abuse between individuals, and multinational groups' power is encouraged and promoted at the expense of human relationships, happiness, survival of communities and ecological balances. It is interesting that the German architect, after having explicitly written about the attempt at approaching, through digital tools, the «compelling beauty of living beings» [5], decides to abandon this definition and names Parametricism the architectural style emerging from this research.

A choice that shifts the focus from an expressive nature referred to the living form—that evokes the need for negentropy, i.e. order and complexity as mentioned by Morin [6] and recently demonstrated by contemporary neurosciences as far as the deep bond between human beings and terrestrial ecosystem is concerned—to the technical, instrumental form, omitting the starting choice, the tendency towards a "natural beauty" that, detached from social and ecological thinking, would reduce everything to an ephemeral hedonism.

Similarly in the definition an aspect is minimized: the choice of "primitive" geometries (more or less smooth lines, curved surfaces, elements of which the spatial shell can be made) that determine the final appearance of a parametric project, that, instead, can take on both the characteristics of Malevich's Architektons and those of a cellular tissue during mitosis. Almost as if it were possible or desirable to bypass the designer's will, which instead remains central also for parametric design—since, paraphrasing Gianluca Bocchi, technology without man is stupid–, so as to emphasize this way the exquisitely technical nature of the latter.

As for the difference between architecture and art, Schumacher seems to suggest that the designer cannot act of his/her own free will because he/she should "inform" the computational tools of the right parameters as regards the activities and the project constraints. So, while the German architect describes the question of productivity, in the manifesto there is no trace of the term "openness", the concept of "urban carpet" and in general the concept of a private–public, outside-inside, anthropocentric—non-anthropocentric interpenetration. There is a neutralization of all cultural and political contents of the project—the critical elements regarding the organization of the contemporary architectural order—and of the rules concerning the external characteristics of the fluid and "internally correlated" language, where all the generative systems are an organic multiplicity, systems of correlated systems, the only objective of which is to ensure the readability of the three-dimensional spatial continuity to make it easily navigable by its consumers.

The character, at least ambiguous, of this formulation is already clear: there is no intention of defining a "parametric" architecture in a literal sense (because through parametric software it is, of course, possible to create any style and then any project including the traditional Mediterranean house with gabled roof), on the contrary it refers to a bio-mimetic language that would be able to organize the growing complexity of contemporary social systems that are increasingly interconnected, dense and dynamic, and, we add, more and more competitive, egocentric and anti-ecological. For these reasons, this deals with the proclamation of a "biomorphic" architecture that, in full contradiction, expresses the post-Fordist capitalist economy, the engine of today's oedipal "liquid" and biodiversity-destroying society.

3 A Total Biomimicry: Renunciation of Complexity

Starting from the "post-factum" rules related to the work carried out in the past (such as "use elastic and non-rigid shapes" or "avoid repetition and standardization"), a checklist of "biological" geometries emerges from nearly thirty years of Hadid's design when the naturalization of the architectural language was instrumental as to a project that aimed at revitalizing and reconnecting parts of the city and therefore was freely inspired by the self-organized structures of living forms (to design them Parametricism was not needed, as shown by Michelangelo's projects of the Florentine fortifications in '500 and Paolo Soleri's designs, or Musmeci's work and Moretti's "parametric" stadium in the 60s). Yet these geometries, deprived of any relational desire, bring back the matter to a mere grammar, which can be used to say everything and its opposite. What therefore remains is the use of complex forms through a technological medium that would seem to guarantee the automatic success of the project, or, at least, the positioning of the project within the "parametric paradigm", as Schumacher himself defines it. The contradiction that results is very similar to that which gave rise to the reproduction of international architecture on a world scale starting from the famous five points for architecture by Le Corbusier: this simplification extends the contradiction to the very meaning of architecture, which this way has nothing to do with the themes of weaving city and landscape, regenerating biodiversity or generating bio-centric communities, and refers instead to a sequence of curves, "blobs" and other bio-digital images, as happens in the Chinese project Galaxy Soho dominating and crushing the humble context using some monumental convexities. These figures result in a machination which, by simulating the lines of the living world, in most computers of young designers inspired by this research, becomes a fetish, the illusion of a new computer-generated corporeity running away from the complexity and ecological, urgent needs of the real world.

So, the computational research chases, inside the computer, an intricate feminine sensual curvilinearity which seems to be an attempt at re-appropriating a missing corporeity/promiscuity, i.e., a missing contact with nature. They can catch a sterilized literalness of the biological dimension, enclosed in a virtual dimension: this refers to the process described by Bauman about the use of social networks, where new generations are trying, often unsuccessfully, to find contact with one another and with the body—as evidenced by the success of the online porn—which appears elusive in a real-world that is increasingly atomised, alienated from the body dimension. So, the results of this research represent a spatiality that, instead of regenerating a lost connection with the body-biological sphere, multiplies the distance between environment and building, megalopolis, and biosphere. It seems a new extremism that to the "all culture" (as a drift of "humanization" the philosopher Roberto Esposito writes about) of historicism opposes an "all nature". As a matter of fact, it deals with the view of nature as a tool, a technique, once again an "object" in the human being's hands replicating its shape in the laboratory, through the most "advanced technology" (in line with the function of technology of a part of the anarcho-capitalist transhumanism), so as to put it at the service of his own individualistic utility. This brings to a construction that is also losing the uncertainty, unpredictability, and chaos that, in nature, contribute to determining the epigenetic landscape from which life emerges with its organized but never "perfect" structures, that are always unique, asymmetrical, rough, hybrid. In this reality as a network of unpredictable possibilities, algorithmic architecture replaces a hyper-deterministic scenario, locked in the mathematical values chosen by the designer, without capturing exceptions, without expressing any freedom and often producing an imitation of the autopoietic structures of the living world, pretending to be a sort of "second nature" [7].

The morphogenetic dream pursues rooms like cells, spaces like organelles able to correlate, open up to be penetrated by light and air through the right exposure; corridors as arteries and capillaries, structures like self-optimized skeletons, integrated to space divisions so as not to "drill" spaces/organs; a mediation space, like a nervous system, keeps everything together, interconnected, without a way out, without providing a coexistence of different solutions as happens in the ecosystem. So, from parametric masterplans (as, for example, the Kartal Pendik Masterplan by Zaha Hadid Architects) a new totalitarianism seems to emerge. It generates a reproduction of closed "perfect" systems, where no indeterminacy differentiates each body (think for example, on the contrary, of the imperfect symmetry of the human body), and where new buildings are indifferent to the languages, morphologies, typologies, chiaroscuro and even to the scale of the spaces in which they are settled.

On the contrary, each wild landscape is surprising because it is full of different characteristics and different unpredictable logic, as well as each living form is always at least in part different from the others. Even a bacterium is only a phenotypic "attempt" because the organic molecules that make it up are not generated by simple digital algorithms but by a complexity of factors and contributory causes—also random and therefore undeterminable—that it is not possible to predict any outcome. There are no unique solutions since life presents itself in a "rhizomatic" multiplicity of unpredictable phenomena in the same context: a form of life can proliferate in different habitats as well as the same habitat also belongs to an immense variety of different life forms as for structure, "seniority" and complexity. It is like an equation in which, even maintaining fixed parameters, there is not one only correct result—that is, "able to live"—but an immense wealth of compossibilities [8] all having an equal organic dignity, all having equal rights to exist in a perspective in which biodiversity is not only a variable but a real value, as well as ethology itself becomes a system of values in which every living being is autonomous yet linked to the fate of all the others.

Finally, even if we could plan a system containing the variable of indeterminacy as a random fracture of some logic, that is, even once an "artificial ecosystem" consisting of a perfectly optimized organic form, it is not certain at all that the human being would feel at ease and choose to live there. The same could happen even in the presence of an exceptional natural environment, with beautiful landscapes, and even intimate and comfortable natural sites like a sea cave or a clearing surrounded by trees. A temporary enjoyment does not imply the will or ability to live in a place without making any change, i.e. without "humanizing" the environment by a "cultural" trace. Because the human species (and many scientists start thinking the same even for other species, respecting due differences) possesses an "external genetic code", handed down through language but even through the environment—architecture—which forms the main framework: an "exoskeletal" culture that can be certainly developed and transformed but cannot be ignored or bypassed. Even this—the set of external factors constituting the cultural code—is part of nature and in recent years, on the other hand, various scientific theories have shown that culture and in general experience, as an event external to the body occurring after birth, for example, a trauma, changes the genetic code itself handing down the change to offspring. This is left entirely out of the parametric/biomimetic system, which instead of triggering a transformation of the existing architectural language, provides for a replacement, ignoring that tabula rasa occurs in nature only on the occasion of tremendous disasters destroying entire ecosystems, bodies and their "cultural trail". Even in this case—think of the impact of the meteorite that, as it is assumed, almost completely destroyed the ecosystem about 66 million years ago—most of the next species probably would preserve a structure very similar to that previous, as if the earth scenario contained in itself rules and information that can shape the living structure. Recent research has highlighted, through studies on different types of samples, that the bony tissue of mammals is nearly the same in the different species as if there had been a topological deformation of one only starting form that modified only the extensive properties while maintaining the geometric relationships between the different parts. That's why the linear transposition of morphogenetic research in architecture is not only literal to a fault—it frequently shifts the focus from the relational issue generating introversion—but falls into the temptation starting "from scratch", excluding the existence of the starting point that is represented of course by the millennial architectural types, rooted, and developed in their epigenetic landscape. In this, it differs from the previous and current research of many other designers that hybridize and regenerate the existing such as, for example, in the BIG's Courtscraper, a typo-topological contamination between a skyscraper and a court. The types, that is to say, would be nothing more than a mnemonic registry, the code (DNA) of architectural species that the Earth's epigenetic landscape—bio-anthropogenic—has shaped with time, in a transformational continuity that was first triggered by the biological matrix (which originally turned primates' physical structure into that of the human) and later continued through the historical process, passing from the mutations generated by "primary" needs to those connected to the more and more complex, up to the "cultural" needs, that weave together biological, social and political necessities. This is the concept behind the chreod Sanford Kwinter refers to [9].

But to translate this concept is essential to broaden the speech to the multiplicity of factors in architecture that make up the landscape conforming to the anthropic space, without locking it up in the everlasting values—and then in the codes and conventions developed over the centuries for psycho-social cultural needs of an earlier era—but even without reducing it to a purely biological process of the elementary forms of life, through a return to prehistoric origins that sets architecture in a "uterine" figuration, designed for a man-fetus with no memory. This shows the relationship between spatial type and geographical context, where geography is defined as a geo-political interlacement that contains both the factors related to the environment and landscape (and hence the natural peculiarities in terms of insolation, ventilation, humidity, etc.) and those related to the economy, politics, uses and customs of the historicised place. Once again, the appeal to the biological shape is not enough to produce architecture, since the latter, even and above all in the complex vision, can only be a hybrid architecture, linked to the present cultural living system of mankind, as a remedy for the self-destructive motion of marginalization from the biosphere [6]. Because space influences and simultaneously stimulates the sensory-tactile sphere of the body and memory.

The "systemic" changes of Thompson's deformations—that in the case of post-deconstructivist architecture are easily reproduced through nonlinear deformers (see modelling programs like Maya)—refer once again to a type of mutation that maintains its continuity, readability and recognition of the original figure, i.e. its geometric relations and maximum proportions, without interrupting the membership to its own epigenetic landscape, as happens in the vast majority of animals and humans, where the characteristic features of the face (two eyes, two cheeks, two ears, forehead, nose, mouth and their disposal) are always the same, making an inter-species empathy possible. This would suggest, instead of trying to start from scratch, a hybridization with the biological language that can graft onto the existing architectural code the characteristics of interactivity, openness, continuity, multiplicity, interconnection, flexibility, and adaptability which are now socially, economically, and ecologically necessary in a symbiotic vision. And this is the reason why we need to aim for something that is simultaneously sensual, that is psycho-somatic, cultural, and socio-political. On the contrary, the bio-mimetic way tries to appeal to a pure imitation of natural shapes and in most cases does not take into account the minimum energy-environmental issues (the renewable sources such as solar, wind, etc.), the bioclimatic essential mechanics (passive exploitation of natural resources to reduce consumptions), the necessary grafts of fauna and flora (to regenerate biodiversity) and does not look after the choice of eco-friendly materials and technologies (which would require the renunciation of double curvature) in favour of research fixed on the language.

From this perspective, the research on 3D printed structures in harmony with the physical characteristics of the materials carried out by Neri Oxman represents the apex of the discourse. Cellulose, chitosan, pectin, and calcium carbonate are among the most abundant materials found in nature and are biodegradable: their choice automatically solves all problems relating to pollution and CO_2 production. Yet, we must ask ourselves what it means "to design a culture attuned to the systems of nature", as Paola Antonelli writes about Oxman's research [10], when human "culture" is just a part of "nature", as all the other species' cultural forms.

Oxman is developing a way to "write" the genetic code of the building or the furniture, whether it is made up of the "recipe" of the biopolymer blend, the geometry of the envelope, the path that the robotic arm will follow to create the component: thus an organism is created capable of functioning in all respects as the exoskeleton of an arthropod, and therefore the final structure assumed by the project belongs to the same anatomical/physiological/aesthetic domain as the "biological" one. In this way it is certainly possible to satisfy the needs related to structural resistance—as bones and vegetable fibres do—or to thermo-hygrometric well-being—insulating membranes, more or less transpiring and transparent, which interact with solar radiation or with the need to look outside—but what remains of human expression that is intrinsically linked to the languages thus far produced and developed by civilization? In other words, also—and above all—in Oxman's material biomimicry there is no longer any trace of what is normally understood as architectural culture: if nature, understood as a non-anthropic domain, as a self-organized place, becomes the "primary customer", paraphrasing her writings, the planning operation necessarily transforms itself into the most radical linguistic and formal tabula rasa. The same attempted by modernists to historical architecture but even more absolutist, since, even in the cruciform skyscrapers of Ville Radieuse, there was an echo of classical tectonics, the formal system that has characterized architecture since the dawn of times. It is a very critical point. Because the human being is a 100% natural and 100% cultural animal [6]: he is born and grows orienting himself through the spatial structures he experiences, and it is in the relationship with those systems that his memory develops, its identity, its ability to feel at home in a given habitat. It could be said, then, that biomimicry replaces the anthropocentric system, where "nature" is simply the "outside" of the anthropic territory, the "other" space to occupy, consume and from which to differentiate in all respects, with an opposite therefore formal—value system, where man is effectively cancelled from the equation. A cultural operation that evokes the dramatic results described in Kafka's Metamorphosis; as if, overnight, we were forced to communicate through a different body and a different language.

On the other hand, as we have seen with the so-called parametric architecture, the insertion of a biomimetic structure—both a piece of furniture among other furnishings in the home environment and a building among other buildings in a city—does not show any will of coexistence but re-proposes the simple juxtaposition. It is always the same separation logic—the Cartesian one between subject and object, mind and body, culture, and nature—which is the very basis of the anthropocentric culture responsible for the ecological catastrophe we are starting to experience. Just look at the images of the Aguahoja pavilion inserted in the museum environment to find this extraneousness.

Yet Oxman's invention is equally valuable because it is capable of showing a new path. Imagining an architecture where structure, spatial partitions, and casings are made with the same material as the shrimp shells is essential to shift the axis of design from anthropocentric ontology—the man who uses and consumes terrestrial materials without worrying about the ecological effects—to a different ontology.

4 Conclusions: Towards a Posthuman Architecture

At a time of profound change, characterized by economic, political, social, and environmental crises—inextricably linked to one another—shaking contemporary societies, it is certainly necessary to critically rethink the role of architecture and architectural research in terms of today's transformations. It seems evident that it is unthinkable to continue reproducing and representing the same traditional spatial model of settlement as if nothing had happened and nothing was happening, proposing again the Western split between culture and nature, mind, and body.

Therefore, in the perspective of a transformation, the problem cannot be to "eliminate" computational technologies from the project.

On the contrary, an architecture capable of supporting and promoting the creation of a community of men and ecosystems—a post-anthropocentric [11] community—as well as any form of art aiming at understanding, inspiring, and transforming today's world, can and must take advantage of the expansion of possibilities provided by 4.0 Industry computational technologies. In other words, the possibility to hybridize, and transform a space that so far has spoken about man, strength, and autonomy of civilization, into a crossbred *post-human* space able to cancel the sense of alienation towards the rest of the planet, to tell of weaving, interdependence and symbiosis. A road open to life but without amnesia, capable of recomposing the ancient contraposition between organic and rational, implementing the code rather than rewriting it from scratch.

An architecture that is capable of evoking the figures of history, talking about our deep need to differentiate and emancipate from the violent necessity of natural life, and simultaneously drawing from the negentropy of our bodies and the other natural structures, telling about our awareness and comprehension of our membership and interdependency with Earth, inserting a revolutionary element: mestizo tectonics, where the trilithon and orthogonal volumes would come to life, taking on mechanical strength, transforming themselves into light Voronoi conformations, paving the way for the use of zero impact materials and making it possible to hold the *two worlds* together [12–17].

This would be a biocentric ontology, in which man neither prevaricates nor cancels himself but establishes a fertile exchange with natural otherness; an architecture that is not just a temple or an exoskeleton but a hybrid space, a contaminated language, a meeting place.

References

1. Argan, G.C.: Introduzione a Frank Lloyd Wright. Metron n. 18 (trans. Coppola, M.)
2. Coppola, M., Caffo, L.: L'architettura del postumano. Domus n. 1016(2)
3. Coppola, M., Bocchi, G.: La maniera biomimetica. D Editore, Roma (2016)
4. Bauman, Z.: Modernità liquida. Laterza (2011)
5. Schumacher, P.: Digital Hadid: landscapes in motion. Birkhauser (2004)
6. Morin, E.: L'anno I dell'era ecologica. Armando Editore (2007)
7. Esposito, R.: Pensiero vivente. Einaudi (2010)
8. Deleuze, G.: La piega Leibniz e il Barocco. Einaudi (2004)
9. Kwinter, S.: Un discorso sul metodo (translation by Lucio Di Martino). In: Explorations in Architecture Teaching, Design, Research, Reto Geiser, Birkhäuser, Basilea, Boston, Berlin (2008)
10. Antonelli, P., Nurckhardt, A.: The Neri Oxman Material Ecology catalogue. MoMA (2020)
11. Braidotti, R.: Postumano. DeriveApprodi (2014)
12. Barberio, M., Colella, M.: Architettura 4.0 Fondamenti ed esperienze di ricerca. Maggioli Editore (2020)
13. Perriccioli, M., Rigillo, M., Russo Ermolli, S., Tucci, F. (eds.): Design in the Digital Age. Technology Nature Culture | Il Progetto nell'Era Digitale. Tecnologia NaturaCultura, Maggioli Editore (2020)
14. Coppola, M.: Visioni biocentriche-Neri Oxman e l'architettura sintonizzata sui processi della natura. Bioarchitettura n. 133 (2022)
15. Amirante, R., Coppola, M., Pone, S., Scala, P., Chirianni, C., Lancia, D., Pota, G.: Reloading architecture. Metamorfosi (Sep. 2021)
16. Coppola, M.: Architettura della complessità. In: Ceruti, M. (ed.) Cento Edgar Morin. 100 firme italiane per i 100 anni dell'umanista planetario, p. 153. Mimesis (2021)
17. Coppola, M., Cresci, P., Caffo, L., Velardi, B.: Verso un'architettura postumana. Spazi, figure, linguaggi e tecniche per l'antropocene. In: Melis, A., Medas, B., Pievani, T. (eds.) Catalogo del Padiglione Italia «Comunità Resilienti» alla Biennale Architettura 2021. Ediz. italiana e inglese, vol. 1: Catalogo della mostra

Open-Source for a Sustainable Development of Architectural Design in the Fourth Industrial Revolution

Giuseppe Gallo and **Giovanni Francesco Tuzzolino**

Abstract Ten years after the first conceptualisation of Industry 4.0, we took part in the largest ITC experiment ever conducted. When during the quarantine, governments chose digital as the exclusive means for education and work, revealing its inclusion limits. Issues that also affect architecture, and in the perspective of the sustainable development of our role, force us to think about inclusivity, starting with the tools we use. When considering the fragmented panorama of software, it is possible to make a distinction according to a gradient going from proprietary to open-source. The latter guarantees the greatest inclusivity and is a requirement for architectural design to continue to develop within the horizon of research. As described in our article, open-source is already alive and present in contemporary architecture, and its contributions can promote quantitative and qualitative turning points. There is a clear tendency to distrust open-source tools: a condition that, in the perspectives stimulated by the industry 4.0 enabling technologies, risks placing architects in an eccentric position on the project. Based on these observations, our article reconstructs the diffusion of open-source tools and formats, outlining the contributions and possibilities ensured by an effective knowledge exchange: a condition necessary to keep the architecture as research, shared and comparable. Interviews with architects with extensive experience in digital tools enrich the article in a path that highlights problems caused by proprietary software, and the solutions promoted by designers who are already aware of the need for open-source tools in the AEC industry.

Keywords Open-source · CAD · Digital architecture · Digital tools · BIM · Inclusion

United Nations' Sustainable Development Goals 8. Decent work and economic growth · 10. Reduced inequalities · 9. Industry, Innovation and Infrastructure

---

G. Gallo (✉) · G. F. Tuzzolino
Dipartimento di Architettura, Università degli studi di Palermo, 90133 Palermo, Italy
e-mail: giuseppe@giuseppegallo.design

The global quarantine of 2020 and the response of governments and institutions that have chosen the digital as an exclusive tool to convey relationships, education, and work have highlighted the limits of inclusivity that the digital paradigm brings with it [1]. A condition that concerns architecture, both as a response to human needs and as a design practice that increasingly evolves using digital tools. While digital tools allow the project to reach a capacity for technical analysis and prediction that is unprecedented in the history of analogue architectural design. It is important to remember that an instrument is never neutral to the purposes for which we use it [2], and it always leaves its mark on what we produce with it, constraining the process. With the development of Industry 4.0, this seemingly invisible trait will take on an increasingly important weight in the design and production of architecture.

In the early 2000s, when CAD tools had not yet reached their current diffusion, we imagined a future in which architects would become aware authors of increasingly advanced tools [3]. To date, this has only happened in part in the few studios that can customise or create their own tools thanks to computer skills that most architects do not have. Everyone else has become de facto consumers, often unaware, of products developed by companies specialised in software for architectural design [4]. These companies feed a market animated by a commercial narrative that is geared towards economic purposes transversal to architecture.

Today, the paradigm of Industry 4.0 promises to overcome the separation between design and production that characterised architecture in the modern era, expanding the scope of digital methods and processes through automation. This extends the architect's responsibilities, in his capacity as a human-centred designer, as a producer or simply as a consumer of tools, and requires him to make an inclusive effort at every level, starting with the software used to design architecture.

A condition that is only possible through the development and adoption of open-source tools, which are the only ones capable of keeping the evolution of the architectural project within the research horizon, removing the risk of transformation into a blind application of methods, further marginalising the role of the architect in society.

In this paper we reconstruct the relationship between architectural design and open-source software, describing the often invisible but crucial contributions that open-source formats, programmes, libraries and cultural approaches bring to architecture, guaranteeing the degrees of freedom necessary for the coherent development of our profession. This is also possible thanks to the interviews [5] with architects of great experience in digital architecture: Steven Chilton, Director of SCA, Daniel Davis, former Director of Research of WeWork, Aurélie de Boissieu, London Head of BIM Grimshaw Architects, Xavier de Kestelier, director of Hassell, Al Fisher, Head of computational development of Buro Happold engineering, Harry Ibbs, Europe design technology director of Gensler, Arthur Mamou-Mani, director of Mamou-Mani studio, Edoardo Tibuzzi, director of AKTII and Pablo Zamorano, head of computational design of Heatherwick studio.

1 From Free Software to Open-Source

We encounter the term open-source more and more often in the specialised press and the general media. Opinion makers, journalists, politicians and experts declare the importance of developing an open-source policy based on the free exchange of knowledge: a very current paradigm that is finding a growing number of supporters in academia and also in architecture [6].

A seemingly recent phenomenon, which truly dates back to the beginnings of information technology, when researchers conceived the first computer programs as scientific products, like mathematical or physical research, with scientific experiments, hypotheses, methods and results. Similarly, authors distributed software sharing its source code, so that others could carry out evaluations and develop further applications of the methods described.

This initial phase come to an upheaval when software attracted market interest, with the establishment of commercial groups aimed at the development and distribution of packages for specific purposes. It is the birth of commercial software and licences, defined according to jurisprudence that has evolved over decades to protect products that fall under trade secrets and distributed as executable files for a fee. Thus, also thanks to extensive and systematic marketing activities, commercial software has reached an ever-increasing number of users, attracting investment and developing innovative functions and features at a pace that free software cannot support. This contributes negatively to the reputation of free software, which is perceived as unreliable and of poor quality.

We owe the term open-source to an informal group of developers belonging to the free software movement. They were aware of the ambiguity of the term free software, and proposed a repositioning aimed at improving the perception of open-source collaborative products and methods [7]. It is possible to classify free software according to the possibilities its licence allows: starting with public domain software, not protected by any copyright: anyone can create a new version of the program to which to declare rights.

Something not possible with open-source licenses, which however allow the software distribution in its textual code, including organisation, functions and parameters. These license, therefore, permit anyone to modify the original program and distribute a newly edited version according to rules defined upstream.

The first solution that goes in this direction is proposed by Richard Stallman, founder of the GNU project, who introduced the Copyleft practice in 1985. Copyleft provides additional protection for the distribution terms set by the author of a package while allowing for modifications and preserving the same rights in any derivative product [8].

In 1989 Stallman himself, together with the Free Software Foundation, drafted the first GPL, the General-Purpose License. A license that aims at the legal protection of free software and allows the free distribution of the original programs and the subsequent versions created by their modification. The license, expanded and articulated over the years, is now widely used to guarantee open-source software. It is based on what Stallman [9] and the Free Software Foundation define as the four basic freedoms of software users:

1. The freedom to run the program, for any purpose;
2. The freedom to study how the program works, and change it so it does the computing as the user wishes;
3. The freedom to redistribute copies so users can help others;
4. The freedom to distribute copies of modified versions to others.

These freedoms are not limited to use and distribution, but also provide for the possibility of modifying the program in all its parts. As set out in freedoms number one and three, which make the source code sharing mandatory.

2 Project Culture from the Reims Cathedral to the Linux Bazaar

Computer science and architecture have a much closer connection than we usually think. Nowadays, architects look to IT companies as models to replicate their successes. In the second half of the twentieth century, when software engineers were looking for models that could solve the software crisis, they looked to architectural design culture and tradition. Among them was Fred Brooks, pioneer of software engineering and author of the famous book *The Mythical Man-Month* [10], where he describes software development as an exercise of complex interrelationships. For Brooks, there are several similarities between architecture and software development. The latter is divisible into two distinct phases: essence, the mental conceptualisation of the software, and accident, the software implementation or construction phase of the program.

The essence represents defining the conceptual structure of the system and is the most important part of the IT development process, which the author relates to architecture.

As an example, Brooks compares ideal software development to the construction site of Reims Cathedral (Fig. 1). Most European cathedrals result from combinations of elements from different periods, built according to different styles and influenced by external changes. In this large series, Reims Cathedral is a rarity, built at different times by several generations that followed the original project and produced a fortunate architectural coherence [11]. Similarly, the approach to software development must strive for uniformity, avoid inconsistencies due to the work of different actors, and not add features that, even if valid, would make the system uncoordinated and inefficient.

An important cultural turning point in Brooks' model is related to the best-known and most recognised open-source software: Linux. Conceived in 1991 by Linus Torvalds, then a university student, it is a powerful alternative to proprietary operating systems such as Microsoft Windows and macOS.

Fig. 1 Notre dame de Reims, photo Reno Laithienne

Today, Linux is the basis of many applications created by multinational IT companies, research institutions and data centres. We use it every day when surfing the web and on every Android OS: created by Google on a Linux kernel. The success of Linux is undisputed from both a technical and commercial point of view, so much so that multinational companies such as Microsoft, Amazon, Facebook and Google invest millions of dollars every year in its development [12]. The cultural evolution that this operating system has brought with it lies above all in the openness and sharing that the project has maintained from the very beginning and which it still keeps thirty years after its birth.

Inside his *The Cathedral and the Bazaar* [13], Eric Raymond, already part of the informal group to which we owe the term open-source, retrieves the image of Reims Cathedral, suggesting for Linux that of a large and confusing bazaar. An environment teeming with projects and approaches that even contradict each other, from which only a miracle would have allowed the development of a stable and coherent operating system.

On the contrary, Raymond admits that the bazaar method worked very well for Linux, strengthening the process, rather than throwing it into chaos, proceeding at a speed that was difficult for those who wanted to build cathedrals. The merit of Torvalds, says Raymond, is that he has made the most of a virtual community without geographical boundaries, extending a solitary activity like programming thanks to the intelligence and power of a community. By anticipating Web 2.0, Linux represents an important turning point for open-source software, since then expanded in every field. Today, this different cultural approach contributes to the success of epoch-making initiatives such as Wikipedia, which is based on the open-source platform

MediaWiki. But also OpenOffice, a free alternative to the Microsoft Office package, collectively developed by a worldwide community that today includes thousands of authors [14].

3 Open Source and Architectural Design Software

The success of Linux, Wikimedia and other initiatives have confirmed the strength and scalability of open-source projects, which today have touched every industry, paving the way for new approaches that are now applied in all fields of knowledge.

Open-source diffusion is rather limited in the AEC industry, where the main digital tools adopted by architects are often proprietary software, created by a few large companies specialised in the design and development of packages specifically tailored to our sector. This is true for BIM programs and digital modelling software used since the early stages of projects: programs covered by trade secrets and protected by proprietary licenses. Less when we turn our gaze to advanced computational tools, often integrated into commercial applications and that are increasingly the result of open-source initiatives shared by groups of heterogeneous users.

3.1 Three-Dimensional Modelling Software

The predominant role played by a few software houses or even single programs is not limited to our industry, but has gained considerable weight in architecture. If between the late 1900s and the early 2000s many architects considered Autodesk AutoCAD the ideal software to develop their projects, today, with the spread of the BIM paradigm, architects, engineers and builders are increasingly opting for Autodesk Revit. A program that, despite all the alternatives, is on the verge of becoming the standard in the global architecture industry [4]. This circumstance brings with it the danger that architects will find themselves in an eccentric position to the project. While it is almost inconceivable today to approach architectural design without the aid of digital tools, the commitment of BIM, set by various governments around the world, translates into a constraint on the use of Autodesk Revit.

As evidence, consider an event that shaped the history of architectural software in 2020: the open letter to Autodesk about the rise in Revit costs. Signatories include international firms such as Zaha Hadid Architects, Grimshaw Architects, Rogers, Stirk, Harbour + Partners, Mecaano Architecten, Diller Scofidio + Renfro, MVRDV and other firms that alone make up an important part of the Autodesk Revit market. In the 5-page letter [15], the studies lament an increase in licencing costs of over 70% in five years, stagnation in software development and Autodesk's inability to understand the dynamics of the AEC industry.

Regardless of the content and the subsequent response from Autodesk, which has activated initiatives to share objectives and gather feedback, the letter alone is a clear signal of how commercial software is affecting our profession, and it shows the powerlessness of architects in addressing the software problem. This happens despite several three-dimensional modelling software currently available on the market under an open-source license. The longest-running of these is BRL-CAD, one precursor in the CAD family. Initially developed within the US Army, it is available for free in both executable and source code form since the early 2000s, allows three-dimensional modelling, and also integrates rendering tools [16].

Another feature-rich tool is FreeCAD, which was released in 2002 in an open-source format for mechanical engineering. It integrates not only two-dimensional and three-dimensional modelling functions but also advanced finite element analysis (FEA) methods that are rarely available in commercial CAD programmes. Most importantly, it falls into the category of BIM software, as it allows for a proper parametric description of architectural models and also supports the IFC format [17].

Among the champions of open-source digital modelling, we should therefore mention Blender, a programme born in 1994 and developed for digital animation, like the most famous Autodesk Maya. Unlike the package used by protagonists of the first digital turn, such as Greg Lynn and Zaha Hadid Architects, Blender is not proprietary software and it is available for free. In the nearly thirty years since it was first released, it has grown to include advanced features peculiar to other commercial programmes. These include NURBS, subdivision surface methods, physics simulation techniques, rendering packages, virtual reality engines, and an indefinite number of utilities that the user community has developed and distributed for free [18].

3.2 File Formats

The transmission of information has always been central to architecture as we know it. Traditionally, architects have represented their projects through technical drawings and documents on paper to share them with clients, engineers and builders. With the proliferation of CAD, designers have translated the same information into digital dimensions and although there were no reference formats at first, the widespread use of Autodesk AutoCAD has made DWG files a de facto standard in AEC. Autodesk introduced DWG, an abbreviation for the English Drawing, as a binary file for storing two-dimensional drawings with the first version of Autodesk AutoCAD. It is a proprietary, licenced file format owned by Autodesk for the management and operation of the various applications developed by the company. With the presentation of DWG, Autodesk also released DXF, Drawing Exchange Format, developed as a solution for data exchange between Autodesk AutoCAD and other programmes.

While Autodesk has never released technical specifications useful for developers of other companies to integrate DWG functionality into other programs, with DXF, the company publishes substantial documentation [19]. However, the success of this extension is limited, probably because of the lower possibilities of representing objects and solids that have enriched the different versions of DWG files. So much so that all CAD programs developed over the years support DWG as the main file exchange format. This is possible, not because Autodesk decided at some point to make the format characteristics public, but because a non-profit consortium, the Open Design Alliance [20], engaged in the reverse engineering of the format, creating open-source libraries available to anyone who wants to interpret and create DWG files within programs not developed by Autodesk. The enormous diffusion of Autodesk AutoCAD and the forced opening of the DWG format by ODA will contribute to the diffusion of the format, which is now available on any digital modelling tool.

Among the companies that develop programs for architectural design, one of those that have shown greater attention to open-source practices so far is certainly McNeel and Associates, which distributes Rhinoceros 3D with a proprietary license but has kept the 3DM format open, sharing the source code through the Open Nurbs project [21]. An initiative aimed at developers of CAD programs, who can rely on the company's format for the creation and management of many geometries including NURBS, B-rep, meshes, point clouds and SubD objects, store data and manage them through programming languages such as Python, JavaScript and C #. The Open Nurbs initiative became decisive in the success of the 3DM format, allowing programmers and designers the development and distribution of applications and utilities often published with open-source licences.

To testify to the growing attention towards open-source, and the need for a neutral format for sharing information and interoperability between programs, we mention the establishment of the IFC format, Industry Foundation Classes, proposed as a standard for the BIM software [22]. The author of this format, registered as an ISO standard since 2013, is Building Smart, an organisation founded in 1994 and which today brings together some of the major AEC software companies, together with a large group of multinational technology and construction companies.

3.3 Software Libraries

A sometimes underappreciated contribution to software is that of libraries: sets of functions or data structures that allow programmers to reuse code written by others and thus extend the functionality to their projects. As an example, consider some of the most complex software ever developed, such as social networks: systems that combine hundreds of different functionalities, perceived by the public as monolithic entities, but which could not function without libraries created by third parties. Instagram, one of the main promotional vehicles for contemporary architecture firms, uses almost 30 open-source software libraries, without which it would not be possible to view images and videos on our mobile devices or even simply register on the platform [23].

Even 3D modelling packages integrate third-party libraries, such as mathematical models, that allow them to solve geometric problems, from describing basic elements such as points and lines to more complex ones, such as curved surfaces. Among these, a very special place in the history of contemporary architecture is that occupied by the splines, without which the protagonists of the first digital turn would not have been able to conceive the continuous organic forms that characterise their works.

The recent history of splines, and T-splines in particular, highlights the problems and toughness of using proprietary software within the design process. This technology, developed in 2004 by a US start-up company, has proven successful in several firms interested in designing organic shapes. For about 6 years, the company distributed the library as a plug-in for the Rhinoceros 3D and SolidWorks packages. This happens until 2011, when Autodesk acquires T-Spline Inc., first eliminating the plug-in for SolidWorks and then discontinuing the upgrade and sale of the t-spline plug-in for Rhinoceros 3D, preventing the activation of new licenses to keep the t-splines only on Autodesk Fusion 360 [24].

The inability to activate new licences for Rhinoceros 3D has caused many problems within architectural firms, that until that moment have used the t-splines on Rhinoceros 3D to rationalise the shapes designed in the concept phase with subdivision surface methods on other packages such as Autodesk Maya and Blender. This has negatively impacted workflow in design studios, which, with the new versions of Rhinoceros 3D from 6 onwards, have sought alternative tools. We directly observed these setback when we took part in some meetings of the BIM group of the Zaha Hadid Architects studio in 2019. At one of these meetings, the team of architects tried to address the problem, as the t-spline tool was part of their internal workflow and the limited number of licenses owned by the studio, together with the inability to activate new licences, led to many slowdowns within the practice. They could find a temporary solution thanks to the availability of McNeel and Associates, who provided them with an ad hoc version of Rhinoceros 3D 5 with some features of the newer version.

Problems like this would not occur if the t-splines libraries were under an open-source licence that allowed users to view and modify the source code, take part in their development and make it common knowledge. A decision made by Pixar, a company that has shaped the history of digital modelling, with the first release of Opensubdiv in 2012 [25]. This open-source library for subdivision surfaces was developed together with Microsoft and other volunteers, who collaborated on a method to use the efficiency of parallel computing to produce organic surfaces starting from meshes.

Some ten years after its initial release, the library is available for architects and designers to use in some of the leading 3D modelling programs such as Autodesk Maya, Autodesk 3ds Max and Blender.

3.4 Visual Programming Languages and Open-Source Culture

The category of software where a cultural evolution in the relationship between open-source culture and architecture is most evident is undoubtedly that of visual programming languages. Environments that allow defining complex algorithms through a graphical node interface and that, even when distributed under a proprietary license, allow communities to extend the software capabilities by including features developed by others.

This openness to other actors has been crucial to the spread of Grasshopper 3D, which, as Pablo Zamorano, Steven Chilton and Xavier de Kestelier testify, has established itself among visual programming languages for architecture and design, thanks to its community gathered on Food4Rhino. The site hosts a group of hundreds of thousands of users who have collectively developed over eight hundred plug-ins and components in about ten years. On the website it is possible to find products of different quality and breadth, from components that perform a single function, to real packages with dozens of different utilities for architecture, engineering, robotics, digital manufacturing and much more [26]. As proof of how much the opening towards open-source is primarily a cultural phenomenon, it is important to note that almost 90% of the plug-ins developed are free to use, and most of them are also open-source software. While the success of Grasshopper 3D is certainly due to its community and the incredible number of open-source tools developed independently by users, McNeel and Associates released it under a proprietary license. This permit the company a relative control over the evolution of the central application needed to develop new components.

A limit exceeded by Dynamo, a direct competitor of Grasshopper 3D in the category of visual programming languages, which was developed by Ian Keough and released under an open-source license in 2015. The package, now also integrated within Autodesk Revit, allows designers to adopt algorithmic strategies according to a syntax similar to that of Grasshopper 3D, and is rapidly enriching itself with new components that users can create and share directly through the program interface. A feature that facilitates the flourishing of new shared computational strategies within the community. Today the number of packages created and shared by users had exceeded 500 units [27], and we also observe how several developers already authors of Grasshopper 3D components have translated them to be used on Dynamo.

3.5 *Simulation and Optimisation*

The methods of computational simulation and optimisation in terms of topology and environmental sustainability mark the clearest difference between the architectures of the first digital turn and contemporary approaches. As architects today, we are aware of the urgency of adopting these strategies, part of the collective effort in contrast to global heating. What we are probably less aware of is the fact that it is still almost impossible for architects to develop some of these strategies without open-source tools.

One of these, well known by many contemporary architects, is Kangaroo Physics, a plugin for Grasshopper 3D developed by Piker [28]. With over half a million downloads, it is one of the most popular among the VPL plugins, because it allows interactive simulation, optimization and constraint solving. The program, totally open-source, permitted many users to adopt form-finding strategies similar to those experimented with by Frei Otto, favouring the development and dissemination of innovative approaches not easily obtainable through commercial packages [29].

Another popular open-source plug-in available for both Grasshopper 3D and Dynamo is Lunchbox, a collection of computational tools that includes functions for managing data and geometries, but also for the rationalisation of architectural forms, to favour the interoperability, and adopt generative methods. The tool, created and distributed by Proving Ground, has over 600,000 users around the world and architects use it to define facades and divide them into regular panels according to grids with square, rhomboid and triangular modules [30].

Among the open-source tools available to all Grasshopper 3D users interested in developing architectural sustainability strategies, it is worth mentioning Ladybug tools, an application created in 2012 by Roudsari [31]. As an architect and IT developer, he felt the need for a tool to simplify the modelling activities, and speed up the architectural workflow, then slowed down by repetitive operations. The plugin also aimed to solve interoperability problems between different instruments and integrate tools for environmental analysis. 2013 is the year of the first release. It now includes thirty different components for integrating environmental data, solar radiation models and levels of brightness of the building environment. Subjects of great importance to achieve sustainability within the architectural project, and that any designer can today address via this powerful tool, close to becoming a standard within the contemporary architectural workflow. After several releases, the application has evolved, also integrating Honeybee, another package aimed at connecting architectural models with tools for the simulation and energy validation of buildings such as Radiance, Energy Plus and others.

Ladybug is an excellent example of how the opening towards open-source contributes to the improvement and dissemination of new shared design approaches. While before its release, architects could conduct similar strategies only through commercial tools, purchasable at sometimes prohibitive costs. Since its first distribution, over 300.000 users downloaded the package, making it one of the most used tools in sustainable architectural design. Several developers and architects have also taken part in the project, collaborating in the development of new features. As Chris Mackey, who, on the occasion of his thesis in architecture at the Massachusetts Institute of Technology [32], developed a thermal comfort analysis tool now part of the plug-in. Or as it recently happened on the occasion of other tools, developed by volunteers, and which now allow the connection between Ladybug and other advanced software useful for data management and to extend the environmental simulation to entire portions of a city.

The need for a different understanding of the architectural artefact through simulation practices has led designers to get closer and closer to physics and its computational methods. Similar tools are often created within major international universities and released under an open-source license. This is the case of Open FOAM, Open-source Field Operation And Manipulation, a numerical simulation application released by Henry Weller and Hrvoje Jasak in 2004 [33]. The tool, created for the solution of continuum mechanics and fluid dynamics problems, is currently adopted by some large architectural firms to predict wind effects on buildings and optimise comfort in public spaces.

In our process that collects the main open-source tools adopted by contemporary architectural designers, we have followed a path that starts from generalist CAD tools, up to programs created for resolving specific problems and the deepening of the architectural artefact in environmental terms and optimization. A vast and heterogeneous ecosystem populated as much by free programs developed by small groups as by large corporations specialised in creating programs for architecture: companies that at different levels have adopted open-source strategies and related business models.

The relationship between open-source and architecture, however, does not stop with programs created by companies or informal groups. It becomes even more solid when the architectural firms directly address the development of tools based on internal needs. An increasingly frequent eventuality, as confirmed by Aurelie de Boissieu, who in describing the internal practice of her studio, tells how frequently architectural designers, BIM technicians and developers work together in defining new tools sewn around their internal workflow. This software is often born as scripts composed of a few code lines of code, structured to automate processes, link different programs together, or correctly carry out internal praxis. As witnessed by the interviewee, these scripts frequently evolve until they become real packages, available to those who cannot understand computer languages.

This is also possible thanks to Python, an open-source text programming language that has become one of the most widely used languages in architectural firms thanks to its ease of use. Developed in 1991 by Guido van Rossum, the language is now integrated into the main architectural modelling programmes, from Grasshopper 3D to Autodesk Revit, and has thus become a fundamental tool for the establishment of design strategies aimed at environmental sustainability.

3.6 Interoperability

The multiplicity of tools available for architectural design and the impossibility of solving the complexity of the project with a single software has presented architects with an issue that has caused not a few delays in the studios: interoperability. A subject that, despite the commitment of various actors and the creation of IFC, has not led to the definition of a standard capable of integrating the algorithmic components of the architectural project into an interoperable BIM model. This means that to develop algorithmic processes and optimisation of architectural forms, several studios today rely on Rhinoceros 3D and Grasshopper 3D during the early stages of the project and subsequently translate the model according to the BIM paradigms on Autodesk Revit. A transformation that results in the loss of algorithmic information. Despite the new possibilities offered by integrating Dynamo into Autodesk Revit, the node logic and the algorithmic syntax do not find a direct counterpart in any of the currently available standards.

This discontinuity creates many difficulties for designers, as confirmed by Pablo Zamorano, who is familiar with the common problems in contemporary architectural practises and also testifies to how his firm has found an effective solution to the problem thanks to Flux. An application developed as part of the research programmes of Google X Laboratories, a Google offshoot that deals with energy and economic sustainability issues. The goal of Flux was to solve interoperability between software, enabling remote collaboration via cloud systems, and data management between actors taking part in the design process. It created a dialogue between packages, converting models and information according to the intrinsic logic of the different programmes.

Unlike other data sharing systems based on file transfer, Flux users had to install it as a plug-in on each software to be connected, it then acted as an exchange node between programs and enabled the sharing of projects, analyses and other data in real-time. It instantly linked Rhinoceros 3D/Grasshopper 3D and Autodesk Revit/Dynamo pairs with each other and with Excel, Autodesk Autocad, and Sketchup [34]. Since 2010, the year in which the project began, the research group has grown, detached from Google, became a startup, getting 8 million dollars in funding in 2014, and developing a large collection of useful interoperability tools for architectural workflow [35].

Utilities, which as Zamorano recalls, have become part of the tools adopted by designers in many offices, where architects converted their workflow to operate with that system. Despite this, the company interrupted the Flux project in March 2018, bringing back the architectural firms to confront again with problems already solved: a case incredibly similar to the one that has already occurred with t-splines.

Given the clear need for an interoperability solution for architectural design, other initiatives filled the void left by Flux, developing applications and utilities aimed at facilitating the exchange of information. Among the various platforms recently released, one of the most complete is undoubtedly Speckle, a data exchange system started by Dimitrie Stefanescu at the University College of London in 2016. The project took the form of a start up at the beginning of 2020, proposing as its main product a system for sharing information between professionals connected remotely, also allowing the desired connection between the pairs Rhinoceros 3D/Grasshopper 3D, Autodesk Revit/Dynamo and other tools. Speckle is based on a web infrastructure, it enables users to structure highly configurable interoperability systems, supporting the creation of specific data classes created by end-users also integrating structured formats such as IFC. The data transfer system runs on a server installed on each computer, which serves as a node for sharing information and processes, and ensures the correct functioning of the system in case the other nodes in the network are offline by recording any operation on a shared model.

Speckle implements a hierarchical access system, which grants effective control over the processes and actors who collaborate in the project's definition. An integrated utility named SpeckleViz allows the real-time visualisation of the processes shared between different users through the server, collecting each operation and illustrating architectural workflow in real-time [36]. Unlike Flux, which was a proprietary software and requested the subscription of licenses, Speckle is free to use, and it is an open-source initiative. This bodes well for its longevity, less influenced by market fluctuations, and capable of growing based on a community of users and supporters [37].

A turning point in solving interoperability issues between Rhinoceros 3D and Autodesk Revit is certainly the 2019 solution proposed by McNeel and Associates, which displacing the competition introduced under the new features of version 7 of Rhinoceros 3D, the open-source Rhino.inside component [38]. The utility allows the popular modelling package to be used in any other programme, starting with Autodesk Revit, extended to include Rhinoceros 3D's modelling capabilities and Grasshopper 3D collection of plug-ins, ensuring levels of bi-directional interoperability never achieved.

In the interviews, Al Fisher testifies the growing role of open-source within the AEC industry, talking extensively about The Bhom, Buildings and Habitats Object Model, an interoperability platform started by his office in 2018 [39]. With the Bhom, Buro Happold intends to create a shared elastic infrastructure, accessible and modifiable by end-users who can easily integrate new functions and connections thanks to the innovative paradigm represented by the ECS, Entity Component System. This is a different IT structuring paradigm experimented with for the first time in 2007 within the video game industry. While BIM currently follows the model of OOP, a paradigm which allows the users to create entire families of objects from shared properties and functions, ECS adds new degrees of freedom to the system, determining entities based on data that define them, and thus allowing to include new properties and functions as the project needs change [40]. Inside the interview, Al Fisher cites a practical example that helps to understand the usefulness of ECS. If on the occasion of a project at an advanced stage, the lighting specialists want. To change the characteristics of a lighting system, it will be necessary to return to the definition of those objects to vary their properties upstream of the workflow. An activity that the ECS speeds up and simplifies, allowing an easier injection of new properties without incurring problems and viscosities that currently delay the project. The Bhom provide designers new flexibility in terms of interoperability and easier integration of new tools, but also and above all, it represents taking responsibility for a common problem, and trying to solve it in collaboration with other actors in the AEC field through open-source.

4 Limits to the Development of Open-Source Software for Architectural Design

Despite the open-source contributions to contemporary architectural design, although several architects exploit the strength of open source tools for the design and analysis of architectural models, it is palpable a lack of awareness and a substantial distrust of open-source software within the industry.

While many of the interviewees, such as Arthur Mamou Mani and Edoardo Tibuzzi, recognise the contribution of open-source tools, and testify to their use, hoping for a greater shared effort in their adoption. It is also true, as Daniel Davies states, that most architects limit themselves to using proprietary software. According to Aurelie de Boissieu, the limited adoption of open-source tools in our industry is because of the relative novelty that digital represents in architecture. A factor that binds architects to the use of commercial tools, but which the industry will overcome in the coming years.

In April 2019 we took part in a talk on the digitalisation of architecture organised in London, a debate between three important experts in architectural design and digital tools: the aforementioned Aurelie de Boissieu, head of the London BIM department of Grimshaw Architects, Mauro Sabiu of Zaha Hadid Architects and Benedict Wallbank of Viewpoint Construction Software, a multinational specialised in the development of applications for the management of the architectural order and the construction of architectural artefacts [41].

On the occasion of the debate, I asked the three what their opinion was on the role of open-source tools in architecture, engineering and construction, and it is interesting to note how the opinion of these experts varies. While de Boissieu sees open-source as the only solution to free designers from the constraints imposed by development companies. Wallbank, which admits how the open-source paradigm can lead to resolving specific problems, also argues that the broader problems, shared in the global landscape of digital tools for architecture, will be more easily solved by large companies rather than by individual developers or informal groups. Referring in particular to data sharing and the digitalisation of the built environment.

This diversity of opinions is not uncommon, while on the one hand, there is a shared opinion that open-source programs have a role to play, and even are fundamental for the sustainable development of the entire sector. Many think that the small number of individuals involved in open-source software will never reach the reliability of a large specialised company. An opinion shared by Harry Ibbs, who sees the limits of open-source initiatives in the modest size of their groups, a characteristic that makes them less solid, unable to provide continued support, and guarantees that architects receive by multinationals against the payment of commercial licenses. The interviewee compares, as an example, the proprietary software Autodesk Maya, with the open-source Blender. Affirming that the latter is a perfect alternative to the first, able to perform all the functions of a commercial competitor, sometimes even better.

Despite this, Ibbs continues, very few architects adopt Blender. To achieve this, there would need to have in the team a specialist who cannot make the best use of the software, but also to customise its structure and functionality to compensate for the support that Autodesk provides for Maya to anyone who has purchased a licence. Without this kind of support, or someone with specialist knowledge in the studio, architects risk getting stuck and experiencing costly delays in the design process.

This is true for Blender and for dozens of other open-source tools that perfectly fulfil the functions of commercial packages but offer no support. With this in mind, the interviewee said, anyone who wants to encourage the use of an open-source tool for architectural design needs to provide users with reliable support. According to Ibbs, this could come directly from the developer community by creating an ecosystem aimed at both feature development and end-user support.

Ibbs sees another difficulty in the market of software for architecture. Whereas until the end of the 1990s there were only a few companies specialising in the production of digital tools, these companies have now become real giants.

So if it is relatively easy even for a small group to propose an application to solve a particular problem, developing a programme like Autodesk Revit, which has released dozens of new features over the years, would require an investment of tens of millions of euros. An investment into which additional funds for marketing activities would have to be added. Similar budgets are hardly available for small informal groups. Even if they produce innovative products, they are often acquired by large companies that can alternatively develop similar functions and integrate them into commercial tools.

5 Conclusions

Although not all architects are aware of it, the relationship between open-source software and architectural design is already alive and well and is likely to develop further in the coming years. As several interviewees testify, architects mainly work with proprietary software developed by a limited group of specialised software houses. But, when the architectural project reaches the complexity of environmental sustainability, topological optimisation and simulation, the tools used to develop it are mostly open-source software.

The open-source culture is spreading in architecture thanks to ever-increasing proximity to disciplines such as computer science and physics, areas where open-source initiatives have been alive for years, proving their reliability and capturing important market shares. This cultural evolution is evident in the success of Grasshopper 3D, proprietary software that would not have achieved the popularity it has over the past 15 years without the help of a community of thousands of developers willing to share the source code of their plug-ins. Similarly, and at a different pace, some software houses that base their business on proprietary software are approaching open-source, starting with formats and ending with the release of fully open-source applications.

Meanwhile, individuals and small groups of architects-programmers are busy solving specific problems, developing effective software and orienting their business models towards open-source. These small groups certainly cannot compete with the software giants in developing complete modelling programs. However, there is no shortage of solid packages capable of offering solutions similar to those of proprietary software and delivering results of the same quality, as with Maya or FreeCAD, in which it is possible to invest, either economically or through direct collaboration.

Buro Happold's The Bhom initiative is not only a sign of openness towards open-source directly promoted by a professional firm, but an important precedent for the future of architecture. This is because it represents a willingness to face the problems common in the AEC industry with the declared intention of sharing processes and results.

To avoid problems and delays like with the DWG format, T-splines and Flux, or even worse in the future, AEC professionals and especially architects need to develop a greater awareness of the weight that tools have in our field. This means that we need to perceive our role differently as consumers and, where possible, producers of architectural design software. Alternatively, we can think of sending letters of complaint to the companies that develop the products we use every day.

References

1. Gallo, G.: Digital and quarantine. In: Milocco Borlini, M., Califano, A. (eds.) Urban Corporis Unexpected, pp. 284–290, Anteferma edizioni, Conegliano (2021)
2. McLuhan, M., Fiore, Q.: The Medium is the Message. Penguin Books, New York (1967)
3. Ceccato, C.: Integration: master, planner, programmer, builder. In: Soddu, C. (eds.) Proceedings of Generative Art conference, pp. 142–154. Generative Art, Milano (2001)
4. Gallo, G., Tuzzolino, G.F., Wirz, F.: The role of Artificial Intelligence in architectural design: conversation with designers and researchers. In: Soellner, M. (ed.) Proceedings of S.Arch 2020, the 7th International Conference on Architecture and Built Environment, pp. 198–206. S.Arch, Tokyo (2020)
5. Ratti, C., Claudel, M.: Open Source Architecture. Thames & Hudson, London (2015)
6. Gallo, G.: Architettura e second digital turn, l'evoluzione degli strumenti informatici e il progetto. Doctoral dissertation, University of Palermo, Palermo (2021)
7. Raymond, E.S.: Goodbye, free software; hello, open source. http://www.catb.org/~esr/open-source.html. Accessed 21 Apr. 2022
8. Heffan, I.V.: Copyleft: licensing collaborative works in the digital age. Stanf. Law Rev. **49**(6), 147–155 (1997)
9. Stallman, R.M.: Free Society: Selected Essays of. Gnu Press, Boston (2002)
10. Brooks, F.P.: The Mythical Man-Month. Addison-Wesley, Reading (1975)
11. Thurner, N.: Proiettili d'argento nella rete, Frederick Brooks: un punto di partenza tecnico per una riflessione filosofica sulla natura del software e delle architetture digitali. In: Ciastellardi, M. (ed.) Le architetture liquide: dalle reti del pensiero al pensiero in rete, pp. 145–153. LED Edizioni Universitarie, Milan (2009)
12. Asay, M.: Why Microsoft and Google are now leading the open source revolution. https://www.techrepublic.com/article/why-microsoft-and-google-are-now-leading-the-open-source-revolution/. Accessed 26 Apr. 2022
13. Raymond, E.S.: The Cathedral and the Bazaar. O'Reilly Media, Sebastopol (1997)
14. Gamalielsson, J., Lundell, B.: Sustainability of Open Source software communities beyond a fork: how and why has the LibreOffice project evolved? J. Syst. Softw. **89**, 128–145 (2014)
15. Letters to Autodesk. https://letters-to-autodesk.com/. Accessed 23 Apr. 2022

16. Castro, H., Putnik, G., Castro, A., Fontana, R.D.B.: Open Design initiatives: an evaluation of CAD Open Source Software. In: Putnik, G. (ed.) 29th CIRP Design Conference 2019, vol. 84, pp. 1116–1119. Elsevier, Póvoa de Varzim (2019)
17. Logothetis, S., Valari, E., Karachaliou, E., Stylianidis, E.: Spatial DMBS architecture for a free and open source BIM. In: International Archives of the Photogrammetry, Remote Sensing & Spatial Information Sciences, vol. 42, pp. 467–472. ISPRS, Ottawa (2017)
18. Tan, G., Zhu, X., Liu, X.: A free shape 3d modeling system for creative design based on modified catmull-clark subdivision. Multimed. Tools Appl. **76**(5), 6429–6446 (2017)
19. Lu, J.D.: Analysis on AutoCAD DXF file format and the 2nd development graphics software programming. Microcomput. Dev. **9**, 101–104 (2004)
20. Dwg Toolset. https://www.opendesign.com/solutions#dwg-toolset. Accessed 26 Apr. 2022
21. Open Nurbs initiative. https://www.rhino3d.com/opennurbs. Accessed 21 Apr. 2022
22. ISO 16739-1:2018 Industry Foundation Classes (IFC) for data sharing in the construction and facility management industries. https://www.iso.org/standard/70303.html. Accessed 26 Apr. 2022
23. Libraries We Use. https://www.instagram.com/about/legal/libraries/. Accessed 23 Apr. 2022
24. Mings J.: Yes, Autodesk is Finally Ending T-Splines. https://www.solidsmack.com/cad/yes-autodesk-finally-ending-t-splines/. Accessed 26 Apr. 2022
25. Nießner, M., Loop, C., Meyer, M., Derose, T.: Feature-adaptive GPU rendering of Catmull-Clark subdivision surfaces. ACM Trans. Graph. **31**(1), 16–37 (2012)
26. Apps for Rhino and Grasshopper. https://www.food4rhino.com/browse?. Accessed 26 Apr. 2022
27. Kensek, K.: Visual programming for building information modeling: energy and shading analysis case studies. J. Green Build. **10**(4), 28–43 (2015)
28. Piker, D.: Kangaroo: form finding with computational physics. Archit. Des. **83**(2), 136–137 (2013)
29. Senatore, G., Piker, D.: Interactive real-time physics: an intuitive approach to form-finding and structural analysis for design and education. Comput. Aided Des. **61**, 32–41 (2015)
30. Cubukcuoglu, C., Ekici, B., Tasgetiren, M.F., Sariyildiz, S.: OPTIMUS: self-adaptive differential evolution with ensemble of mutation strategies for grasshopper algorithmic modeling. Algorithms **12**(7), 141 (2019)
31. Roudsari, M.S., Pak, M., Smith, A.: Ladybug: a parametric environmental plugin for grasshopper to help designers create an environmentally-conscious design. In: Wurtz, E., Roux, J.J. (eds.) Proceedings of the 13th International IBPSA Conference, pp. 3128–3135. IBPSA, Chambery (2013)
32. Mackey, C.C.W.: Pan Climatic Humans: Shaping Thermal Habits in an Unconditioned Society, Doctoral Dissertation. Massachusetts Institute of Technology, Cambridge (2015)
33. Chen, G., Xiong, Q., Morris, P.J., Paterson, E.G., Sergeev, A., Wang, Y.: OpenFOAM for computational fluid dynamics. Not. AMS **61**(4), 354–363 (2014)
34. Afsari, K., Eastman, C.M., Shelden, D.R.: Cloud-based BIM data transmission: current status and challenges. In: Salah, M., Abu Samra, S., Hosny, S. (eds.) Proceedings of the International Symposium on Automation and Robotics in Construction, pp. 213–235. ISARC, Auburn (2016)
35. Lunden, I.: Flux Emerges From Google X And Nabs $8M To Help Build Eco-Friendly Buildings. https://techcrunch.com/2014/05/06/flux-the-first-startup-to-spin-out-of-google-x-nabs-8m-for-its-eco-home-building-platform/. Accessed 20 Apr. 2022
36. Poinet, P., Stefanescu, D., Papadonikolaki, E.: SpeckleViz: a web-based interactive activity network diagram for AEC. In: Chronis, A., Wurzer, G., Lorenz, W.E., Herr, C.M., Pont, U., Cupkova, D., Wainer, G. (eds.) Proceedings of the 11th Annual Symposium on Simulation for Architecture and Urban Design, pp. 419–428. SimAUD, San Diego (May 2020)
37. Speckle 2.0: Vision & FAQ. https://speckle.systems/blog/speckle2-vision-and-faq/. Accessed 23 Apr. 2022
38. Rhino.Inside. https://www.rhino3d.com/inside. Accessed 22 Apr. 2022
39. The Buildings and Habitats object Model. https://bhom.xyz. Accessed 20 Apr. 2022

40. Alatalo, T.: An entity-component model for extensible virtual worlds. IEEE Internet Comput. **15**(5), 30–37 (2011)
41. Watch our talk on the digitalisation of architecture with Zaha Hadid Architects Grimshaw and Viewpoint. https://www.dezeen.com/2019/04/25/knauf-digitalisation-architecture-talk-livestream/. Accessed 26 Apr. 2022

Educating the Reflective Digital Practitioner

Ioanna Symeonidou

Abstract The chapter discusses the processes of a designer's reflection-in-action when employing simulation-based design tools. It revisits Donald Schön's seminal work on the Reflective Practitioner, considering the current technological mileu where simulations of physical or environmental behavior educate future architects on how to reflect in action during the design process, rather than analyzing and modifying their design a posteriori. Schön argues in favor of the idiosyncratic element in design decision making which is based on practice. Hence, the digital practitioner of our times develops an intuition and knowledge that derives from the exposure to simulations and computational tools. The chapter will expound on processes of experiential learning in architecture and discuss the findings and experiences from architectural studio case studies that employed computational tools for form-finding to provide real-time feedback on the behaviour and geometry of the projects. The curriculum aimed at combining teaching strategies, digital media and design processes towards the objective of educating the reflective digital practitioner of the future.

Keywords Algorithmic design · Physics simulation · Donald Schön · Architectural education · Design computation · Form-finding

United Nations' Sustainable Development Goals 4. Quality Education · 9. Industry · Innovation and Infrastructure · 12. Responsible Consumption and Production

1 Introduction

Architectural education has always followed an experiential learning approach, which in older times took the form of apprenticeship near the master builder, and gradually evolved into a systematic study in design studios, which until today are considered as the signature pedagogy of architectural education, placing a strong

---

I. Symeonidou (✉)
Department of Architecture, University of Thessaly, Pedion Areos, 38334 Volos, Greece
e-mail: symeonidou@uth.gr

© The Author(s), under exclusive license to Springer Nature Switzerland AG 2024
M. Barberio et al. (eds.), *Architecture and Design for Industry 4.0*, Lecture Notes in Mechanical Engineering, https://doi.org/10.1007/978-3-031-36922-3_9

focus on physical models, tacit knowledge, hands-on experimentation and several cycles of reflection and modification of the student projects. In his seminal work "*The Reflective Practitioner*" Donald Schön argues that practitioners develop "*reflective conversations with the situation*" [1]. Architectural education has a long tradition of craftsmanship and 3D models and a strong connection to the means of industrial production. The BAUHAUS pedagogy was a response to the 1st Industrial Revolution, introducing workshops for industrial production of artifacts ranging from furniture to ceramics, metal working and weaving, all of which aimed to train students with regards to the construction methods of that time. Gropius highlighted the necessity to educate '*a new generation of architects in close contact with modern means of production*' [2], exemplifying the benefits of standardization and mass production. Similarly, the 4th Industrial Revolution is marked by the increasing interconnectivity of tools and processes, the automation of production methods and mass customization. According to BAUHAUS pedagogy, Josef Albers would affirm that students should develop a "*feel*" for materials and processes. Similarly, Schön considered primordial that designers are to develop a "*feel*" for situations as the sine qua non of design practice. He argued that there is a shift from "*problem solving*" to "*problem setting*" which would nowadays become of crucial importance if seen through the prism of Industry 4.0, as the current generation of architects have in their disposal computational tools that accelerate design thinking and can compute a large number of parameters in a simulation-based design process. The question that arises here, is how do we prepare future architects for the challenges and opportunities of Industry 4.0? Furthermore, how can we instigate concepts of reflective practice when we shift from physical models to digital simulations? The chapter aims to address the importance of educating the reflective digital practitioner of the future, to discuss contemporary design education protocols and examine the way students develop a "*feel*" for digital media in concurrence with the ideas of Schön about reflection in action. Although the understanding of computational tools for simulations evidently falls within the spectrum of "*hard*" knowledge of science and scholarship, this chapter wishes to explore the "*soft*" knowledge of artistry and unvarnished opinion with regards to simulation-based design processes, inquiring the epistemology of digital practice, opting for an insight into the professional knowledge of competent digital practitioners and enriching contemporary design teaching protocols with tacit knowledge and experiential learning.

2 The Legacy of Donald Schön

Donald Schön, professor of education and planning at MIT, has often been a reference for architecture educators [3–12]. Schön is known for developing the concept of reflective practice and contributing to the theory of organizational learning. In his book "*The Reflective Practitioner - How professionals think in action*", Schön inquires the epistemology of practice. He considers that a competent practitioner should not only be able to solve a problem but emphasizes the designer's ability to construct a problem. Problem setting and framing of the context is of crucial importance for Schön, and this can directly translate to computational design simulations, where the actual parameter setting is more important than the solver itself. Schön researches the epistemology of practice that is inherent to intuitive design processes, the spontaneous response of an experienced practitioner. As he explains *"We are often unaware of having learned to do these things; we simply find ourselves doing them. In some cases, we were once aware of the understandings which were subsequently internalized in our feeling for the stuff of action"*.

In his writings he describes the difference between *"knowing-in-action"* and *"reflection-in-action"*: Knowing-in-action is *"….the repertoire of routinized responses that skilful practitioners bring to their practice"*, gained through training or experience [1]. He expounds that if we operate outside our normal routines, outcomes are not as expected—surprises, uncertainty or non-understanding occur. Therefore, we need to *"reflect"* on our actions, on the spot, so we can still have an impact on the outcome. *"Our spontaneous responses to the phenomena of everyday life do not always work. Sometimes our spontaneous knowing-in-action yields unexpected outcomes and we react to the surprise by a kind of thinking what we are doing while we are doing it"*, a process he calls *"reflection-in-action"*. The reflection *"... has a critical function, questioning and challenging the assumptional basis of action, and a restructuring function, reshaping strategies, understanding of phenomena, and ways of framing problems"* [1].

Based on this assumption, a designer should be trained not only with regards to the technical requirements and programming skills that are needed to create a simulation tool for architecture, but this knowledge should target the ability to reflect in action within the dynamic context of a simulation. An experienced digital designer will intuitively know which parameters of the simulation need fine tuning, without necessarily being able to explain why. An analytical mindset would turn to science or engineering to understand the influence of parameters on the result of a simulation. However, a design practitioner without engineering or scientific background, after several iterations of experimentation with simulation-based design tools, will be able to intuitively predict the behaviour of the simulation and drive the design outcome. Within a digital simulation environment, a designer can reflect in action, by varying parameters, combining and recombining a set of figures *"within the schema which bounds and gives coherence to the performance"* [1].

During the educational activities presented in this chapter, the students engaged in simulation-based design processes, in a highly experiential approach, where their choices and *"spontaneous responses"* highly depended on the material or software at hand. Both in the digital as well as in the physical environment an elastic material would form different geometric configurations if stretched in a certain way, whereas a steel rod would spring back within certain range of bending and obtain a permanent deformation if the elastic limit is exceeded, therefore changing its behaviour and the experimentally obtained geometries. Similarly, in cases of doubly curved surfaces that comprise of planar parts, a planarization algorithm will behave differently depending on the size of panels and their anchorpoints, giving rise to a range of solutions that are feasible and can be digitally manufactured, capitalizing on the potential of Industry 4.0. The interconnectivity between machinery and design processes, has not yet been fully exploited by contemporary designers, but there is already evidence of a broad range of projects that adopt strategies of performance-oriented processes with the aim to reach informed design solutions. In this realm Schön's reflection process, with the necessary adaptations for the current technological milieu, offers a huge potential for architects to envision new ideas, solutions and theories:

> Depending on the context and the practitioner, such reflection-in-action may take the form of on-the-spot problem-solving, theory-building, or re-appreciation of the situation. When the problem at hand proves resistant to readily accessible solutions, the practitioner may rethink the approach he has been taking and invent new strategies of action. When a practitioner encounters a situation that falls outside his usual range of descriptive categories, he may surface and criticize his initial understanding and proceed to construct a new, situationspecific theory of the phenomenon [1].

3 The Challenges of Digital Design Media in Architectural Education

Never before in the history of architectural education was there such a drastic shift in the way we design and make, and furthermore in the way we think. The next generation of architects, which are currently being trained in architecture schools, is a completely different one, there is a big discontinuity taking place. Marc Prensky refers to them as **digital natives**[1] while their educators are undoubtedly the last generation of **digital immigrants**.[2] We are therefore passing through a moment of *"singularity– an event which changes things so fundamentally that there is absolutely no going back"* [13], and this is marked by the arrival and rapid dissemination of digital design tools.

Digital technologies have changed the nature of architecture; this was inevitably reflected on architectural education. *"Educators in engineering schools are experiencing a new pressure to change the way they teach design-related courses in order to equip their students to interact with CAD/CAM/CAE systems and have a knowledge of their fundamental principles"* [14]. This transition is currently occurring in a rather unplanned manner, as new tools are quickly adopted by schools around the globe, as a chain reaction and without proper curriculum planning and organization. Technological advancements took place so fast and architecture schools were the early adopters of computational media, therefore, learning how to integrate new tools in design, almost on the go, without prior strategic planning. The rush to engage with digital media was further accelerated by the market needs and the digital culture in all media of our everyday lives. Within this transition phase several

---

[1] Digital Natives refer to the students today, as they are all "native speakers" of the digital language of computers, video games and the Internet [13].

[2] Digital Immigrants learn—like all immigrants, some better than others—to adapt to their environment, they always retain, to some degree, their "accent," that is, their foot in the past [13].

educational experiments have taken place, and new design opportunities appeared. Currently, in architecture schools around the world, both students and educators have already gone through several years of experimentation with digital media, some of which have proven to be particularly useful for the advancement of education and architecture in general. With regards to digital design media in particular, the use of simulation tools for design represents only a very small percentage within the possibilities offered by digital media. However, it is an indicative example where research and production tend to agglomerate towards two extreme cases. On one hand there are the engineers that are fluent with setting up simulations, and even programming routines from scratch but with little or no design training at all. On the other hand, there are the competent designers who lack background knowledge in fields of science or applied science who may use readily available simulation tools but are not always capable of understanding the underlying parameters so as to frame the design problem. The current generation of computational designers respond to this knowledge gap, bridging the two extremes, by either working in multidisciplinary groups or by opting for dual degrees and specializations. Edgar Schein in his book "*Professional education: some new directions*" discusses the three components of knowledge, with each component relying and building upon the previous [15]. He distinguishes between science (ex. math, physics) applied science or engineering (ex. computation, engineering) skills and attitudinal component (ex. computational designer) that is based on the previous but eventually is responsible for delivering a project. More specifically for simulation-based design processes, this would require an understanding of the underlying mathematical and physical principles. But is this always possible for design students? Could we possibly bypass the lack of background knowledge by employing apprenticeship models like those proposed by Donald Schön?

Although a lot has been said and written about the education of architects, very little has been said about the preparation of teachers of architecture [16]. It is interesting to observe current trends in architectural education, looking at established theories like constructionism [17, 18] and experiential learning which have been embraced by architectural educators, and speculate how new technologies and digital media can be integrated in the architectural curriculum, exploiting the new possibilities offered by the use of simulation modeling towards novel design processes that boost creativity and innovation.

4 Jumping in at the Deep End

Introducing students without scientific background to simulation-based design processes might seem like throwing the students in at the deep end and letting them "*sink or swim*". Several educators would claim such approach to be stressful for the students as it is forcing them to act outside their comfort zone, however Schön would claim that "*the student must begin to design before she knows what she is doing*" [19]. However paradoxical that may sound, it proved to be more time-efficient and resourceful than traditional academic learning, that would request a robust knowledge of the underlying science prior to experimentation. Students are often unaware of the knowledge that is already there and find themselves acting as a reflective digital practitioner, learning and reflecting in action. As Schön would explain, through reflection a student can *"surface and criticize the tacit understandings that have grown up around the repepitive experiences of a specialized practice, and can make new sense of the situations of uncertainty or uniqueness which he may allow himself to practice"* [1].

The three design studio projects presented in this chapter are on one hand systematic, based on traditional learning theories, incorporating contemporary technology, software, employing new methods of knowledge documentation, and on the other hand chaotic with regards to the use of digital media and algorithms in architecture, facilitating the acquisition of tacit knowledge about emergent behaviors, multi-criteria optimizations, structural simulations and agent-based systems. By revisiting the theories of Schön, we are now able to take advantage of the technology to seamlessly transfer design ideas to projects. The three case studies presented in this chapter are drawn from design studio courses taught by the author. In all cases students had no previous experiences with computational design. All three studios employed simulations that were undertaken in Kangaroo Physics engine developed by Piker [20], as it is a relatively simple interface and students can intuitively draw parallels between the material feedback and behavior that would occur in the digital and the physical world. Kangaroo comprises of a collection of algorithms that computationally simulate some aspects of real-world physical behavior of materials and objects, acting as vehicles for design experimentation and innovation. Simulation based design processes and form-finding fall within a well-established design methodology, where dynamic models are influenced by internal or external parameters. In the case of a physics simulation, the architectural form is obtained as a negotiation of forces, environment and constraints. All of the above are eventually design parameters that drive the design towards a state of equilibrium were the design intent and the design criteria are met. Hence, Schön's persistence to shift the weight to problem setting rather than problem solving becomes more topical than ever. The simulation will eventually reach one or more solutions that optimize form and performance for certain parameters. What drives that design however is the way the problem is set and not the solution itself. Design decisions are reflected in the way the problem is set, and the hierarchy of parameters within the simulation. Surprises and unexpected results do occur, due to the emergent character of simulation-based design models. Simulations of physical forces for architectural form generation are increasingly gaining ground in architectural education as there is a broad selection of computational tools readily available that allow quick experiments to be conducted. The integration of simulation tools within the process of architectural design with digital media will be seen in the three case studies that follow.

4.1 Tensile Structures and Dynamic Relaxation

The studio which took place at Graz University of Technology, introduced students to the design of tensile structures through analogue and digital form-finding (Fig. 1). Being the basis of most design-oriented simulations, spring-particle algorithms have become a great design tool for architects; they allow experimentation and computational form-finding of a great variety of forms, ranging from catenaries and gridshells, to membranes and cable nets.

Spring-particle systems are based on lumped masses (particles) which are connected by linear elastic springs. Each spring is assigned a constant axial stiffness, a rest length, and a damping coefficient. Springs generate a force when displaced from their rest length. Each particle in the system has a position, a velocity, and a variable mass, as well as a summarized vector for all the forces acting on it.

The students used dynamic relaxation which is "*a computation modeling, which can be used for the form-finding of cable and fabric structures [...] The system oscillates about the equilibrium position under the influence of loads. The iterative process is achieved by simulating a pseudo-dynamic process in time*" [21].

There are many benefits of structural form-finding using spring-particle systems. It embeds criteria of structural optimization from the first stages of the design process, while it assists architects to increase their intuitive understanding of the structural behaviour of geometrically complex forms. While traditional architecture and engineering aims at the structural optimization of an existing form, a dynamic form-finding system can lead to a real-time discovery of structural form [22] encouraging the morphogenesis of optimized structures rather than a post-design optimization. Understanding the association between geometry and material behavior, the elastic properties of membranes or computational spring meshes and the obtained form, leads to a '*synergetic approach to design integrating form, structure, material and environment*' [23].

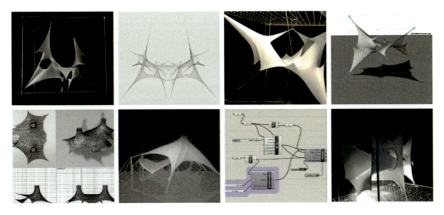

Fig. 1 Physical models and digital simulation models for tensile structures from the studio at Graz University of Technology

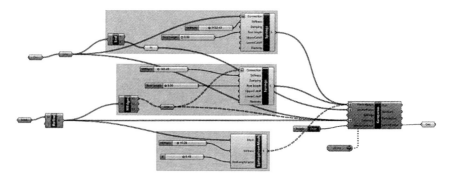

Fig. 2 Typical parametric definition including simulation of spring-particle systems, setting the rest-length of springs, variable stiffness values and anchor-points for the simulation in Kangaroo Physics Engine

In an attempt to mimick the physical behaviour of a material system, we translate physical properties into mathematical equations that generate the geometry in the computational enivironment. Thus an elastic textile can be represented by a spring-particle system (Fig. 2), translating mesh vertices to particles and mesh edges to springs, a system of points and lines (Fig. 3). Having obtained an understanding of the forces acting upon the models through manual model making and observation, the students were able to build their own grasshopper definitions, compare the results to the physical models and fine tune the process.

In all of the student projects, there had been several cycles of design generation and modification, both in analogue and digital environment, in order to reach the desired geometry. What the students percieve as trial and error proved to have an enormous educational impact. The process they followed is what Kolb would describe as "Do-Observe-Think-Plan" [24]. Through this iterative process the students were able to take design decisions, and reflect in action as Schön describes, for example for achieving the tensioning of a membrane, they would increase the stiffness of the computational springs or decrease their restlength, when the simulation would not give the desired result, they would devise alternative solutions such as stabilizing it with edge cables of different stiffness, etc. During the process, students would observe the result and reflect upon it, comprehend which actions benefit their model and extract the knowledge from this observation. These realizations would help them further plan their digital experiments based on the new tacit knowledge they acquired.

As Piker explains "*one great advantage of physically based methods is that we have a natural feel for them, and this intuitive quality lends itself well to the design process*" The biggest benefit of form-finding in Kangaroo, is that the designer can embed rapid simulation in early design stages, enabling reflection in action and driving the design towards informed and optimized solutions. In Piker's words, "*through the application of real-world physics we can make computational tools that really work with us to design in a way that is both creative and practical*" [20].

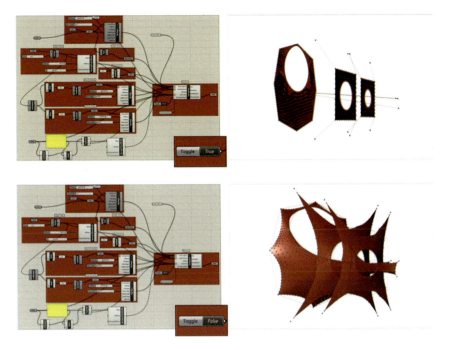

Fig. 3 Computational model before and after the dynamic relaxation of spring-particle systems

The experience highly relied on a learning-by-doing process, understanding the correlations between problem setting and design result. The students would conceptualize and understand the forces in play, extract valuable conclusions about the amount of tension and its repercussion on geometry and realize the role of reinforcements and edge cables.

4.2 Bending Simulation as a Morphogenetic Tool

The design studio about computationally simulated active bending took place at Graz University of Technology. It introduced students to the design of elastically bent rod and surface structures through analogue and digital form-finding. The aim of the workshop was to utilize bending behavior as a design tool for formal experimentation [25]. A system of bending rods, just as any other form-found structure tries to minimize its energy to span between the given borders. Eventually, the material system settles in a stable configuration.

Though form-finding is regarded as an intuitive process and experimentation with analogue media leads to tacit knowledge of materials, there is a considerable knowledge gap in the mathematical and computational understanging of the geometry of bending. Architect Marten Nettelbladt has been experimenting with the geometry of bending and has created an important online resource of experiments and observations concluding that all elastic deformations follow the same principals and therefore the same geometry [26]. Recent developments in *Kangaroo physics engine* can provide a very good approximation of bending geometry, and for this reason this tool proved to be a great resource for simulating the geometry of elastically bent elements, particularly when more bent elements interact resulting in tridimensional spatial curves (Fig. 4). Kangaroo works in an iterative way calculating the interaction among predefined forces, springs, bending resistance, pressure and gravity, and their repercussion on the geometry, until a stable form is reached.

Comparing the results obtained from the simulation to the physical models, there is great precision when simulating a single bending rod or a small set-up of interacting rods. The digital experiments become more demanding when several interacting rods are in play, adding to the complexity of the system [25, 27]. At the same time the physical form-finding experiments revealed some unpredictable results that emerged from the self-organizational capacity of the system to regulate and distribute forces to reach equilibrium.

For the simulation of bending behavior, as Achim Menges explains *"there are no simple mathematical equations which can describe the entirety of a surface defined by the equilibrium of applied force – be it tensile, compressive, or pressurized"* [28]. Having acquired some intuitive understanding of bending behavior through the analogue experiments, the students were able to reflect in action while experimenting with computational simulations of bending rod behavior.

During the studio the students used the recently incorporated hinge components, which means that apart from the standard spring-particle parameters like rest length,

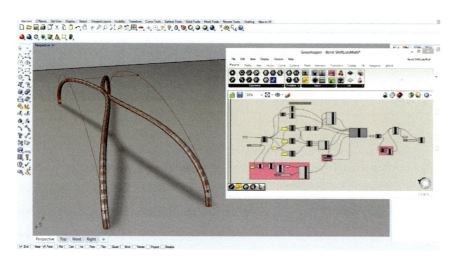

Fig. 4 Digital form-finding model of two rods constraining the tangency of the rod axis to be parallel to Z at start and end points and a connection point in their mid-length. Rods self-organize through dynamic relaxation and obtain a common tangency near the connection point

spring force, damping, they would define a rest angle for the hinges to be computed during dynamic relaxation. For the case of elastically bending rods, the initial geometry is represented by a curve which is divided into a finite number of line segments which will be introduced as springs with a fixed length and high spring stiffness (for avoiding changes in the total length of the rod). The rest angle (the angle between the tangents of two consecutive line segments) is set to 0, so that the natural tendency of the rod is to spring back to its initial straight condition. As the overall length of a rod is not negotiated during the simulation, it is important to set the start and end conditions of the rod. The designer would decide the degrees of freedom at the two ends of the rod, in the case of pinned connection, the end point of the spring would be anchored to fixed xyz coordinates, whereas for simulating a fixed connection we would require to constrain the start tangent. This would geometrically mean to set the first two control points of a degree 2 curve in the desired direction, or in the case of spring-particle systems with hinges, to set the position of the first two particles as fixed for the desired tangent direction. The knowledge gained from the physical models was directly fed into the design decisions in the digital environment. Having mastered the bending behavior of a rod or a group of rods, the students experimented with bending planar surfaces (Fig. 5). Similar to the previous case, the aim was not to increase the overall area of the surfaces through bending, as it would happen with membranes and tensile structures. The obtained geometries were shapes of single curvature, however in often cases there were some unpredictable results, with the surface bending inwards or outwards when the start tangent was not constrained. Nevertheless, the students were able to understand the behaviour of the simulation and modify the parameters accordingly, usually in real-time, without necessarily restarting the simulation, they were able to reflect in action when the obtained result was diverting from their expectations.

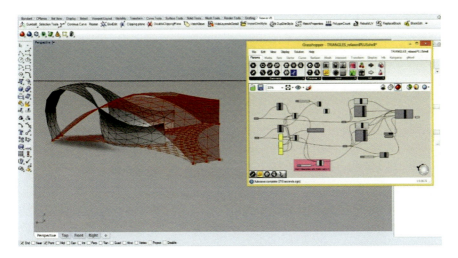

Fig. 5 Digital form-finding model of a surface element that bends in two directions

4.3 Planarization Algorithms for Design Rationalization

The studio Digital Tectonics which took place at the University of Thessaly, introduced students to simulation-based design processes for design rationalization and construction optimization. Some geometric problems, like the discretization and paneling of freeform surfaces with flat elements [29] cannot be easily solved through a mathematical approach. In several cases, a simulation-based strategy is necessary [30]. With the aim to design architectural structures that embed fabrication criteria and cost-effective design decisions, students set up a simulation that implements planarization algorithms in Kangaroo, as a design decision tool, that would drive the overall shape of their constructions towards feasible solutions.

Students had to decide the approximate shape and size of the panels, and their distribution on the freeform surface. They would have to decide the anchorpoints, in this case the edges of the panels that would connect to the ground (Fig. 6). During the simulation the rest of the vertices would move, until they find a stable position when the requirement for planar panels is met. To visualize this, a surface from planar curves is created for each panel. When the condition of planarity is met, the surface is created, when the vertices snap out of plane the surface will not be created. Therefore, the visual hint of planar subsurfaces that are parametrically created from boundary curves, is indicative that the simulation has found at least one of the possible solutions. During the design process students would try several different panel configurations, shapes, boundaries and volumes. If the system is overconstrained, it will not reach a solution with planar panels, hence the simulation does not always result in valid solutions, it may occur that some polygons due to their constraints and acnhorpoints never become planar. Such event would require either the modification of the overall geometry, or the shape and density of the panels. Alternatively, opting for a variable panel density based on the curvature may lead to an optimized configuration.

The students' intuitive understanding about the discretization of double curved surfaces into polygonal flat panels led to the variation of panels, Among the parameters that influenced the simulation result were the polygon sizes, anchorpoints, adjustment of the overall height of the shell (Fig. 7). The aim to design a complex tridimensional form that can be constructed from flat panels led to intuitive design decisions, and real-time design modifications during the simulation. As there were plenty of cases that could not be solved for construction with planar elements, the simulation helped define the optimized cell size. Students could reflect in action and modify the design parameters so as to obtain the desired result.

It was interesting to observe that after several iterations of Experience, Observation, Conceptualization, Experimentation, also known as Kolb's Cycle,[3] the students started to develop a capacity for reflection in action within the digital environment. Even if they hadn't been capable of explaining the reasons why they adjusted the parameters of the simulation, the design decisions they took were at all times coherent to their design intent, and surprisingly enough the modification of parameters of the

---

[3] Kolb's Cycle comprises of 4 stages: 1. Concrete Experience 2. Reflective Observation 3. Abstract Conceptualization 4. Active Experimentation.

Educating the Reflective Digital Practitioner

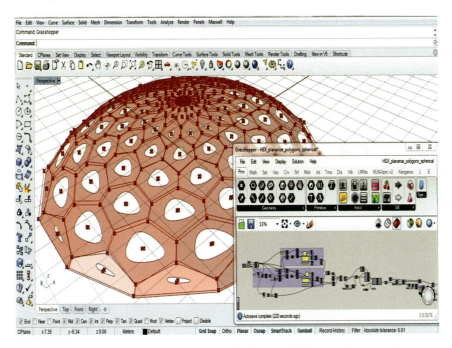

Fig. 6 Semispherical shell structure comprising of polygons with planarization constraint

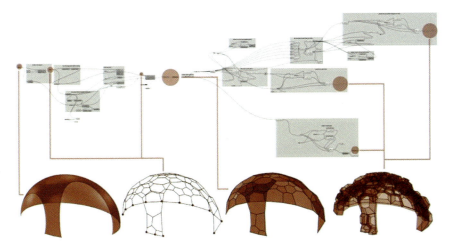

Fig. 7 Simulation-based modeling of a shell structure employing planarization algorithm (Kangaroo Physics Engine)

simulation was almost always in the right direction to obtain geometry that was optimized for fabrication. Beyond doubt, the students had obtained tacit knowledge about the simulation by experimenting through practice. The simulation in turn would educate the designer which design decisions would benefit the design with regards to the criteria of planarization and feasibility of fabrication.

5 Conclusions

In an era of transition from manual to computerized and further on to computational design, the ideas of Schön seem to be evergreen. In our days the computer has become the main tool for sketching and drafting, and very often a tool for thinking. We have only seen a small fraction of the great potential that computation has to offer to the design disciplines. We are currently witnessing a growing interest in emergent design technologies and processes, where experimentation gives rise to new design opportunities. Yet, how do we explore unknown grounds? How do we train future designers to reflect in action, to be alert and creative at the same time, to seize the design opportunity? Is it always possible to be in full command of all new digital tools? Or are we seeking to develop our capability to adapt to the new challenges regardless of previous computation skills? The designer's ability to reflect in action becomes of crucial importance for dynamic digital design processes. Simulation-based modeling offers immediate visual feedback to the designer, a modification in design parameters can inform the design in real time. While theory, science and technology courses will answer the questions about "knowing that", the involvement of students in hands-on exercises and simulations of physical forces will train them with regards to "knowing how" [31]. Dewey in his writings about "*Experience and Education*" [32] polarizes the two extremes, traditional and progressive education, referring on the one hand to structured and disciplined education and on the other hand to unstructured, flexible, student-led process. While the first one lacks an holistic approach and is overly focused on content rather than process, the second is based upon a very weak philosophical basis of freedom in education. Dewey therefore suggests that educators should move beyond such "*paradigm wars*" and instead of contrasting theory with experience, employ a "*theory of experience*". According to Dewey, past experiences influence the future experiences in the sense of "*continuity*" while the notion of "*interaction*" takes the theory one step further by examining how past experiences interact with the current situation. Computational tools marked a new epoch in the design disciplines. Industry 4.0 calls for new education models, that are more open ended and adaptive. There is no clear answer about how digital media and simulation modeling should be incorporated in architectural education. Simulations of physical forces and interactions (stretching, bending, planarizing) are difficult to build from scratch, but at the same time for a designer that uses a readily available tool like Kangaroo, the process and design outcome is self-explanatory in the sense that the designer can observe the dynamic model oscillate around its stable position until it reaches equilibrium. The design studios presented in this chapter had exactly this goal, to experiment with readily available computational tools and train the students' ability to reflect in action within a digital environment. Simulation-based modeling affords us the opportunity to envision computational design tools of the future and learn to adapt to change, to learn through experience and develop new design tools and methodologies. As Keeton and Tate explain, experiential learning is the *"Learning in which the learner is directly in touch with the realities being studied. It is contrasted with the learner that only reads about, hears about, talks about or writes about these realities but never comes into contact with them as part of the learning process"* [33]. This type of learning is the link between the classroom and the real world. In an era of abundance of information through the worldwide web, the students need to develop skills of filtering the received information, but also of combining existing knowledge.

While the traditional academic set-up helps students to develop skills related to symbolic and perceptual aptitudes, behavioral and affective abilities are better fostered through active involvement of the student, through hands-on activities. A lot has been said about our ability to retain knowledge; Edgar Dale published a visual model of knowledge retention, the Cone of Experience, a visual model for classifying mediated learning experiences.[4] According to Dale, our ability to retain knowledge relates to the type of instruction; the Cone of Experience exemplifies *"Direct, Purposeful Experiences"* over *"Visual and Verbal Symbols"*. This is in line with the famous Chinese proverb attributed to Confucius 450BC *"Tell me, and I will forget. Show me and I may remember. Involve me, and I will understand"*.

It is important to highlight here, that not any computational design experience can lead to learning, only experience that is reflected upon can yield new knowledge. It is the duty of the educator to provide the circumstances where an educational experience produces learning. Michael Polanyi had always supported the importance of *"tacit knowledge"*, the special type of knowledge that cannot be put into words, the experiential knowledge usually related to creative disciplines, associated with the actual praxis of design. Polanyi's belief that *"we can know more than we can tell"* [34] may be the way to describe the students' early experience with simulation modeling in architecture and Schön's legacy may form the basis for educating the reflective digital practitioner of the future.

Acknowledgements The work presented in this chapter is based on doctoral and post-doctoral research projects of the author. The educational experiments took place at Graz University of Technology in Austria and the Department of Architecture of the University of Thessaly in Greece. The author would like to acknowledge the work of the students that participated in the design studios, and express gratitude to colleagues and student assistants for their valuable input.

---

[4] There have been several variants of Dale's Cone of Experience, some of which are known as the Pyramid of Knowledge, offering percentages of the learner's ability to retain knowledge. However, such adaptations of the original model are not based on scientific research and are falsely presented by researchers as evidence about knowledge retention. The original model by Dale as presented in the book is not aiming to be presented as the outcome of scientific model, it is suggested as a conceptual, visual model.

References

1. Schon, D.A.: The Reflective Practitioner: How Professionals Think In Action. Basic Books, New York (1984)
2. Gropius, W.: Scope of Total Architecture. Collier Books (1962)
3. Al-Qawasmi, J., Velasco, G.P.V. de: Changing Trends in Architectural Design Education. csaar (2006)
4. Dowdle, D., Ahmed, V.: Teaching and Learning Building Design and Construction. Routledge (2013)
5. Dutton, T.A.: Voices in Architectural Education: Cultural Politics and Pedagogy. Praeger, New York (1991)
6. Harriss, H., Widder, L. (eds.): Architecture Live Projects: Pedagogy into Practice. Routledge, Oxon, New York, NY (2014)
7. Nicol, D., Pilling, S. (eds.): Changing Architectural Education: Towards a New Professionalism. Taylor & Francis, London, New York (2000)
8. Salama, A.: A theory for integrating knowledge in architectural design education. ArchNet-IJAR: Int. J. Arch. Res. **2**, 100–128 (2008)
9. Salama, A.M.A., Wilkinson, N.: Design Studio Pedagogy: Horizons for the Future. ARTI-ARCH (2007)
10. Schön, D.A.: Toward a marriage of artistry & applied science in the architectural design studio. J. Arch. Educ. **41**, 4–10 (1988)
11. Waks, L.J.: Donald Schon's philosophy of design and design Education. Int. J. Technol. Des. Educ. **11**, 37–51 (2001)
12. Webster, H.: Architectural education after Schön: cracks, blurs, boundaries and beyond. J. Educ. Built Environ. **3**, 63–74 (2008)
13. Prensky, M.: Digital natives, digital immigrants part 1. On Horiz. **9**, 1–6 (2001)
14. Lee, K.: Principles of CAD/CAM/CAE. Prentice Hall, Reading, Mass (1999)
15. Schein, E.H.: Professional Education: Some New Directions. McGraw-Hill, New York (1972)
16. Weaver, N., O'Reilly, D., Caddick, M.: Preparation and support of part-time teachers Designing a tutor training programme fit for architects. In: Nicol, D., Pilling, S. (eds.) Changing Architectural Education: Towards a New Professionalism. Taylor & Francis, London, New York (2000)
17. Ackermann, E.: Piaget's constructivism, Papert's constructionism: what's the difference. Futur. Learn. Group Publ. **5**, 438 (2001)
18. Harel, I., Papert, S.: Constructionism. Ablex Publishing, Norwood, N.J. (1991)
19. Schon, D.A.: Educating the Reflective Practitioner: Toward a New Design for Teaching and Learning in the Professions. Jossey-Bass, San Francisco, Calif (1987)
20. Piker, D.: Kangaroo: form finding with computational physics. Arch. Des. **83**, 136–137 (2013)
21. Lewis, W.J.: Tension Structures: Form and Behaviour. Thomas Telford (2003)
22. Symeonidou, I.: Flexible matter: a real-time shape exploration employing analogue and digital form-finding of tensile structures. Int. J. Archit. Comput. **14**, 322–332 (2016)
23. Oxman, N., Rosenberg, J.L.: Material-based design computation: an inquiry into digital simulation of physical material properties as design generators. Int. J. Archit. Comput. **5**, 26–44 (2007)
24. Kolb, D.: Experiential Learning: Experience as the Source of Learning and Development. Prentice Hall, Englewood Cliffs, N.J. (1984)
25. Symeonidou, I., Gupta, U.: Bending Curvature: Design Research and Experimentation: Analogue and Digital Experiments. LAP LAMBERT Academic Publishing (2012)
26. Nettelbladt, M.: The Geometry of Bending. Förlag Mårten Nettelbladt, Stockholm, Sweden (2013)
27. Symeonidou, I.: Analogue and digital form-finding of bending rod structures. Presented at the August 18 (2015)

28. Menges, A.: Behavior-based computational design methodologies. In: Proceedings of the 31st Annual Conference of the Association for Computer Aided Design in Architecture (ACADIA), Banff, Alberta (2011)
29. Eigensatz, M., Deuss, M., Schiftner, A., Kilian, M., Mitra, N.J., Pottmann, H., Pauly, M.: Case studies in cost-optimized paneling of architectural freeform surfaces. Adv. Arch. Geom. **2010**, 49–72 (2010)
30. Jiang, C., Tang, C., Tomičí, M., Wallner, J., Pottmann, H.: Interactive modeling of architectural freeform structures: combining geometry with fabrication and statics. In: Block, P., Knippers, J., Mitra, N.J., Wang, W. (eds.) Advances in Architectural Geometry 2014, pp. 95–108. Springer International Publishing, Cham (2015)
31. Ryle, G., Dennett, D.C.: The Concept of Mind. University Of Chicago Press, Chicago (2000)
32. Dewey, J.: Experience and Education. Free Press, New York (1997)
33. Tate, P.J., Keeton, M.T.: Learning by Experience-What, Why, How Morris T. Keeton, Pamela J. Tate Editors. Jossey-Bass (1978)
34. Polanyi, M.: The Tacit Dimension. University Of Chicago Press, Chicago (1966)

Teaching Digital Design and Fabrication to AEC's Artisans

Maurizio Barberio

Abstract This contribution describes an operative research activity within the teaching of digital design and fabrication to Architecture, Engineering and Construction (AEC) artisans. The didactic approach described arises from the lack of academic paths thought for AEC's artisans, highlighting the reason why this aspect is relevant for both the AEC and the artisanal fields. In particular, the article reports a research project carried out by two artisans who attended the C.E.S.A.R. Course, an annual university course organized by the Politecnico di Bari in collaboration with Les Compagnons du Devoir, a historic French professional association. In particular, the research project concerns the study and the digital transposition using digital design and fabrication processes and tools of the "Bridge over the Basento River" designed by Sergio Musmeci.

Keywords Digital fabrication · Digital crafting · Architectural didactics · Digital artisans · Basento Bridge

United Nations' Sustainable Development Goals 8. Promote sustained · Inclusive · And sustainable economic growth · Full and productive employment · And decent work for all · 9. Build resilient infrastructure · Promote inclusive and sustainable industrialization · And foster innovation · 12. Ensure sustainable consumption and production patterns

1 Introduction

The paper aims to describe an educational experience of designing and building a prototype using digital fabrication techniques and its consequent didactic implications. The experimentation is based on the didactic method of learning by doing and experiential learning, where knowledge is transmitted not only through lectures but also and above all through proactive and laboratory experimentation. This approach

M. Barberio (✉)
Dipartimento di Meccanica, Matematica e Management, Politecnico Di Bari, 70124 Bari, Italy
e-mail: maurizio.barberio@poliba.it

has already been adopted by the author, with other colleagues, and with architectural students during the design workshops in the third and fourth years. The method is particularly useful when it is necessary to acquire skills in complex and interrelated topics such as digital design and fabrication, architectural geometry, and form-finding. The case study of Musmeci's Basento Bridge was chosen because it is suitable for bringing all these aspects together and giving students an overview of such aspects. In particular, the paper refers to the teaching of digital design and fabrication to professional artisans without a specific background in AEC design topics, which is an uncommon didactic case study for architectural schools' agenda. The advanced digitalization of the craftsman operating in the AEC sector (both inside factories and building sites) is essential to reach the new 4.0 standards of the fourth industrial revolution, also when they are employed in SMEs, or they are self-employed.

2 Background: Digital Design and Fabrication Within the Academic Context

Digital fabrication is intended as the manufacturing process in which the physical model is produced through machines controlled by the computer starting from a digital model. Between the 1990s and the 2000s, the widespread use of computers in architecture changed the way buildings were designed and built. In 1992 the Gehry Partners LPP studio created a fish-shaped pavilion to be placed on the Barcelona seafront. The three-dimensional IT model was obtained starting from a study maquette. The surface thus generated was then utilized to perform the structural analyses and to obtain all the building components. For the first time, the production and assembly of the components of the structure were completely directed starting from the digital model [17: 8]. Kolarevic and Male-Alemany [13] emphasized that, following Gehry's example, the architects understood that the information of the digital model could be used directly for fabrication and construction, thanks to the use of numerical control machines. Kolarevic stated that the most interesting potentiality of integrating digital fabrication in the architecture practice is to revitalize the close relationship that once existed between architecture and construction: "By integrating design, analysis, manufacture, and the assembly of buildings around digital technologies, architects, engineers, and builders have an opportunity to fundamentally redefine the relationships between conception and production. The currently separate professional realms of architecture, engineering, and construction can be integrated into a relatively seamless digital collaborative enterprise, in which architects could play a central role as information master builders, the twenty-first century version of the architects' medieval predecessors". In this direction, other authors [6] stated that the contemporary designer can be defined as a novus architetto adaucto; in other words, an "expanded designer" who possesses new (robotic) arms which allow him to cut and shape the materials according to his direct requirements (almost) without any external mediation, paradoxically like the architect-master (or master builder) of the past. Anyway, as for architects and engineers, there is no evidence to exclude artisans from this important change. In fact, like during the Middle Ages, when stonemasons directed the construction of the cathedrals, being effectively responsible for how they were built, nowadays artisans can take part in the design and construction choices of the contemporary building sites, if adequately prepared and skilled. It is a fact that digital fabrication brings a significant change in Architecture, Engineering and Construction (AEC) industry, particularly in the planning and execution phases. As a result, scholars have already highlighted that it is expected that current construction roles will evolve, and new roles will be created: the responsibilities of the construction workers will shift from unsafe and hard conditions to safer and less labour-intensive, such as monitoring and control automated processes by transferring their know-how to the robotic systems [5]. In the absence of specific academic training paths for artisans who operate in the AEC sector (that we call "AEC'S artisans"), the acquisition of digital competences is left to the resourcefulness of individual workers or to the companies where they work. The birth and diffusion of the Internet have contributed to creating a pervasive digital culture (makers culture), that has allowed us to fill formative gaps casually and informally. Lee [16], analysing the "maker mindset" and its implications for education, has defined this mindset as playful, asset-and growth-oriented, failure-positive, and collaborative. Some scholars believe that the Fab Labs could potentially challenge the structure of society in the coming years because with their diffusion, knowledge is no longer statically placed in universities, companies, or research centres, but it is increasingly moving towards the creation of a fluid and adaptive network able to informally spread knowledge and innovation [17]. Despite this view, it is undeniable that the academic context had a crucial role in the birth of the maker's movement: the first digital fabrication laboratory (Fab Lab) has been founded at MIT in 2001. Again, at MIT in 1998, Neil Gershenfeld - director of the Center for Bits and Atoms - inaugurates the course called "How to make (almost) anything". As a computer scientist, Gershenfeld conceives an interdisciplinary course, in which students can learn how to use CNC machines of industrial derivation to develop fully functional experimental prototypes [10]. Afterwards, several scholars studied the relationship between didactics and digital fabrication, especially in architecture schools. For example, a 2-year course called "File-to-Factory Digital Fabrication" has been launched at the University of Nebraska-Lincoln in the early 2010s. The course goal was for students to synthesize various disparate architectural assemblies and materials with the file-to-factory digital fabrication process to understand the making architecture [11: 22]. A mix of classes and lab periods has allowed the students a better understanding of digital design and fabrication processes and the production of physical prototypes [11: 29]. Another didactic model, called "Digital Design Build Studio" has been organized in both individual activities (first part of the studio) and group work (second part); for the final part one project has been selected and developed further, to test ideas on a 1/1 scale, to allowing "the studio to fit in the existing curriculum but also allows for an investigation and research that goes beyond the regular design studio setting" [21: 201]. In a study about engineering education, Sheppard et al. [22] proposed the categorization of laboratory instruction into three levels. As summarized by Celani [4: 476], the first level concerns novice students that must follow the instructor's directions strictly, step by step, to reach the desired results, which will demonstrate a concept. Next step concerns intermediate students, that must do some exercises to understand the mathematical description of the theory. The last step consists of developing laboratory simulations that illustrate the same phenomenon, in which advanced students can validate the concepts learnt by testing them with different parameters and conditions. Celani states that "as digital fabrication labs become more common in architecture schools and are assimilated by design instructors, they can promote changes in architectural education, allowing students to become closer to the production process and to have a better control over building parts and materials" [4: 480]. Furthermore, Celani affirmed that digital fabrication laboratories have the potential of promoting experimental methods in architecture together with a scientific approach, which is the basis of contemporary architectural practice [4: 480]. In fact, in the last decade, several digital fabrication laboratories have transformed their pioneering explorative research into a scientific activity, with the design and fabrication of full-scale models (that we may also call "proto architectures") that aim to demonstrate the goodness of an empirical hypothesis. Anyway, these advanced research activities are not usually accessible at the undergraduate level of education, but they are thought for master's and Ph.D. students, generally enrolled at Institutes of Technology and Polytechnics. To overcome this limitation in the so-called "post-digital era", Figliola [9: 35] proposes the inclusion of modules relating to computational design and digital fabrication in educational programs starting from the first university education cycle. As stated before, a similar approach was developed by the author's research group during architectural design studios held during the 3rd year course (out of 5 years degree program) at Politecnico di Bari [7, 8], in which a "learning by designing" approach was adopted both in the realization of scaled models of building components, realized by using digital design and fabrication tools, and in the architectural design of the whole buildings, which embedded those components into the overall design. Stavric et al. [24] underline that the teaching approach for learning digital design and fabrication should be based strongly based on geometry, mathematics, programming, hardware computing and material behaviour. Again, the translation of digital models into physical ones is one of the cores of the teaching activity. Anyway, other scholars argued that the introduction of digital fabrication and design in upper primary and lower secondary schools poses several issues related to the contradiction between a curriculum-based and highly goal-oriented school setting and an experiment-based and highly explorative maker culture. The study revealed that teachers were not technologically or methodologically prepared for an educational program that did not align with the structure of conventional training, because the explorative nature of digital fabrication challenged the authority of the teachers and jeopardized their feeling of being in control [23: 46]. In any case, the education on digital design and fabrication seems nowadays essential for all kind of students, especially for those will start an academic path into STEM disciplines or for those that will work into companies related to manufacturing or engineering. Numerous architecture schools all around the world have incorporated digital design and fabrication coursed into their degree programs. Regardless of the education path of each, Gershenfeld believes that the digital revolution in manufacturing will allow people to produce objects and machines on demand, allowing the birth of new hybrid professionals, named "makers" or "digital artisans".

3 A Didactic Approach for AEC's Artisans Within the Academic Context

In [3] Richard Barbrook and Pit Schultz coined the term "digital artisan" in their Manifesto for describing who works within hypermedia, computing, and associated professions. Even if the definition is not specifically related to who commonly can be defined as "artisan", their Manifesto does not preclude the inclusion of them, because the authors intended to celebrate the Promethean power of the digital artisans' labour and imagination to shape the virtual world. They imagine that digital artisans will build the wired future through their own efforts and inventiveness by hacking, coding, designing, and mixing. Thus, the introduction of such topics into traditional teaching programs poses different challenges but it could also be an important opportunity for developing a holistic approach and developing critical thinking skills. Anyway, the use of the Internet to share knowledge openly and fluidly within the makers' context is one of the reasons which is not easy to establish proper academic paths to transfer knowledge from universities to qualified workers who need to update (or create) their digital skills. In other words, it is easier for them to search informal didactic resources (articles, blogs, tutorials, etc.) rather than start an academic formative path. There are two reasons for it: the first lies in the lack of academic coursed dedicated to AEC'S artisans; the second lies in the fact that often a professional diploma is not a sufficient requirement for being accepted in a traditional academic course. These challenges can be overcome by establishing innovative formative partnerships between professional associations and academic institutions, joining their efforts to support workers and companies in the so-called "lifelong learning". Regarding the relationship between the AEC industry and the artisanal field, there is a need to formulate a didactic approach adequate to train artisans (who may be employed both by manufacturing and construction companies) in a way that they can be part of a holistic framework where design, fabrication, and construction aspects are seamlessly linked together. The recent development of the Industry 4.0 imposed the development of the homologous "Architecture 4.0" in which designers (architects and engineers), artisans, workers contractors, suppliers and construction companies share the same language and the same processes [2]. Lanzara [14] states that the involvement of the academic world plays an important role in the process for the improvement of collective awareness towards a multidisciplinary collaborative ecosystem, by sharing advanced activities to support training and entrepreneurial activity of students or artisans, and for developing a digital conscience.

4 The Theoretical Models Adopted for Teaching Digital Design and Fabrication to Artisans

The operative research described in this paper is representative of the experience that the author has accumulated over the years at the C.E.S.A.R. Course (Cours de Enseignement Supérieur en Architecture et Restauration), held annually since 2015 at the Politecnico di Bari. The uniqueness and the novelty of the course stand in the fact its goal is to create and train a professional profile who can create a closer connection between the restoration site manager (architect or engineer) and the various specialists involved in the study, protection, restoration, management, and enhancement of the architectural heritage, also adopting contemporary tools, such as parametric and digital modelling software or digital fabrication techniques. Inside the overall didactic scheme of the C.E.S.A.R. Course, the classes held by the author about digital design and fabrication are essential for training the new generations of "digital artisans". The digital design classes concern the understanding of the different levels of interaction between the designer and the digital environment, also explaining the differences that exist among the various 3D modelling techniques (Table 1). Digital fabrication classes followed a similar structure that those regarding digital design themes. They are summarized in Table 2. They are concerned essentially with the relationship between the design outputs allowed by using different digital fabrication tools and techniques. In other words, the intention was to transfer the design thinking underlying the different projects who take advantage of digital fabrication processes.

The course is incardinated on the use of the NURBS-based modelling software Rhinoceros. This software has been adopted not only for its user-friendliness but also for its versatility which allows transforming the software into a powerful platform, capable of easily embedding different 3D modelling techniques (NURBS, mesh and subd modelling) thanks to both its native features, above all, using specific plug-ins (for example Grasshopper for parametric modelling, VisualARQ for Bim, etc.).

Table 1 Digital modelling strategies topics of the course

Modelling strategy	Modelling typology
Direct modelling	Solid
	Parametric solid (semi direct)
	Polygonal
	NURBS
	Sub-d
	VR modelling
	Digital sculpturing
Non-direct modelling	Procedural
	Parametric-associative
	Computational
	BIM

Table 2 Digital fabrication strategies topics of the course

Fabrication strategy	Fabrication typology
Subtractive fabrication	Cutting of flat elements
	Cutting of volumetric elements
	Carving of volumetric elements
Bending	Bending of rigid elements
	Bending of flexible elements
	Bending of flat elements using a cutting pattern
Formative fabrication	Digital weaving
	Stretching of elastic material
	Thermoforming
Additive fabrication	Material extrusion of monolithic objects
	Material extrusion of discrete assemblies
	Binder jetting of monolithic objects
	Binder jetting of discrete assemblies
	Additive formworks

Furthermore, Rhinoceros allow the investigation of the three levels of interactions aforementioned: direct modelling, parametric-associative modelling, and computational modelling. Direct modelling refers to the use of modelling software through a consequential but static process. In other words, the digital model is manipulated directly by the user, but any additional modification makes it impossible to go back. In this type of modelling, it is therefore important to preserve the fundamental steps of the modelling process, to be able to return to an earlier phase of the process. It is a typical design process in which we start from the global geometry up to the definition of all the details. A change in the initial global geometry determines the need to restart the modelling process from the beginning. Parametric-associative modelling refers to the use of a parametric modelling software or application for defining the digital model. In this case, the designer concentrates on defining the logical consequentiality of the various steps, which can proceed from the overall geometry to the detail or vice versa. Thus, the designer does not directly generate the digital model, as in the previous case, but generates a parametric "code", i.e., an algorithm governed by some fundamental parameters that define the geometry. For computational modelling, we mean the use of a programming language (embedded or not in a parametric or modelling software) for the definition of an interactive model, in which the designer can simulate various types of phenomena characterized by high conceptual or geometric complexity. Also, in this case, we can proceed from the global geometry up to the detail or vice versa.

The course goal is not only to transfer the artisans some 3D modelling skills but to provide critical thinking to understand the theoretical differences between the different levels of interactions and their use to achieve different fabrication results.

This is because the modelling strategy to be undertaken cannot follow predetermined paths but will be influenced by the need of the specific case. Summarizing these differences, direct modelling is indicated in all cases in which is possible to generate the 3D model easily and at the same time it is not yet possible to define the project parametrically, due to the uncertainty on the road to be taken. This is particularly useful in the initial study phases of the forms. Parametric-associative modelling can be useful when the project is still in an exploratory phase, but it is already possible to define some parts of it from an algorithmic point of view (for example, the tessellation of a vaulted system that is not too complex). In this way, different solutions for a specific design aspect can be examined more easily. This type of modelling can be useful even when the complexity of the project is not so high that it must necessarily use more sophisticated computational tools. The increasingly widespread dissemination of parametric-computational strategies in the design field has made it possible to apply new operative models of computer origin also in the fields of architecture and engineering.

In general, it is possible to state that the computational and parametric design can follow two models: top-down and bottom-up. Both models have been theorized in the field of computer science and are used as strategies for writing parts of program codes. In top-down models, the starting point of the design process is represented by the formulation of a general systemic idea, from which all the sub-problems that compose it follow. The model provides the progressive finishing of all parts as they are designed, and new elements are added to the system. In bottom-up models, the starting point of the design process is represented by the detailed definition of individual elements of the system, which are subsequently connected and interrelated to each other, up to the definition of the overall system.

5 Operative Research: The Digital Design and Fabrication Transposition of the "Bridge Over the Basento River" by Sergio Musmeci

In this section, top-down operative research (final exam) carried out by two student-artisans is described. This project is described as an example of the application of the didactic approach described before, concerning the study and the digital transposition using digital design and fabrication techniques of the shape of the "Viadotto dell'Industria" (Industry Viaduct), commonly known as "Ponte sul Fiume Basento" ("Bridge over the Basento River"), designed by Italian engineer Sergio Musmeci and built between 1971 and 1975. Musmeci's Bridge has been chosen for different reasons: firstly, the project is a unique engineering (and architecture) masterpiece, and it is also a clear example of what is possible to achieve when the architectural shape is completely linked to structural behaviour aspects. Plus, the project has been originated by generating different physical models, as described afterwards, and it is generally considered one of the precursors of contemporary digital form-finding techniques. Lastly, the bridge presents a non-Euclidean, complex shape, suitable to be used as a case study to train the students in advanced and parametric modelling, digital fabrication, and architectural geometry.

The operative research method was based on the learning-by-doing approach. Thus, a sequence of tasks has been assigned to the students to allow the acquisition of knowledge on the chosen topic proactively and progressively:

1. Historical investigation of the case study and understanding of its cultural value for architecture and engineering.
2. Investigation of design and form-finding strategies to obtain the overall shape of the bridge.
3. Critical evaluation of the design and form-finding outputs and the model-making feasibility.
4. Definition of the final digital design and fabrication process.
5. Critical evaluation and description of the design issues and improvements.

6 Historical and Cultural Research

Basento Bridge is one of the best examples of a shell structure built during the XX Century in which physical models have been used to determine its optimal shape. Even if physical models have always been used in architecture for different reasons, like representing the project, studying its proportions, its structural behaviour, etc., the models used for searching the optimal shape of a given structure are of more recent introduction. It is important to underline that not all scale models can be used for structural purposes. The phenomena or structural behaviours that can be scaled linearly, concern the linear dimension of a structure, the funicular form of a vault, of a dome or a shell, and the stability of a masonry structure subject to compression only [1]. In the Sixties Sergio Musmeci used a form-finding technique originally developed by Otto and Rasch [18], to determine the initial form of the structure of the Basento Bridge. The technique consists of the immersion of metal profiles of the desired shape in soapy water, and it has been used to research the shape and to start the initial calculation processes [15]. Musmeci continued the research by building a neoprene model that allowed the study of the tensions in the two perpendicular directions. Subsequently, a methacrylate model of two spans of the bridge was then built on a scale of 1:100 to verify the correspondence of the form to the design program and was subjected to elastic tests that allowed a first partial control of the calculation forecasts. Finally, before the construction of the bridge, the Superior Council of Public Works requested the construction of a scaled-down (1:10) model made of micro-concrete for loading tests [19: 17–24], a technique already used by Eduardo Torroja in 1933 for the project of the colossal dome of the Algeciras market in Spain. Later, different analogue form-finding techniques have been translated into the digital environment, especially in the last decades.

7 Investigation of Design and Form-Finding Strategies

Among the various digital form-finding techniques developed, like Dynamic Relaxation, Force Density Method, and Thrust Network Analysis, students investigated the use of the Particle-spring system for the investigation of the bridge's shape. As the name suggests, the Particle-spring system is composed of a set of particles connected by a system of springs: the particles represent the points where the mass is concentrated, and the springs are schematized as elastic lines connecting two points. Applications of this form-finding system within computational design have been developed by Kilian and Ochsendorf [12] conceiving CADenary, and Daniel Piker who developed Kangaroo Physics, a particle-spring tool available inside Grasshopper, the visual programming language of Rhinoceros [20]. Considering this background, a particle-spring form-finding technique has been used during the course to train students to understand the relationship between architectural geometry and structural optimization and behaviour. Kangaroo 2 has been initially used for trying to recreate in the digital environment the form-finding process utilized by Sergio Musmeci. The simulation consists of creating a basic flat mesh placed on the XY plane, which represents the membrane on which to apply the form-finding process. On the base mesh, the designer defines the anchor points that will remain fixed while the other points (particles) are free to move according to the resistance of the elastic lines (springs) that connects the various particles. However, in this case, the process provides that some anchor points will no longer be on the XY plane, but they are moved on the Z axis to give the bridge the actual arcuate shape (Fig. 1). Students were asked to replicate the form-finding process described to evaluate the feasibility of physical model fabrication, considering also different materials and production methods.

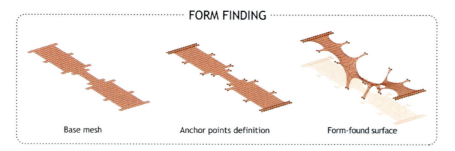

Fig. 1 Particle-spring form-finding workflow

8 Critical Evaluation of the Design and Form-Finding Outputs

Although the output model was fine in its pedagogical value, it was not for its geometrical properties. The form-found model was quite different from the actual shape of the bridge. This difference is because the actual Musmeci's bridge is not a funicular compressed-only shape but, instead, it can be approximated by a tensile minimal structure. For this reason, it has been decided to realize a more accurate 3D model analysing the laser-scanned survey carried out for the restoration study of Musmeci's Bridge. In this case, a mix of basic and advanced modelling has been used to achieve the result. The multiple modelling approach has been encouraged by the author because in this way students had the chance to be aware of the different possibilities that can choose to accomplish the fixed goal. It is important to note that the research goal was not to recreate a surface perfectly identical to the original. Instead, the main interest was to use digital processes and tools to study how to evocate the bridge's shape by taking advantage of the bending properties of a typical material available in a FabLab, like thin plywood, dividing the whole shape into small pieces. This implies the development of a comprehensive computational strategy (although simplified due to the didactic nature of the experiment), from design to fabrication.

9 Definition of the Final Digital Design and Fabrication Process

The author guided students during the development of the whole process, which is formed by several steps (Fig. 2):

- Recreation of the bridge's shape by extracting the fundamental curves from the survey, using them for creating the base NURBS polysurface of the bridge.
- Conversion of the discontinuous NURBS polysurface into a mesh model by creating an ultra-simplified network of quad meshes (coarse mesh).
- Subdivision of the previous mesh using the Catmull-Clark algorithm by simultaneously pulling the obtained mesh onto the base polysurface.
- Extraction of transverse and longitudinal mesh edges (u and v directions).
- Creation of the continuous NURBS surface by a network of curves using the ordered lists of mesh edges of the previous step.
- Study and test the tessellation pattern shape, the material type, and its physical properties (like bending).
- Population of the continuous NURBS surface according to the chosen pattern.
- Testing on a smaller part of the whole prototype the chosen pattern and material behaviour.
- After validation, production of the final model (all the pieces need to be numbered and oriented).

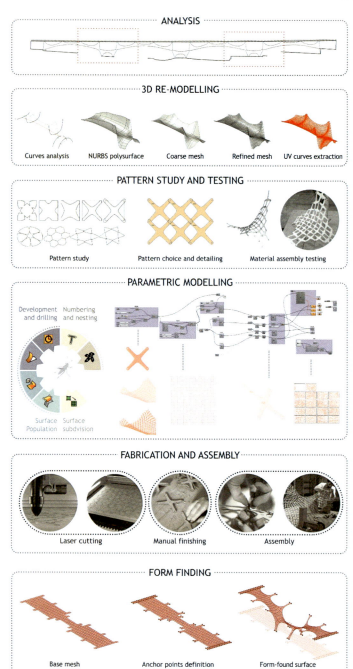

Fig. 2 Project workflow

- Nesting of all the pieces into the defined sheets of material.
- Fabrication through laser cutting tools.
- Final assembly.

10 Critical Evaluation and Description of the Design Issues and Improvements

Some considerations on the consequences of the didactic value of the learning-by-doing approach need to be highlighted. The first assumption of the research was to build a complex surface using only small flat elements. At the start, students intuitively came up with the idea of triangular modules because they are always flat. Anyway, students experimented several challenges testing triangular tessellation, as for example, the problem of the junctions between each element that converges in a point. They tried to solve this issue by adding a soft leather part to each end, but the solution was expensive, difficult to realize and inelegant. After abandoning triangular tessellations, they started to experiment with quadrangular patterns, especially studying the relationship between material bending properties and the pieces' shape. Soon they discovered that using quad pieces allowed a much cleaner and more efficient division of the surface, a better data order into Grasshopper, and a great bendable of the modules if constituted by 4 branches (Fig. 3).

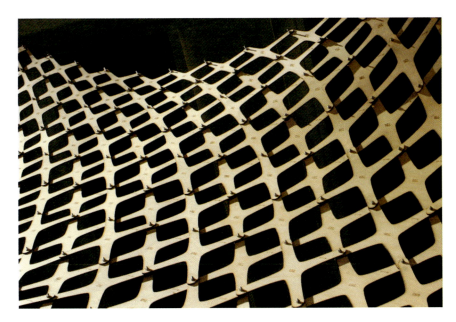

Fig. 3 Details of the assemble plywood elements

Finally, the used pattern was chosen for its aesthetics, but also and above all for its shape: the four branches that make it up are narrow, which has improved the flexibility of the modules. In this way digital and physical models have been conceived together, one influencing the other and vice versa. In fact, after the design phase, students needed to make the first prototype to assess the reaction of the material to the double curvature. The goal was to push the limits of the material as much as possible. For that, they modelled a surface like the bridge one but with a stronger curvature. After having cut all the plywood pieces, they assembled the structure by starting flat. They soon noticed that as they added more pieces, the overall shape began to form due to the tension established by the bent pieces of wood. Indeed, the fact of forcing the parts to be aligned with respect to the screw holes forced the structure to find its final shape (Fig. 4).

In the end, the final model was formed by 980 unique pieces of plywood, fabricated by means of a laser cutter. Each piece has been overlapped and fastened by bolts with to the adjacent one. The model is held on itself, using nylon threads that keep it under tension across its width. The final model has been suspended in the air at the atrium of the Architecture Department of the Politecnico di Bari: the "Flying Musmeci" prototype is ready to intrigue the next generations of students (Fig. 5).

Fig. 4 View of the finished model recalling the shape of the Basento Bridge

Fig. 5 Picture of the suspended model inside the atrium of the Politecnico di Bari's Architecture Department

11 Conclusions

Through the operative research presented, the proposed didactic approach shows a possible learning path for the education of professional artisans. The framework used for guiding the didactic experience of the students suggests that establishing academic paths on the critical use of digital design and fabrication tools could be a feasible way for enhancing the digital awareness and skills of artisans employed in the AEC industry, with benefits for the entire chain. It can be considered also a reference for new lifelong learning didactics for reskilling operations for experienced artisans who need to gain new abilities required by the labour market. Lastly, it's possible to state that the same approach may experiment also for the undergraduate student's curriculum (i.e., bachelor's degree) because they have similar general knowledge of AEC verticals compared to artisans, especially in the first year of studies.

Acknowledgements The research project was carried out in 2020–2021 during the academic Course named C.E.S.A.R. (Cours de Enseignement Supérieur en Architecture et Restauration) at the Politecnico di Bari, where the author has taught digital design and fabrication starting from 2015. The author thanks the students of the Course, Louis Gibault, and Timothée Michel for having carried out the operative research presented in this paper. The author thanks also the Director of the Scuola di Specializzazione, Prof. Monica Livadiotti, and the Course Coordinator and Supervisor Prof. Giuseppe Fallacara for the scientific support. A special thank goes to the associations Les Compagnons du Devoir et du Tour de France and the Fondation de Luxembourg for funding the student's scholarship and to Politecnico di Bari's FabLab and the company Romeo Srl, for supporting the realization of the physical models of the Course. Video of the project can be watched on YouTube: https://www.youtube.com/watch?v=31TH20mWMdA.

References

1. Addis, B.: Toys that save millions-a history of using physical models in structural design. Struct. Eng. **91**(4), 12–27 (2013)
2. Barberio, M., Colella, M.: Architettura 4.0. Fondamenti ed esperienze di ricerca progettuale. Maggioli, Santarcangelo di Romagna (2020)
3. Barbrook, R., Schultz, P., The digital artisans manifesto. Imaginary Futures (1997). http://www.imaginaryfutures.net/2007/04/16/the-digital-artisans-manifesto-by-richard-barbrook-and-pit-schultz/. Accessed 10 May 2022
4. Celani, G.: Digital fabrication laboratories: pedagogy and impacts on architectural education. In: Williams, K. (ed.) Digital Fabrication, Nexus Network Journal 1. Springer-Birkhäuser (2012)
5. de Soto, B.G., Agustí-Juan, I., Joss, S., Hunhevicz, J., Habert, G., Adey, B.: Rethinking the roles in the AEC industry to accommodate digital fabrication. In: Skibniewski, M.J., Hajdu, M. (eds.) Proceedings of the Creative Construction Conference, pp. 82–89 (2018)
6. Fallacara, G., Barberio, M.: An unfinished manifesto for stereotomy 2.0. Nexus Netw. J. **20**(3), 519–543 (2018)
7. Fallacara, G., Barberio, M., Colella, M.: Learning by designing: investigating new didactic methods to learn architectural design. Turk. Online J. Educ. Technol. (Special Issue for IETC 2017), 455–465 (2017)
8. Fallacara, G., Barberio, M., Colella, M.: Con_corso di Progettazione. Learning by designing. Aracne Editrice, Rome (2017)
9. Figliola, A.: The role of didactics in the post-digital age. AGATHÓN | Int. J. Arch. Art Des. **3**, 29–36 (2018)
10. Gershenfeld, N.A.: Fab: The Coming Revolution on Your Desktop–From Personal Computers to Personal Fabrication. Basic Books, New York (2005)
11. Hemsath, T.L.: Searching for innovation through teaching digital fabrication. In: Gerhard, S. (ed.) Future Cities [28th eCAADe Conference Proceedings], pp. 21–30 (2010)
12. Kilian, A., Ochsendorf, J.: Particle-spring systems for structural form finding. J. Int. Assoc. Shell Spat. Struct. **46**(2), 77–84 (2005)
13. Kolarevic, B., Male-Alemany, M.: Connecting digital fabrication. In: Klinger, K.R. (ed.) ACADIA 22-Connecting Crossroads of Digital Discourse, pp. 54–55. Ball State University (2003)
14. Lanzara, E.: Generative design strategies for customizable prototypes. Academic research and entrepreneurial education. In: Garip, E., Garip, S.B. (eds.) Handbook of Research on Methodologies for Design and Production Practices in Interior Architecture, pp. 68–93. IGI Global, Hershey (2020)
15. Magrone, P., Tomasello, G., Adriaenssens, S., Gabriele, S., Varano, V.: Revisiting the form finding techniques of Sergio Musmeci: the bridge over the Basento river. In: Cruz, P.J.S. (ed.) 3rd International Conference on Structures and Architecture Conference (ICSA2016), pp. 543–550 (2016)
16. Lee, M.: The promise of the maker movement for education. J. Pre-Coll. Eng. Educ. Res. (J-PEER) **5**(1), 30–39 (2015)
17. Naboni, R., Paoletti, I.: Advanced Customization in Architectural Design and Construction. Springer International Publishing, Cham (2015)
18. Otto, F., Rasch, B.: Finding Form: Towards an Architecture of the Minimal. Edition Axel Menges, Stuttgart (1996)

19. Petrizzi, C.: Sergio Musmeci a Potenza: il ponte e la città. Basilicata Regione Notizie **104**, 17–24 (2003)
20. Piker, D.: Kangaroo: form finding with computational physics. Archit. Des. **83**(2), 136–137 (2013)
21. Gernot, R.: The digital design build studio. In: Luo, Y. (ed.) CVDE 2011: Cooperative Design, Visualization, and Engineering, pp. 198–206. Springer, Berlin, Heidelberg (2011)
22. Sheppard, S., Colby, A., Macatangay, K., Sullivan, W.: What is engineering practice? Int. J. Eng. Educ. **22**(3), 429 (2007). https://www.webofscience.com/wos/woscc/full-record/WOS:000238371800004?SID=EUW1ED0CCDNuGwWkVJmtWYO13PtrS
23. Smith, R.C., Iversen, O.S., Veerasawmy, R.: Impediments to digital fabrication in education: a study of teachers' role in digital fabrication. Int. J. Digit. Lit. Digit. Competence **7**(1), 33–49 (2016)
24. Stavric, M., Wiltsche, A., Tepavčević, B., Stojaković, V., Raković, M.: Digital fabrication strategies in design education. In: Conference 4th ECAADe International Regional Workshop: Between Computational Models and Performative Capacities, pp. 139–14 (2016)

The Corona Decade: The Transition to the Age of Hyper-Connectivity and the Fourth Industrial Revolution

Alexandros Kallegias, Ian Costabile, and Jessica C. Robins

Abstract The COVID-19 pandemic continues to profoundly affect the world socially and economically. The quarantine and isolation strategies adopted globally have advanced online trade to a new level, as people are finding new ways to provide products and services from home. Several digital tools are gaining popularity and delivery services are ramping up production to meet the increased demand. This paper analyses the current situation considering the impact of COVID-19 in technology and society. The first part of the analysis consists of historical connections between epidemics and technological progress. The paper charts the impacts these have had on society and where they have come to define each industrial revolution. The second part of the analysis explores the different strategies to contain the coronavirus and protect economies. Comparisons between countries are developed through available data and displayed in charts. Furthermore, the paper demonstrates the impact of the strategies on social lives and the economic shift from physical to online. It explores the creative adoption of platforms and technologies that are driving the new revolution. As a case study, it also focuses on "the field of architecture and reviews the case of the live data collection process that is made after the erection of an edifice. Through a practice-based project, it speculates on enhancing the energy performance of a building via applying computational techniques to the collected data. It describes the system of an ad-hoc sensory device that gathers energy data from a building as a different option from existing HVAC systems. Considering COVID-19's high impact on society, drastically altering the way the market operates, we suggest this moment as the true beginning of the Fourth Industrial Revolution, bringing with it a new historical narrative.

Keywords Real-time · Data-conversion · Architecture · Computation · AI and machine learning

A. Kallegias (✉) · I. Costabile
Liverpool University, England, UK
e-mail: alexandroskallegias@gmail.com

J. C. Robins
Lancaster University, England, UK

United Nations' Sustainable Development Goals 11. Make cities and human settlements inclusive, safe, resilient and sustainable

1 Introduction

Several technology developments from the previous decade have gained enormous market strength as a result of the COVID-19 situation, and major digital enterprises are now expanding even further. In the AEC industry, it became evident through COVID-19 that there is an increasing dependence on technology to accomplish sustainability goals, with technology serving as a catalyst for much-needed innovation. These goals are related to a variety of urban production but also to climate change challenges. Architectural technology is driving the way toward adaptable design and building systems, from material science to equipment, tooling, and software development. With the use of robotic techniques and decision support systems, it is essential to consider technology beyond the mere collection of physical components generated via tests and assembled to create new goods. There is also experimental thinking that enables technological progression. Together with the AEC industry, several digital and online platforms have benefited from the lockdown, as a swarm of users and consumers have appeared online. Society, economy and culture are being abruptly reshaped and online trade has finally become mainstream. Here, we announce the Fourth Industrial Revolution and a new decade marked by the 'corona' disruption, brought on by a pandemic that has completely changed the system.

2 Epidemics and Industrial Revolutions

For design and technology, the consequences of this pandemic are fundamental. It has forced humanity to stop climbing the ladder of prototype technology and to go directly to the world of stable digital design. As documented in history, unprecedented events have instigated the First, Second and Third Industrial Revolutions. The First Industrial Revolution began around the 1760s and was mainly characterised by steam and water power, along with advances in mechanised factory systems [1]. Although the causes of this revolution are various [2], this was preceded by a series of epidemics caused by diseases such as the plague [3], which have driven societies to invent new ways to operate. For instance, the Black Death (1348–1350) killed between a third and half of the European population, and for Britain and Italy, this resulted in a rise in wages [4]. Workers earned three or four times subsistence [5] and marriage ages increased, since there were more female employment opportunities, particularly in animal husbandry [6].

The Second Industrial Revolution generally corresponds to the period of 1870–1914 [7], marked by the development of public utilities such as gas, water and sewage systems, along with railroad and telegraph networks, further leading to electrical power and the invention of the telephone. The large number of epidemics happening throughout this revolution, such as the third plague pandemic (1899–1940) in Europe [8], several cholera pandemics and the flu pandemic (1889–1890) [9], are seen as contributors to this revolution and have motivated technological developments. Dirty and contaminated water was discovered to contribute to the spread of diseases, and this made improvements in public utilities and sanitation vital for survival. Furthermore, horses in urban communities contributed to the transmission of diseases, which is argued to have sped up their replacement by automobiles in the early 1900s [10].

The Third Industrial Revolution began in the 1950s [11], highly accentuated by the invention of semiconductors, leading to the invention of computers and culminating with digital tools and the internet. These developments came after more series of pandemics, such as the Asian flu (1957–1958 [12]) which caused millions of deaths worldwide, closing factories and causing a global recession. Epidemics, as such, have supported the need for automating factories and inventing ways to make machines operate more independently of humans.

Certainly, pandemics as well as wars and other conflicts, provoke socioeconomic consequences. The direct connection between disease and the First and Third Revolutions can be argued to be tenuous, as there is a much stronger link to the other conflicts that preceded them. However, it is clear that the Second Industrial Revolution was propelled by the need for improvements in public health due to diseases. In a similar way, the isolation strategy of COVID-19 throughout the world has had a great impact on the way countries operate. With the virus rapidly spreading worldwide, several countries have enforced quarantine and closed their borders, establishing rules of social distancing and self-isolation. This has forced people to find ways to work, consume, study and socialise through the internet, creating economic chaos in the outdoor market and hastily transforming our ways of thinking, acting and being.

The arrival of the Fourth Industrial Revolution has been announced before by many historians and technologists [13–15]. Yet, there was no event like the pandemic that could accelerate technological advances and immediately turn the market over. The cause of such a revolution is not the coronavirus per se; all this was expected to develop to this stage. However, the pandemic has caused an abrupt market change from physical to online, where society has had to completely reconfigure itself. The educational sector, for example, has had to rethink how to educate children and adults through dedicated technology platforms that are easily accessed from homes. The demand from businesses, now operating from homes, for easy communication, has been pushing further technological development favouring conferencing software relatively unheard of prior to the pandemic and massive 5G adoption, something that had been under question due to the use of hardware developed by the Chinese State Telecoms Company and scare stories around spying [16]. This certainly marks a revolutionary change. For this reason, now we believe we can officially declare: The Fourth Industrial Revolution has begun, and its starting point is 2020, the 'Corona Decade' (Fig. 1).

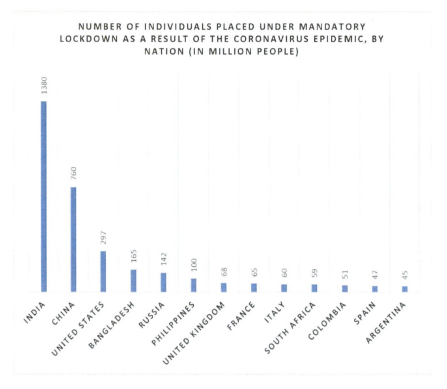

Fig. 1 A graph that depicts the size of coronavirus lockdowns

3 Philosophy of a Transformative Chaos

Our society has suddenly become isolated and many have seen their jobs in danger. The solutions for those who are not key skill workers (e.g. medical or social services) seem limited to two main options: Work at home online or work for the warehouse and delivery industry. Who are the people losing their jobs? All workers who require physical structure and dealing directly with customers, which includes a great part of the arts and entertainment sector (artists, galleries, performers, etc.), the education sector, sales representatives, hospitality, the whole tourism sector including airlines and hotels, personal service workers such as decorators, plumbers, hairdressers. The list continues extensively, addressing all industry sectors [17]. If there is dependence on mobility, the job is at risk. A report from the UN estimates that up to 24.7 million jobs could be lost [18]. However, society carries on, with services and products that require to be delivered. Therefore, there will always be those who will keep old patterns alive. We are living through a dramatic reshaping of society consisting of online work and study. If drones were more ready to be utilised for delivery, human activity could be even more suppressed.

For those people forced to go the route of an online business, skills for acquiring specific tools and managing online marketing are indispensable. This has provoked new technological demands; faster connection services (leading to rapid 5G adoption), faster computers and webcams. Those items, among others, increased their market value following the crisis. As an example, the shortage of protective masks was soon followed by a shortage of webcams [19]. The education sector suddenly relies on this technology, and thus thousands of teachers around the globe will benefit from investing in new digital equipment. Yet, to compete in a saturated online market, high-quality media and internet speed are crucial, which means consuming equipment and services that are not always affordable.

The chaos opens the curtains on the 20 s decade; announcing a drastic shift in our common way of living. The steep increase in internet traffic is further proof. It is evident that this data has increased exponentially since the first quarantines. Digital streaming platforms such as Netflix and Disney Plus are limiting high definition content in Europe so that bandwidths can cope with increased demand [20, 21]. We are now noticeably more financially dependent on the internet and its digital tools. Cash is history. It holds the risk of virus contamination and, now, more complicated, physical, distribution. The total shift online has been predicted many times over but until now had been slow to come about. It has been a key strategy of capitalism: Multinational online companies such as Amazon taking over physical shops, supermarket chains improving their delivery services, online banking, and the arrival of 5G. Despite the predictions, the change hadn't quite happened, until now. The transition has taken place with no warning and all digital resilience has been suddenly lost. One example concerns the elderly population, those who did not have a connection to the online world, who still paid bills by visiting the bank and post office. By unfortunate coincidence, the elderly population seems to be the most affected by this virus. These are thousands of senior citizens who are getting excluded from the digital turn of society as they no longer comprehend how it works. In isolation, people who have ignored digital platforms such as WhatsApp and social media see little choice but to reconsider adopting new ways of communication. All of the extra expenses that can be attributed to the digital shift of home supply, compared to all jobs at risk, point to a global recession [22]. However, it goes beyond that, we are seeing a societal shift to a single economic model of online work, especially for the tertiary industry. The coronavirus situation will be a temporary one as solutions and vaccinations are brought forward, but the changes are bound to remain. Who will invest in an offline business after such an event? This virus is going to be contained or eradicated soon, however, the risk of similar threats in the future remains and this calls for more preparation globally. Whereas the digital dependence can be reversed after recovery, it is unlikely it will return to the same state it was before the pandemic (Fig. 2).

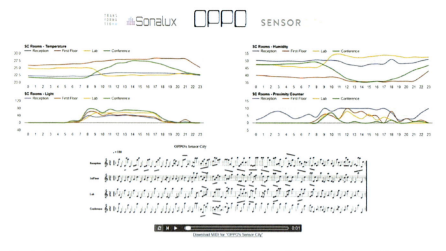

Fig. 2 OPPO: the web-based version translating live data from four sensors in four different rooms onto a musical score

4 Global Cocoons and the Worldwide Transition

As many countries adopt quarantine strategies and the global society adapts to online solutions, we see a stage of change and transition. The worldwide strategy dilemma has divided the priorities between economy and health. This decision can drastically change the current economic position of all affected countries. Governments that better respond with economic tools to protect employment, income and demand, who are providing effective methods, guarantee a better economic future after the outbreak. However, prioritising the economy is not a straightforward option, especially for countries with public health systems that cannot cope with large or concentrated influxes of cases, as a weak public health system during a time of chaos can also profoundly impact the economy.

What is essential here is to include in the mitigation strategies the strengthening of the digital infrastructure. Quarantining the population means an increase in the use of digital tools, from production to consumption. Although countries have been adopting different measures, most countries in the world have encouraged remote working and the closing of schools. Financial support for online working has been provided in Italy and Japan [18]. In other cases like Greece, a proposal for developing digital skills—customer service, technical support, etc.—was considered at the onset of the pandemic but eventually not applied [23]. Nevertheless, this makes the need for training and transition more evident, moving individuals from not using or using some digital tools to becoming efficient in operating daily chores and vocational responsibilities remotely and online.

Media reports data [24] that shows a great part of the world's population has entered a lockdown state, where the tendencies are to socialise and work online. As the event prolongs for longer than a month, the likelihood of having a new structuralisation of socio-economic patterns on a global scale is high. However, it is important to consider that a restructure is not possible for all parts of the world at once. Countries or regions that lack digital infrastructure and technology have more difficulties confronting COVID-19, and as such, many countries will not participate in such a transition as described here. Yet, this event reinforces investments and developments in digital technology, which then leads to the same paradigm shift in harder to reach regions. It is important to remember that the previous industrial revolutions did not exist in all parts of the globe concurrently [25]. However, in the current scenario changes are happening in all industrialised regions through several means of communication and international relations.

We, here, take into consideration the concept of 'revolution' as the hasty and precipitous restructure of a society, which happens for almost urgent reasons. Countries that responded with immediate measures for encouraging online working and that already had a high-end digital structure in place are likely to be transiting towards revolution. According to the Harvard Business Review, countries on the frontline of digital business are the US, the UK and the Netherlands [26]. The list ranks countries for their digital capabilities such as banking, retailing, media and skilled freelancers. These are countries that can function through a wide range of online services and have efficient delivery services. At the bottom of the list, we see Indonesia and Russia, countries that struggle more to find digital solutions and thus will not operate similarly in this historical shift (Fig. 3).

Fig. 3 Customizable sensory devices for the OPPO system

5 Hyperconnectivity and Possible Futures

During this period of rapid transition brought by the Fourth Industrial Revolution, workers will become more accustomed to flexible working. Through the realisation that workers no longer need to waste time on the daily commute, and perhaps a greater appreciation of their homes, there is a high possibility of having more creative pursuits and a return to the old adage from the Second Industrial Revolution: 8 h for sleep, 8 h for work and 8 h for play. A change to the week/weekend dichotomy is another future scenario. Even prior to the house-isolation measures, services such as groceries or product deliveries were already in place throughout both the weekdays and weekends with normal working hours. However, these services have now received a significantly higher demand, consequently changing the habits of many shoppers and attracting new customers. This has substantially increased the number of people connected to the online market.

As the world is counting the weeks in the work-from-home setup, there are certain changes in human behaviour regarding work and leisure to be considered. These changes relate to aspects of flexibility, overtime, management and motivation. We are seeing that people tend to keep a more flexible and loose working schedule. This is to cope with responsibilities at home as they occur (home-schooling children, career responsibilities, household emergencies). While they may compensate with the time gained from not commuting, this, in turn, requires certain skills of self-management. This also requires a new approach to arranging meetings, making deadlines and tracking excess working times in the hierarchical levels of different professions. Online services are now becoming more frequent and accepted. As a result of the imposed quarantine measures and the new rules for supermarkets, which include long queues for grocery shopping, more consumers are registering online [27]. It is likely that after the pandemic crisis people will continue to online supermarkets for groceries. Video conferencing tools have become crucial since the beginning of quarantine and there has been a boom in previously underused platforms such as Zoom [28]. Without cinemas, restaurants, cafes and clubs for socialising, online platforms such as cloud clubbing and film/music sharing have been in higher demand [29]. Digital tools and skills are becoming critical for career development in all sectors. In the entertainment business, for instance, many artists have experimented with online performances, and funding is being offered to create computer-based experiences. Online platforms have expanded their capabilities for audience interaction and have improved features to assist digital concerts. It is possible the new decade will have an increased number of YouTubers as the profession becomes more valued and competitive. The digital culture of micro-celebrities will rise further and will be levelled with other media channels. The online shift has also increased interest in joining VR platforms to experience virtual tourism through online galleries and museums [30]. There has been a surge in connectivity as more people have joined the connected world since the start of the pandemic; we can say now we are 'hyper connected'.

There is another consideration to be made. The limitation of personal services during quarantine and the longer stays at home have an impact on domestic affairs.

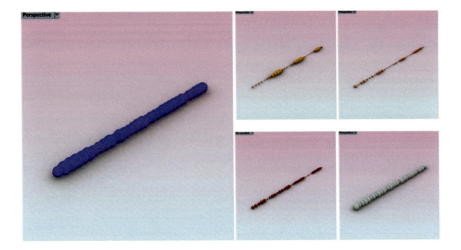

Fig. 4 These are energy data image captures that have been transformed into an interactive graphical set of spheres that follow the relevant audio levels based on the obtained data

As people settle into life at home, we see a return to old skills; craft and repair projects that have been growing dusty on shelves will suddenly look appealing once all the streaming content has been consumed. The internet is replete with sites to help us to repair and create, and older relatives might delight in the chance to demonstrate cooking, sewing or woodworking skills, for example (while possibly learning about video making themselves). In Europe, as this crisis is developing during the start of spring, many people will make attempts to grow their own food in light of the supermarket shortages and inevitable supply chain disruption. This has also coincided with parents being forced to educate their children. Parents are now looking for easy, home-based activities, which will result in a generation of people with traditional skills that were almost lost to the generations before them (Fig. 4).

6 The Architectural Practice

As previously described, COVID-19 has a huge influence on nearly every business on the planet. The architectural field has been no exception. As soon as the epidemic struck, initiatives were either cancelled or postponed until a later date. Every significant epidemic throughout history has resulted in a large-scale architectural shift. While architectural projects may have been cancelled in the early months of the pandemic, they could be crucial in the long run in combating COVID-19.

After contact, viruses and illnesses have been shown to survive and grow on a variety of surfaces across an interior setting. Many organizations are adding temporary "contact-less" features throughout their offices and services, so many of these features are likely to become commonplace in the future. Ventilation systems that remove potentially polluted air, automated doors, and elevator buttons linked to smartphones, voice-controlled lighting, and a slew of other contactless technologies might become the norm in interior areas across the country. Those who are reluctant to change and regard these new trends as transient may be left behind, while those who comprehend the changes and their influence on society drive architecture ahead.

7 The Industry 4.0 and Data

Together with COVID-19, the architectural field was already pushing its boundaries with technologies such as Artificial Intelligence (AI) and the Internet of Things (IoT). Considering the effect of the "contact-less" safety measures against the coronavirus within the AEC industry, there is an opportunity to apply AI and data-driven systems in order to boost the potential of computational design. This is particularly effective when it comes to monitoring the in-use energy performance of an edifice and computers play a crucial role in a better understanding of the energy levels of a building after its construction.

Prior to the use of computers and software for design, the practice of architecture developed by depending on experienced architects' design judgments and assumptions. Architects create and accumulate building design knowledge via their work. Their knowledge progressively transforms into a personal intuitive design technique. Most other architects are unaware of this particular intuition since the gathered data is kept in their heads. Eventually, the building and its package drawings are all that is left as tangible evidence of that knowledge.

The use of computers has sped up the process of creating and documenting data in architecture. Around the mid-1960s, computer-aided design (CAD) became popular, making it easier to store both graphical and non-graphical project data. Architects were able to optimize their productivity by switching from manual drafting to a more integrated and automated generation of package drawings in the late 1990s, thanks to technological breakthroughs in software programs brought on by the Third Industrial Revolution. Traditional design procedures were also automated, allowing architects to organize their workflow with other engineers around a single basic digital model. This fundamental model serves as a primary structure for collecting and arranging the project's data by combining the varied perspectives of all project partners. This is also referred to as Building Information Modelling (BIM).

When examining how organizations in the AEC industry handle data, it becomes evident that companies now have digital design tools that create and document information in a complete and interoperable manner while constructing a structure, thanks to advances in computing technology. The essential aspect of this improvement is that construction knowledge can now be transferred without relying entirely on personal experience. BIM has been evolving since the 1970s and it is currently becoming a prerequisite for the delivery of public buildings in a number of countries, including the United Kingdom, and architects are using it for private projects as well.

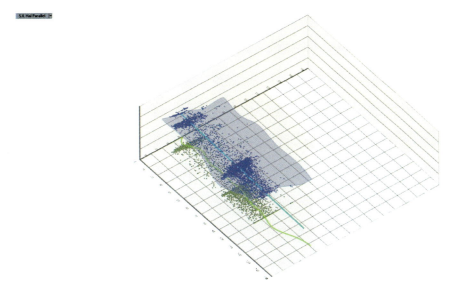

Fig. 5 A 3D machine learning graph of energy data from Sydney Jones' library reception area

Companies who are first in line to take advantage of Industry 4.0 and the Internet of Things' ability to capture in-use performance data will expand even faster. It is critical to design policies that link performance analytics apps to benefits, duties, and liabilities in this case. Aside from rules relating to industry interoperability, there is also a demand for targets for designing and managing buildings using in-use energy data. There is an opportunity here to address the UN's position in the building industry in terms of its sustainability goals. With live energy data, aside from the automated HVAC system, there is the possibility of more creatively conveying the building's energy levels. Such an initiative is described in the following case study (Fig. 5).

8 Ad-Hoc Energy Gathering System

The position of the UN's goals toward a brighter future, in addition to fighting climate change, is also to make cities and communities more sustainable. Examining the energy levels of a structure is a step in achieving that aim. The UK's Secretary of State for Business, Energy and Industrial Strategy has also said that the Net Zero Strategy would be used to achieve that aim. They indicate in their white paper that they will seek suggestions on regulatory measures to increase building energy efficiency. Here, the flow of project data is an essential part of this efficiency. Certain protocols (IFC, COBie) have been established in the AEC sector to simplify the flow of project information among the many collaboration stakeholders. These transfer procedures appear to place a strong emphasis on the stages that span conception through delivery. However, there is no feedback loop in place once the project is completed to optimise efficiency or monitor changes in usage. Projects will be established and managed more effectively with the availability of performance data combined with the goals of regulating bodies and AEC industry data validation in the design process. This is when design computerization comes in handy. Having real-time updates of a building's physical properties, in addition to its fixed digital representation, will change the way architects develop and deliver their projects. This will be a BIM-driven transition.

As a case study, a system was designed to enable the creative and practical collecting and sharing of energy data while recognising how the Internet of Things technology includes changing real things into smart devices. The technology, which is part of a project called Oscillating Personal Places Occurrences or OPPO, uses an audio-visual medium to transform raw numerical data into valuable information. The goal here has been to improve our understanding of a building's energy behaviour by using working tools from both the fields of architecture and music. While we humans are visual beings, our cognitive ability to hear is greater than that of sight. Our hearing sense is not only the quickest, but it also has the ability to calculate minute variations in sound strength. With this in mind, and because communication is so important when it comes to transmitting information, the project examined the visual and audible translation of numerical energy data.

The OPPO project entailed the development of sensor boxes and an Internet of Things system that can collect data (light, temperature, humidity, and proximity) from a building and convert it to another, editable format. It was installed from February 2020 to July 2021 at two libraries simultaneously at the University of Liverpool and investigated how a large collection of energy data may be used in the predictive modelling process using artificial intelligence techniques. Specifically, the *LunchBox* software plugin was used to employ a method known as regression analysis. This method of prediction is widely utilised, and it has a lot in common with machine learning. The approach, based on Adrien-Marie Legendre's least squares method, displays a curve that best fits a set of data points, such as a set of energy values, on a Cartesian system by minimizing the sum of the offset points from the projected curve. By extending the curve, which displays the relationship among specific data points, it is possible to get a sense of the future relationship of the points that are still to be generated.

The project used computation to develop novel methods for gathering and processing real-time energy data. The speed with which these complex data sets were acquired was very fast and large databases formed quickly. These databases will be impossible to manage with standard data processing software. However, these vast amounts of data could be used to solve a number of previously unsolvable challenges in the industry. Big Data, or large, complex data sets, could help the architectural profession with information visibility and process automation. When Industry 4.0 and additional BIM levels are implemented in small increments, new working methodologies and rules for accessing, validating and processing open data in a standard format will emerge.

9 Conclusion

Through this paper, we have proposed the COVID-19 event as a historical mark. One that is reshaping society by motivating a number of industrial sectors to move to fully online operations and changing the face of work for millions. We have compared this to previous industrial revolutions, elevating the current crisis to a catalyst for a fourth revolutionary phenomenon; an event that has the potential to characterise the whole next decade. The current state of quarantine of several countries has been seen as a transitional phase which can have profound socio-economic consequences and a large impact on global society. Estimated changes in society have been expressed in possible futures, as the digital market and technology continue to be impacted by the pandemic. The decade of the 2020s will greatly differ in many aspects from the previous two, spawning a whole new cultural generation that functions through live virtual interaction. As an example of such interaction, it was demonstrated how in the architectural practice there is an increasing interest and benefit for having a system that can keep a virtual record of the building information gathered as well as display it remotely. As a result, rather than a segmented survey, a broader perspective is possible. The immersive outputs make it easier to interact with data using various representation approaches. The case study's ambition has been to show a way of adding another layer of data to cross-reference a physical entity rather than creating a model capable of precisely predicting outcomes. Moreover, data that are collected using open-source sensor technology such as OPPO, can be subsequently translated using design tools for use in a more creative visual and audio communication context.

10 Discussion

Besides the architectural case study, the aspect of harvesting data and storing them holds a lot of implications. It can have benefits in promoting a more efficient way of understanding our world but it can also become a tool used for partial purposes. This delicate balance between the two sides became clear when measures against the coronavirus were created based on personal data. Hence, the steps that governing bodies consider taking to tackle this pandemic can also endanger public privacy and new concerns will continue to emerge as governments seek to prevent future pandemics. However, there is the potential for research into skills acquisition and actions towards self-sufficiency during this time, which can be complemented by research into a phase in experiments in governance and social welfare models.

Acknowledgements This article takes the view of technologists members of the Transformation North West cohort working with the industry in the UK'S North West. We acknowledge the sponsorship of the AHRC and NWCDTP as a contribution to our doctoral research.

References

1. Encyclopaedia Britannica. Industrial Revolution. (2019). Available from: https://www.britannica.com/event/Industrial-Revolution
2. Deane, P.M.: The first industrial revolution. Cambridge University Pres (1979)
3. Zuckerman, A.: Plague and contagionism in eighteenth-century England: The role of Richard Mead. Bull. Hist. Med. **78**(2), 273–308. The Johns Hopkins University Press (Summer 2004)
4. Allen, R.C.: The British industrial revolution in global perspective: How commerce created the industrial revolution and modern economic growth. Available from: https://www.ehs.org.uk/events/assets/AllenIIA.pdf
5. Allen, R.: Why was the Industrial Revolution British? VOX. (2009). Available from: https://voxeu.org/article/why-was-industrial-revolution-british
6. Voigtländer, N., Voth, H.–J.: How the west "invented" fertility restriction. Am. Econ. Rev. **103**(6), (October 2013). Available from: https://www.aeaweb.org/articles?id=https://doi.org/10.1257/aer.103.6.2227
7. Mokyr, J.: The second industrial revolution, 1870–1914. In: Valerio Castronovo (ed.) Storia dell'economia Mondiale. pp. 219–245. Laterza publishing, Rome (1999). Available from: https://pdfs.semanticscholar.org/d3fc/63c43a656f01f021fb79526d9ba3b25f6150.pdf
8. Bramanti, B., Dean, K., Walløe, D. and Stenseth, N.C.: The third plague pandemic in Europe. The Royal Society. Proceedings of the Royal Society B. Biological Sciences. (2019). Available from: https://royalsocietypublishing.org/doi/https://doi.org/10.1098/rspb.2018.2429
9. Kempińska-Mirosławska, B., WoŸniak-Kosek, A.: The influenza epidemic of 1889–90 in selected European cities—a picture based on the reports of two Poznań daily newspapers from the second half of the nineteenth century. Med. Sci. Monit. **19**, 1131–1141 (2013). Available from: https://www.ncbi.nlm.nih.gov/pmc/articles/PMC3867475/
10. Rosner, D.: "Spanish flu, or whatever it is. . . .": the paradox of public health in a time of crisis. Public Health Rep. **125**(Suppl 3), 38–47 (2010). Available from: https://www.ncbi.nlm.nih.gov/pmc/articles/PMC2862333/
11. Taalbi, J.: Origins and pathways of innovation in the third industrial revolution. Ind. Corp. Chang. **28**(5), 1125–1148 (2019). https://doi.org/10.1093/icc/dty053
12. Jackson, C.: History lessons: The Asian Flu pandemic. Br. J. Gen. Pract. **59**(565), 622–623 (2009). Available from: https://www.ncbi.nlm.nih.gov/pmc/articles/PMC2714797/
13. Min, J., Kim, Y., Lee, S., Jang, T–W., Kim, I., Song, J.: The fourth industrial revolution and its impact on occupational health and safety, worker's compensation and labor conditions. Saf. Health Work **10**(4), 400–408 (2019). Available from: https://www.sciencedirect.com/science/article/pii/S2093791119304056
14. Schäfer, M.: The fourth industrial revolution: How the EU can lead it. Eur. View **17**(1), 5–12 (2018). https://doi.org/10.1177/1781685818762890
15. Penprase, B.E.: The fourth industrial revolution and higher education. In: Penprase, B.E. (ed.) Higher Education in the Era of the Fourth Industrial Revolution pp. 207–229. Springer (2018). Available from: https://link.springer.com/chapter/https://doi.org/10.1007/978-981-13-0194-0_9
16. Bowler, T.: Huawei: Is it a security threat and what will be its role in UK 5G? BBC News. (2020). Available from: https://www.bbc.co.uk/news/newsbeat-47041341
17. Konings, J.: All sectors will feel coronavirus hit but some will recover quicker than others. ING. (2020). Available from: https://think.ing.com/articles/all-sectors-will-feel-coronavirus-hit-but-some-will-recover-quicker-than-others/
18. ILO. COVID-19 and the world of work: Impact and policy responses. (2020). Available from: https://www.ilo.org/wcmsp5/groups/public/---dgreports/---dcomm/documents/briefingnote/wcms_738753.pdf
19. Thomas, B.: Where to buy a webcam: These retailers still have stock. (2020). Available from: https://www.techradar.com/uk/news/where-to-buy-a-webcam
20. BBC News: Netflix to cut streaming quality in Europe for 30 days. (2020). Available from: https://www.bbc.co.uk/news/technology-51968302

21. Bisset, J.: Disney plus throttles streaming quality amid coronavirus outbreak. (2020). Available from: https://www.cnet.com/news/disney-plus-throttles-streaming-quality-amid-coronavirus-outbreak/
22. Elliot, L.: Prepare for the coronavirus global recession. The Guardian. (2020). Available from: https://www.theguardian.com/business/2020/mar/15/prepare-for-the-coronavirus-global-recession
23. Poluzou, K.: Finally in distance training, regular reinforcement of EUR 600 scientists [in Greek]. ERT (2020). Available from: https://www.ert.gr/%CE%B1%CF%84%CE%B1%CE%BE%CE%B9%CE%BD%CF%8C%CE%BC%CE%B7%CF%84%CE%B1/telos-stin-tileka tartisi-kanonika-i-enischysi-ton-600-eyro-stoys-epistimones/
24. Buchholz, K.: What share of the world population is already on COVID-19 lockdown? Statista (2020). Available from: https://www.statista.com/chart/21240/enforced-covid-19-lockdowns-by-people-affected-per-country/
25. Engerman, S.L.: The industrial revolution in global perspective. In: Floud, Johnson (eds.) Cambridge Economic History of Modern Britain, vol.1: Industrialisation, pp. 1700–1860. Cambridge University Press, Cambridge (2004). Available from: https://warwick.ac.uk/fac/arts/history/students/modules/hi153/timetable/wk4/engermanobrien.pdf
26. Chakravorti, B., Chaturvedi, R.V.: Ranking 42 countries by ease of doing digital business. Harv. Bus. Rev. (2019). Available from: https://hbr.org/2019/09/ranking-42-countries-by-ease-of-doing-digital-business
27. Martin R. Coronavirus UK: Will supermarkets announce more online delivery slots? Metro. (2020). Available from: https://metro.co.uk/2020/03/20/will-supermarkets-announce-online-delivery-slots-12431081/
28. BBC News: Coronavirus: Zoom under increased scrutiny as popularity soars. (2020)s. Available from: https://www.bbc.co.uk/news/business-52115434
29. Cuthbertson, A.: How to watch films with friends online under coronavirus lockdown. Independent (2020). Available from: https://www.independent.co.uk/life-style/gadgets-and-tech/news/watch-movies-stream-online-group-netflix-party-youtube-coronavirus-lockdown-a9426586.html
30. Coates, C.: Virtual reality is a big trend in museums, but what are the best examples of museums using VR? Museum Next (2020). Available from: https://www.museumnext.com/article/how-museums-are-using-virtual-reality/

Quasi-Decentralized Cyber-Physical Fabrication Systems—A Practical Overview

Ilija Vukorep and Anatolii Kotov

Abstract Building an effective cyber-physical system is difficult due to the overall complexity of technologies on their own, challenges with the interaction between all parts of workflow and applicability issues. This makes the real-world application of complex cyber-physical systems only available to the big industry parties or high-tech startups, leaving small and medium-sized businesses behind. Therefore, the democratization of such applications is a reasonable goal to achieve. Cyber-physical systems of this kind are in the experimental stages and incorporate robotics, IoT, materials science, visual and 3D-scanning techniques, and machine learning (ML) tools all under the heading of Industry 4.0. This paper covers specific practical approaches in several application fields using robotics and IoT. We use custom-built hardware and software setups together with standard frameworks and the MQTT protocol for different applications. Due to the practice-driven approach, the paper will illustrate both positive and negative effects. We describe the use of our systems in several case studies: Multi-layer automated robotic concrete spraying using ML, IoT spatial awareness via sensors (such as Lidar, Kinect), robotic multi-axis milling and quasi-autonomous robot movements. A unifying issue is a decentralized approach of modular IoT elements that we grouped to achieve specific tasks. The paper illustrates how these elements exchange data and communicate, and all services are controlled on each computer instances, connected to an IoT network to ensure a high level of system stability.

Keywords Cyber-physical system · Digital fabrication · Industry 4.0 · Decentralized digital services

United Nations' Sustainable Development Goals 9. Industry Innovation and Infrastructure · 12. Responsible Consumption and Production

---

I. Vukorep (✉) · A. Kotov
Chair of Digital Design Department, Brandenburg University of Technology, 03046 Cottbus, Germany
e-mail: ilija.vukorep@b-tu.de

© The Author(s), under exclusive license to Springer Nature Switzerland AG 2024
M. Barberio et al. (eds.), *Architecture and Design for Industry 4.0*, Lecture Notes in Mechanical Engineering, https://doi.org/10.1007/978-3-031-36922-3_12

1 Introduction

Our paper discusses practical approaches to robotics and Internet of Things (IoT) in several application fields. In order to increase the reusability of such cyber-physical setups, a combination of custom hardware and software, as well as standard frameworks and protocols are used for different applications. We implement principles of Industry 4.0 paradigm by using several architecture defining principles such as microservices, applied robotics and ML. There are several examples of similar setups of using industry 4.0 principles in robotic fabrication [1, 2]. Micro-services allow us to isolate key services and machines, enabling us to create decentralized, modular IoT elements that can be grouped together to achieve specific tasks. Furthermore, microservices are used within the software and hardware domains, allowing for better communication and data exchange between different segments. This further enhances the integration of our cyber-physical systems, allowing us to achieve greater levels of automation and efficiency. This concept embodies one of industry 4.0's main strengths, which is the democratization of complex tools.

We are discussing some typical services in detail used in our applications and providing links to some own developed libraries and components on GitLab [3]. Our system is demonstrated in several case studies: Multi-layer automated robotic concrete spraying using machine Learning (ML), IoT, spatial awareness using sensors (such as LiDAR, Kinect), robotic multi-axis milling, and in teaching scenarios. A common theme is a decentralized approach of modular IoT elements that we group to achieve specific tasks. One of the tasks is the concrete spraying case study that is aiming to reduce the weight of produced elements and increase the potential design space through its capability to produce free-form shaped parts. Its production incorporates a wide range of machine control, space observation and data processing segments. The paper will describe how these segments exchange data and communicate with each other. In the quasi-autonomous case study section, we will cover the interaction of a pair of robots, where one robot is performing target action, while the second is following the first one via dynamic observation. Micro-services are used both in a software and hardware sense—both key services and machines are isolated. We start by defining the problem statement, which discusses the use of Robot Operating System (ROS), a leading system for developing cyber-physical systems. The following chapter is about categorizing various types of cyber-physical systems in terms of their internal organization and communication of processes.

2 Problem Statement

There are several systems that can help in developing automated and robotic applications. One of the most popular and robust is ROS—robot operating system, a framework for various robots founded 2007 at Stanford University. It is suitable for heterogeneous clusters with a highly developed messaging system with low latency. Its publish-subscription communication management is even near real-time in the new ROS 2 version. Furthermore, its services run decentralized, so demanding processes can execute remotely. ROS' lively community is providing libraries for all kinds of hardware. Communication with external non-ROS clients is possible (Websockets, mqtt). The ROS-industrial is an extension of ROS and provides open source industrial general and vendor specific libraries in an industrial context. The drawback of such a powerful system is that the implementation can be very complex. In a typical academic scenario, where experiments are performed with light and atypical configurations, setting up ROS with its enormous overhead can be extremely time-consuming. Although ROS covers a big range of hardware and allows implementation of one's own libraries, installing unsupported sensors and actuators inside of ROS proved to be unrealistic in our context of projects.

The setup of a construction-related robotic system requires the creation of several services that can work robustly and independently and have a clear way of interacting. As most robotic setups are different, those services should operate in different constellations. The software can be written in different languages, and the services can run on completely different hardware. With this, to distribute the development of single parts to a wider range of people is easily possible, in our case to students and academic researchers. When working with parametric geometry, such as connecting Rhino 3D Grasshopper results to a robotic setup with many sensory inputs and complex actuator control, this flexibility is especially important. Another important feature is also the ability to work remotely, via VPN, when needed, with visual feedback.

3 State-of-the-Art

In the architectural and construction context, most academic experimental robotic setups are very narrowly and pragmatically tailored to solve some specific scientific problem. The majority of related research papers do not detail their technical setup, but a few of them mention their software equipment and this has been analyzed. Generally, we can divide architecture- and construction-oriented robotic setups into several groups:

1. **ROS as a central unit**. ROS is often used for its integrated and decentralized service platform and communication protocol. In [4] research, a collaborative human–robot construction system is developed around a ROS computational core and its ability to communicate with virtual reality in Unity, sensor data, and the robot. In the work of [5], ROS is exchanging data with MATLAB Robotics System Toolbox for path creation. In [6] ROS is used together with MoveIt for trajectory planning of the robots. In [7] Choreo is used along with ROS-industry and Move-it to plan motion and choreography.
2. **ROS as one of several parallel services**. The work [8] involves modeling in Rhino, kinematic modeling in the COMPAS FAB package, and ROS as a robot controller with communication via Roslibpy. This library was developed

at the ETH Gramazio. Kohler and is based on WebSockets and bridges between services. For [9] GH/Lunchbox ML services, as well as serialized data in JSON format, are implemented.

3. **Decentralized system with heterogeneous communication**. The work [10] uses individual services for scanning (python), ML analysis (Tensorflow, python), modelling (GH/Rhino) and path planning (RoboDK) without a specific communication protocol. In this study, [11] Rhino for modelling, GH to simulate robotic processes, Unity for VR-visualization, DynamoDB at Amazon Web Services (AWS) for data handling and data storage are used and heterogeneously connected.

4. **Decentralized system with specific server communication**. Described in [12] is a robot setup controlled by a self-written python server (XML, TCP/IP Ethernet) that transmits data between services (clients: Python algorithms, Rhino visualization, camera, and robot). In [13] the communication is arranged through Java in Processing and an UDP protocol exchanging data with the Scorpion plugin in Rhino/GH for path creation and transmission to the robot.

These categories each have their own advantages and impact on scalability, modularity, and reusability. A decentralized system with a Message Queuing Telemetry Transport (MQTT) protocol is described in this article based on experiences gained in a variety of scenarios. The protocol serves as a basis for IoT development, as all clients and services in an IoT system communicate via the web. These services will be described in more detail within the following chapters.

4 Components/Services

Our system uses a modular microservices architecture. Most commonly, this term describes the organization of software complexes and information systems. Contrary to a monolithic approach where all the code is merged together, the microservices concept divides a program into several independent components (services) that can run on multiple/divided platforms and have a unified communication vocabulary. If one subsystem fails, the whole system will be less likely to crash, what improves stability. Another advantage is a possible update/change of individual components inside the system without having to rebuild the entire system. There are currently many distributed online systems with millions of users using it. As a basis for further development, this approach was chosen due to the need for stability, modularity (reusability), and scalability.

4.1 MQTT Broker—Infrastructure Server

The core element of this decentralized cyber-physical system is the IoT data exchange protocol (MQTT). In contrast to *XML* over *TCP/IP*, *Websockets*, *UDP* or other previously mentioned communication protocols, is that it can deliver messages with requested quality of service (QoS). This includes fire and forget, at least once and exactly once. QoS can help when connections are unstable or we have critical command execution procedures (as running a robot). It's supplying sufficient speed and it's lightweight. The principle of subscription and publishing over topics is similar to ROS' handling messages. The issues of security around MQTT are not explicitly handled in this article and need special care by using client identification rules. The broker runs in the background inside a network. In our settings, we ran a *Mosquitto* broker (mosquitto.org) on *Raspberry* Pi 4 that automatically activates the service on booting. It can run on any OS system with a known IP address to the clients. For testing purposes, we provide a self-developed MQTT server on localhost running inside GH/Rhino as a component. Furthermore, we wrote a MQTT subscription and publication component for GH/Rhino [3].

4.2 Dashboard

To manage all processes, we could theoretically use any MQTT-browser like MQTT-lens [14] but this will not give us a good overall overview of incoming and outgoing information. For visualization and administration of all processes, a central dashboard is necessary. Fig. 1 is showing an example of such a dashboard with several pages. The best method of a dashboard that is compatible for diverse use cases is to keep it growing as services are being added. This means that services that are not used are automatically disabled in some applications. This can be done by internal checking if services are available (online).

Key features of a dashboard are:

– Robot control, with some predefined robot positions or movements, manually loading robot scripts or others,
– System observation, key vital data of robot TCP position, availability of hardware and program components,
– Visual observation from connected cameras,
– Control of additional machines as scanners, grippers,
– Database connection with data presentation and editing options.

Our dashboard is built with *Python* and *Flask* backend programming framework that can easily be scaled as it uses templates and sophisticated data interaction. These packages can run on Raspberry Pi 4 or any other computer (also together with the broker) in the network and can be reached at the local network.

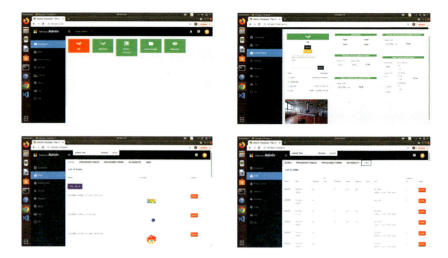

Fig. 1 Dashboard pages: status, robot control, scan data view and info page

4.3 Robot—IoT Interface

One central part of the IoT robotic setup is the connection of the robot to the rest of the system. For this connection, we use an interface service that is transferring incoming data directly to the robot over an Ethernet/Lan connection and vice versa, publishing information like robots joint and TCP positions data. As this interface is embedded in a wireless network to other clients, huge flexibility of robotic control is provided, which means we can control the robot without being directly connected to it.

As every robotic company has got its own I/O solution, this interface has to be tailored individually to robotic families. We tested this setup with *Universal Robots* running the interface on a Raspberry Pi 4 and written in Python. All outgoing information about the robot is read by the IO Modbus protocol of the UR robots and broadcasted over MQTT.

Additional machines connected to the robot are also accessible through this interface as part of the script that is transferred to the robot. This transfer incorporates also some extra features such as comparing the robot position with the target position or relative and absolute movements and some special positions, like home, self-test or tool changing position.

4.4 Data Management

Data management is not organized as a single service but can be found in all other services and processes. In our cyber-physical arrangements, we have four data streams that handles: Configuration files of the setup, incoming (i.e. sensor) data, processed data and logging data. Configuration files are actual settings (home position, special robotic positions, broker addresses, storage credential data) stored as JSON-files. Besides this, we also have environment and tool data stored as Rhino 3 dm-files. Both types are inside the folder structure of the *dashboard*.

Incoming data are files made by sensors and are mostly of high volume. Those are saved in a project folder with a time-stamp and additionally automatically processed or triggered by incoming commands. Those processed files are then saved in additional folders like thumb views of scanning or images, automatically generated scripts or other project specific data. All processed data that are part of closed procedure, are saved in the job-folder. This decentralized database-free approach is working well in closed networks but there are security issues when we expose them in open networks. For this usually some professional data handling services can be used like AWS S3 services that is also handling user-right-management and some data processing through APIs. In our setup we used a *Resilio* Sync—*BitTorrent* protocol for data synchronization as this proved to be fast.

Logging is an important step toward error tracking and statistical analysis. Some of the services in our cyber-physical system are logging their processes. Advisable would be a central logging device, even organized as an own service, that is collecting logging data from other services with the capability of analytic display in dashboard or other viewer.

4.5 Vision

Cameras can be used for tracking movement, security control, position diagnosis or simple documentation purposes. The application will define if camera stream will be recorded or not, if the processed image will be directly processed or high computation is necessary (i.e. OpenCV postprocessing). Any stream data client can be connected to the network and then retrieved from the dashboard. Based on streamed data it's possible to detect movement, risk situations (and use MQTT for signaling) or incorporate customized recognition of elements. In our case we have used the camera for finding edges of the scanning area. For simple low resolution processing, a camera can be connected to a Raspberry Pi 4, which can run OpenCV4, and transfer the data over Wi-Fi. For higher resolution data streams an Ethernet connection is necessary. Remote control of the system is also facilitated by vision.

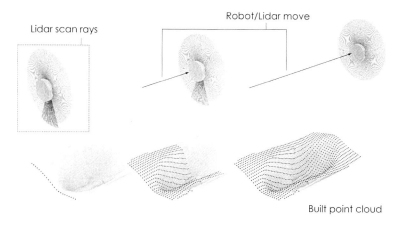

Fig. 2 Illustration of the scanner movement over the observed object

4.6 Sensing

The complexity of the task and models meant that we had to use different scanning systems in our project. We did tests with *Kinect*, *azure Kinect DK* and *RPLidar* scanners. All three have different capacities and options of sensing the environment.

4.6.1 Rotation Lidar Scanner

Rotation lidar scanner like is measuring distance to an object at some specified angle. As it rotates, it can only measure in slices, meaning if we want to measure distances in all 3 axes, the lidar scanner has to move. In combination with a precise movement of a robot we can have a very effective scanner that can, in some cases, make better scans of areas that are obscured, deep objects or with cantilevers (Fig. 2). We connected our RPLidar A2M8 scanner to a Raspberry Pi 4. Together with the UR-robot, both were triggered by an MQTT command [3]. Scanning results grouped with the coordinates of the scanner positions were saved in the incoming data folder and automatically further processed. This data processing usually depends on the application and in our case it consisted of deleting the unnecessary data and transforming the rest into a 3D point cloud.

4.6.2 Azure Kinect SDK Scanner

For certain tasks, such as real-time scanning for object presence and control, a moving lidar is not the best solution. Therefore, a depth camera coupled with a RGB-camera seemed like a good option [3]. We chose an *Azure Kinect SDK* scanner that is connected to a computer with the *Ubuntu OS*. Even though the hardware and software requirements are high, the scanning process worked well because we needed a high level of precision without a high resolution. As the RGB-stream can be read separately, we use the camera also for documentation purposes. Several MQTT commands are running the scanner: RGB-single shot, RGB-stream, hi-resolution scan, low-resolution scan. The results are stored in the incoming data folder and are automatically processed like creating thumb views of the point clouds or detecting objects.

In our case we mounted the scanner above the active zone of the robot, see Fig. 5. In the usage examples described later, the scanner serves for sensing the actual state of robot production. For this, every process step is scanned and the object scan are compared after the process step is finished. Before the process starts, a calibration scan is executed that marks the clean table and initializes the process.

4.7 Point Cloud Processing Service

When scan data is stored in the incoming folder, several application dependent processing steps are necessary. Usually object have to be detected from out of a messy point cloud and for this the following functions are executed: (1) the empty table scan detects a plane, (2) the object scan is been cleaned against this plane, (3) the result is clustered so some flying points eliminated, (4) the outline of the object is found, that all further scans can use for the eliminating unnecessary point data. Although this service depends on the application we wanted to develop, it as a module that can be reused in other applications. The algorithms are triggered by MQTT-commands from the dashboard. Most processing work is done in the background while the dashboard lists the objects in the browser views. Fig. 3 shows a 3D scanned mould after the point cloud cleaning procedures.

4.8 Grasshopper Components

Since the wide adoption and the abilities of Rhino/Grasshopper, it was essential to create viable MQTT components for Grasshopper. During different development phases we also used GH/Python connected modules, but that was not offering enough scalability, since one has to install and configure a Python environment for any new machine. GH components are simpler to use and also better in terms of response of the GUI since they're natively compiled from C# and Grasshopper is built on.NET technology. The components for Rhino 6 and 7 can be downloaded from GitLab [3].

Fig. 3 Mold 3d scan example automatically subtracted from the surrounding environment

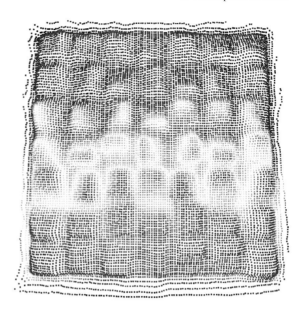

4.9 Path Planning Service

The goal of the path planning service is to provide a viable adaptive trajectory over a given surface keeping in mind execution of additional tasks. Those tasks are the execution of actions or activating hardware connected to the robot depending on path and/or surface properties. The path planning are then combined with action augmentation. This means, in certain positions or areas robot activates connected hardware (e.g. use gripper or run the concrete pump).

A naive path planning approach can be defined as a *zig-zag pattern* based on a given shape (Fig. 4). However, such an approach is ignoring the curvature of the target surface. In this case it is not adaptive, and this can cause problems with complex shapes and where actions require precision. We calculate simple patterns based on the previously mentioned outline form without any usage of additional 3D programs.

An alternative approach would be the use of ML methods with the aim to be adaptive to any form and be compatible with imperfections of the scanned object. We use a *Multi-Objective Optimization (MOO)* strategy in conjunction with other ML methods. These methods are proven to be useful as one of the tools in architectural production [14]. However, to apply those methods, the transformation of several stages of data is necessary. The path planning between different zones can be also formulated with the *Travel Salesman Problem* (TCP), which enables us to find the optimal path. To solve it, we're using internal Grasshopper components with custom Python processing. Thus, we're solving the problem of uniform distribution of the points over the given surface with desired in-between distance (a derivative of action radius and distance between TCP and surface). As a result, a waypoint graph with equal edge length based on the surface of any curvature is generated. Using the

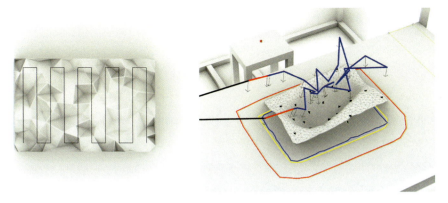

Fig. 4 Left: naive zigzag approach, right: ML approach with the determination of the working area for the path planning algorithm as well as for dynamic hardware control settings (e.g., we stop air & concrete pumps while the robot is travelling outside/between red and blue line, and enable it over islands/inside blue line)

previously generated graph and applying TCP guided local search algorithm, we're able to find the optimal path. In our examples we are using different script generation software like RoboDK (robodk.com) and Robots (github.com/visose/Robots) running in Rhino/GH instances. Those are activated by MQTT commands that sends the number of the job so the identification of the files in the job folder is possible.

To ensure that the robot movement is safe, the path is being simulated in a duplicated virtual environment where the virtual robot and all the robotic movements are synchronously simulated prior to the machine execution. Thus, we're testing that the given path can be executed, that all points can be reached, that there are no singularities and collisions (and signaling an error over MQTT). As a final part of the program, the resulting script is being augmented according to the target setting for acceleration and speed of the robot's linear and joint movements' settings.

4.9.1 Digital Twin

During the implementation of certain use strategies we found it very useful to implement a digital twin of the URs that we use in the lab. This can help us to prevent singularities or collisions with itself, other robots or the environment. The difference between a digital twin and a simulation is that it simulates the setup in real time and with all possible MQTT inputs that is streamed over the *Robot—IoT interface*. Each simulation is using individual UR calibration files, so we can be sure that it's 1:1 to the real setup. The other benefit of using digital twins is the fact that one can develop programs for robots without the physical availability of the systems (e.g. develop robot control system not from the robot lab, but from home office). Later, the algorithms and models tested on digital twins can be propagated onto real hardware. For implementation we are using Python for MQTT-side communication and control, sharing part of the codebase with a *Robot—IoT interface* and RoboDK, all running on a windows machine inside a Rhino/GH instance. The limitations lie in limited sensor (vision and scanning) data or the simulated production process that can be complex.

5 Applications

5.1 Adaptive Robotic Concrete Spraying

Concrete spraying is one of the promising technologies that aim to reduce the weight of produced elements and increase the possible design space through its feasibility in producing free-form shaped parts. Even though thin-formed concrete shells production is efficient for many situations, a variable thickness would increase applications. First situation is resulting from the inconsistent process of spraying. It often suffers from an interruption in concrete feeding that the pump is pushing through the tube. As these interruptions are happening during the robot movement, the applied material unintentionally varies in thickness. The second situation comes from the structural properties of the element that sometimes demands to have ticker or slimmer areas depending on its structural needs.

Using ML methods this system offers different spraying strategies: Using target thickness or matching the target form. The spraying is done automatically using 3d scanning and adaptive robot path planning (Fig. 5). The software subsystem consists of several parts:

- Point-cloud scanning of mould and scanning the processing phases,
- Form comparison/matching—comparison of the given form to target goals,
- Robot path planning component—optimal robot spraying trajectory,
- Robot—IoT interface for execution of desired spraying path, controlling pump and air valve.

While there may be simple forms, the high curvature forms' surfaces present a challenge. This requires a complex path plan strategy as the mechanical movements of the robot should not collide with the mold during the spraying process as well as to follow the goal parameters of the target concrete shape. The outcome of the research can be found [15].

5.2 6-Axis Milling

Traditional CNC-milling is usually limited to milling in 3 axes. However, attaching a milling machine to a 6 axis robot arm allows to achieve much more versatile milling.

Quasi-Decentralized Cyber-Physical Fabrication Systems—A Practical ...

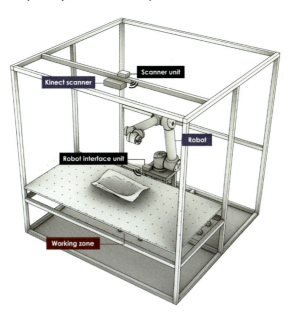

Fig. 5 Hardware and software elements of a cyber-physical concrete spraying system

One of the key abilities is so-called "undercutting", the movement when a robot drill goes under an already existing form without destroying it (Fig. 6).

In our setup the robot was controlled via a Raspberry Pi with our software. The most challenging part of this application was the software that forms the movement trajectories in 6 axis movements, to check if the form is feasible to milling and the robot arm and milling drill will not collide with the milled model itself. Furthermore, there are also issues of visibility of milling paths on the target surface in case of sophisticated target form. Despite the proof-of-work state, developing a full-scale application seems challenging due to the complexity of robot path planning and challenges with fine milling of complex curvatures.

Fig. 6 Left: undercutting maneuver, right: resulting mold

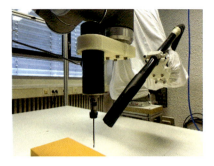 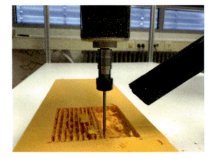

Fig. 7 Left: system setup, right: milling-following-vacuuming

5.3 Milling with Real-Time Relative Robot Movements

Milling produces a lot of dust, which is a problem. In this instance, we are solving this with a robotic IoT toolset. Next to a milling instrument attached to one UR, we added another actor; a new UR with the vacuum nozzle that is designed to follow the milling TCP of the first UR robot to vacuum all dust generated during the milling process.

Hardware setup consists of UR robots each connected with Raspberries. During the implementation of this solution we found the need to create the 1:1 digital twin of the acting robots in the environment due to the fact that it was more convenient to debug controlling algorithms in a digital way. UR1 is milling, while UR2 is following UR1 TCP via constant reading real-time position and axis data (Fig. 7). Using this data UR2 is correcting its position.

5.4 Teaching

Teaching digital design methods in universities often involves the use of robots. The robots are usually housed in cages and separated from the workshop area to minimize injury risks. Several groups of students can work with the robot and develop experimental setups. We found that the possibility to wireless control the robot over MQTT can help in the workflow as there is no need to always re-switch the cable to the robot control unit. Still, as everybody can control the robot from their computer and if the right MQTT topic name and the broker's address is known, this can increase the risk of uncontrolled sending commands to the robot. This type of dangerous interference can be reduced with strict rules in robotic workshops (time-slots for groups) and vision-based recognition to signal, when the robot area is free of people.

6 Results and Future Work

In this paper we described a decentralized cyber-physical setup based on services that can be easily reused for various applications. As the main communication protocol is MQTT with good remote capabilities it can serve as a classical IoT example. Due to security and latency issues, the described setup has still no direct industrial potential but as the system is modular, elements can be evolved towards professional use cases. It is perfect for academic workshops where a lot of task-driven pragmatic and quick development is possible.

All presented services can be improved, especially where real-time is necessary (i.e. controlling safety measures is high risk environments). We noticed 300–500 ms latency when controlling the robot with a gamepad or in the milling-following-vacuuming application example. Also the data management can incorporate industry-grade data handling services such as AWS S3 services, Azure or Google Cloud to avoid exposure of data with a safe user-management. Logging can be incorporated as a single central service together with a good analytic visualization. This setup was tested only with the UR robot family. The interaction with other robots should be further developed and tested. Some of the libraries and GH components are publicly available at GitLab [3].

References

1. Betti, G., Aziz, S., Rossi, A., Tessmann, O.: Communication landscapes. In: Robotic Fabrication in Architecture, Art and Design, Springer International Publishing, Cham, pp. 74–84 (2019). https://doi.org/10.1007/978-3-319-92294-2_6
2. Reinhardt, D., Haeusler, M.H., London, K., Loke, L., Feng, Y., De Oliveira Barata, E., Firth, C., Dunn, K., Khean, N., Fabbri, A., Wozniak-O'Connor, D., Masuda, R.: CoBuilt 4.0: Investigating the potential of collaborative robotics for subject matter experts. Int. J. Archit. Comput. **18**, 353–370 (2020). https://doi.org/10.1177/1478077120948742
3. https://gitlab.com/Digitales_Entwerfen/pub_repository
4. Wang, X., Liang, C.-J., Menassa, C., Kamat, V.: Real-time process-level digital twin for collaborative human-robot construction work. In: Presented at the 37th International Symposium on Automation and Robotics in Construction, Kitakyushu, Japan (2020). https://doi.org/10.22260/ISARC2020/0212
5. Gill, R., Kulić, D., Nielsen, C.: Path following for mobile manipulators. In: Bicchi, A., Burgard, W. (eds.) Robotics Research: Springer Proceedings in Advanced Robotics, vol. 2, pp. 527–544. Springer International Publishing, Cham (2018). https://doi.org/10.1007/978-3-319-60916-4_30
6. Kaiser, B., Littfinski, D., Verl, A.: Automatic generation of digital twin models for simulation of reconfigurable robotic fabrication systems for timber prefabrication. In: Presented at the 38th International Symposium on Automation and Robotics in Construction, Dubai, UAE (2021). https://doi.org/10.22260/ISARC2021/0097
7. Huang, Y., Carstensen, J., Tessmer, L., Mueller, C.: Robotic extrusion of architectural structures with nonstandard topology. In: Robotic Fabrication in Architecture, Art and Design 2018, pp. 377–389. Springer International Publishing, Cham (2019). https://doi.org/10.1007/978-3-319-92294-2_29

8. Ercan Jenny, S., Lloret, E., Gramazio, F., Kohler, M.: Crafting plaster through continuous mobile robotic fabrication on-site. Constr. Robot. **4**, 1–11 (2020). https://doi.org/10.1007/s41693-020-00043-8
9. Ercan Jenny, S., Lloret-Fritschi, E., Jenny, D., Sounigo, E., Tsai, P.-H., Gramazio, F., Kohler, M.: Robotic plaster spraying: crafting surfaces with adaptive thin-layer printing. 3D Print. Addit. Manuf. 3dp.2020.0355 (2021). https://doi.org/10.1089/3dp.2020.0355
10. Nicholas, P., Rossi, G., Williams, E., Bennett, M., Schork, T.: Integrating real-time multi-resolution scanning and machine learning for conformal robotic 3D printing in architecture. Int. J. Archit. Comput. **18**, 371–384 (2020). https://doi.org/10.1177/1478077120948203
11. Ravi, K.S.D., Ng, M.S., Ibáñez, J.M., Hall, D.M.: Real-time digital twin of on-site robotic construction processes in mixed reality 8. (2021)
12. Dörfler, K., Rist, F., Rust, R.: Interlacing. pp. 82–91 (2013). https://doi.org/10.1007/978-3-7091-1465-0_7
13. Elashry, K., Glynn, R.: An approach to automated construction using adaptive programing. pp. 51–66 (2014). https://doi.org/10.1007/978-3-319-04663-1_4
14. Vukorep, I., Kotov, A.: Machine learning in architecture. In: The Routledge Companion to Artificial Intelligence in Architecture, pp. 93–109. Taylor & Francis, London (2021)
15. Vukorep, I., Zimmermann, G., Sablotny, T.: Robot-controlled fabrication of sprayed concrete elements as a cyber-physical-system. In: Second RILEM International Conference on Concrete and Digital Fabrication, RILEM Bookseries, pp. 967–977. Springer International Publishing, Cham (2020). https://doi.org/10.1007/978-3-030-49916-7_94

Latent Design Spaces: Interconnected Deep Learning Models for Expanding the Architectural Search Space

Daniel Bolojan, Shermeen Yousif, and Emmanouil Vermisso

Abstract This work proposes an adoption of Artificial Intelligence (AI)-assisted workflow for architectural design, to enable the interrogation of possibilities which may otherwise remain latent. The proposed design methodology hinges on the "systems theory" consideration of architectural design, expressed in Christopher Alexander's "systems generating systems", offering an alternative to the reductionist and complexity-lacking structure of design processes (Alexander 1968). This logic is pursued through the integration of DL models into an "open-ended" workflow of interconnected Deep Learning strategies (DL) and other computational tools, rather than treating it as a closed "input–output" cycle (single DL model). While a closed cycle risks flattening architectural layers, ending up with a reductionist encoding of design intentions, an open-ended workflow can inquire into an expanded design search space and augment creative decision making. This way, chained model strategies can simultaneously address design intentionality within discrete architectural layers (i.e. organization, composition, structure).This system enables three distinct modes of collaboration: *human–human, human–AI,* and *AI–AI*. Understanding the contribution of human and machine agents within the workflow offers a re-evaluation of designers' processes. The proposed nested workflow reflects the transition from 'expert systems', which rely on hard-coded rules, to 'learning systems', which are inspired by the human brain (DL) (Hassabis 2018). This allows architects to approach design problems which are not fully defined (Rossi 2019) and helps avoid overconstraining the search in creative domains like architecture. Furthermore, the design investigation is strengthened by accessing a search space that is otherwise beyond the designer's reach towards an expanded design creativity.

D. Bolojan (✉) · S. Yousif · E. Vermisso
School of Architecture, Florida Atlantic University, FAU/BC Higher Education Complex, 111 East Las Olas Boulevard, Fort Lauderdale, FL 33301, USA
e-mail: dbolojan@fau.edu

S. Yousif
e-mail: syousif@fau.edu

E. Vermisso
e-mail: evermiss@fau.edu

Keywords Computational creativity · Deep learning · Human–machine collaboration · Architectural design space · Human–machine collaboration · Augmented intelligence

United Nations' Sustainable Development Goals 9. Build resilient infrastructure, promote sustainable industrialization and foster innovation · 11. Make cities inclusive, safe, resilient and sustainable

1 Introduction

1.1 Generative Design Systems

The 4th industrial revolution, or "4IR" [4] is projected to influence a wide range of white-collar jobs in contrast to prior industrial paradigm shifts in history, which directly impacted the blue-collar workforce [5]. This disruption manifests through introducing new problem-solving approaches using, for example, Artificial Intelligence (AI). The design and realization of buildings during the twenty-first century is increasingly informed by computational tools which enhance performance and enable the emergence of a new kind of aesthetics. Data-driven design can address more criteria than before, breaking down levels of complexity which were difficult or impossible through analogue processes alone. Taking advantage of large amounts of available data and corresponding algorithms like Artificial Neural Networks (ANNs), designers can access additional layers of information, which can potentially augment their design thinking. The adoption of AI-assisted working protocols in the architectural domain can enable the interrogation of otherwise latent possibilities. These protocols complement traditional rule-based "expert systems" which use hard-coded rules for problem-solving (input + rules → results), with learning systems (i.e., deep learning models) which learn by example and can generate rules from the ground up (supervised: input + results → rules; unsupervised: input → rules). Instead of depending on pre-programmed answers, learning systems develop solutions from first principles. ANNs are examples of learning systems, inspired by the structure and function of the human brain, as well as the way humans acquire specific types of information. The capabilities of learning systems to generate creative outputs can be attributed to their non-linear and unknown synthesis, and lack of rule determination in the learning process [6]. Expert systems such as algorithmic and parametric models involve rule-based and/or performance-driven algorithms that evolve design as a product of a parametric exploration, with specific parameters and constraints, often to satisfy a set of objective functions [7]. Such deterministic systems require a set of input parameters to examine a feasible search space, offering a reductionist approach to design approach and confining the design space to pre-programmed solutions [8]. By integrating learning systems, a recent second generation of GDSs is emerging, marking a shift in design, towards an expanded exploration of the design space.

1.2 Design Complexity

Existing literature demonstrates that current applications of AI in architectural design occur at the representation level, without addressing other important design layers. The progressing line of research on AI for architectural design remains experimental with a limited scope of investigation, due to its infancy [9]. Current approaches employ a single AI model to tackle multi-faceted design problems. "Design is probably the most complex of human intelligent behavior" [10], therefore, the capability of a discrete AI model to address all design levels is questionable. Rather than using one AI model for a single design task, a new AI-assisted design workflow of multiple connected models can tackle multiple systems. This echoes a "systems theory" consideration of architectural design, expressed in Christopher Alexander's "systems generating systems", offering an alternative to the reductionist and complexity-lacking structure of design processes [1]. In this multi-step workflow, this work initiates a thorough investigation of recent generative machine capability by looking at designer agency and mode of collaboration with the machine.

This work examined a new human–AI collaborative framework ("Deep-Chaining"), where multiple chained deep learning (DL) models were deployed in a sequential logic with feedback from multiple designers to address specific architectural systems and design tasks. Our method is based on task separation to minimize the complexity inherent within AI-assisted workflows. Different strategies were examined: *(1) fully-chained supervised, (2) fully-chained unsupervised and (3) partially-chained hybrid networks*, trained on a specific scale (urban or architectural), and architectural task (e.g. Fig. ground, envelope, structural system, etc.). A case-study was implemented in a workshop format for computational designers/architects. Strategies for chaining results from multiple neural networks enabled assessment of discrete design tasks from human agents when necessary. The primary objectives include assessing the proposed generative framework of human and artificial agency by identifying and evaluating the interactions among the involved agents (Human–Machine; Human–Human; Machine–Machine). In addition, the work interrogates a new type of "Augmented Architectural Intelligence", leveraging the potential of AI-assisted workflows to be computationally creative [11]. The authors have adopted a "process" approach for assessing creativity defined by neuroscience [12], to discuss and evaluate the quality of the proposed strategies and overall framework. This novel framework contributes to innovation in the architectural sector in line with UN Sustainable Development Goal #9 and #11.

The manuscript is structured in seven sections: 1. Introduction; 2. Theoretical framework; 3. State-of-the-art and DL Application in Architecture; 4. Research Methodology; 5. Discussion; 6. Conclusions & future work.

2 Theoretical Framework

2.1 *Intelligence and Creativity*

Our proposal stems from an interest to enhance the designer's creative faculty, so we present here a brief discussion of *intelligence, creativity* and the emerging concept of *"computational creativity"*, in order to frame our subsequent discussion. Architectural design thinking is an inherently creative act. Understanding the relationship between human and computational creativity (AI) is essential to develop AI-assisted computational procedures which encourage creative results while introducing reliability and efficiency.

Our current understanding of intelligence is focused on two properties—which, combined—may help frame the discourse between human and AI: *"Intelligence measures an agent's ability to achieve goals in a wide range of environments"* [13]. This definition identifies two aspects of intelligence, the former referring to task specificity and the latter to context specificity (generality).

In contrast to intelligence, creativity is harder to define, because its study spans across disciplines. Experts identify novelty and utility among manifestations of the creative process: *"Creativity is the ability to come up with ideas or artefacts that are new, surprising and valuable"* [14]; *"Creativity can be defined as an idea or product that is original, valued, and implemented "* [15]. Creative insight is thought to emerge either by sudden perspective shifts or non-incremental solutions reached through logical analysis [12] and can be classified into three types of increased rarity and difficulty: *Combinational, Exploratory* and *Transformational* [14]. Furthermore, everyday creativity (P-creativity: Psychological) is distinguished from less common, and more impactful H-creativity (Historical) in terms of prior precedence.

The recognition of human creativity is slightly more complex, and has been known to fluctuate over time-unlike intelligence, which has a consistent benchmark. Professor Csikszentimihalyi argues that it is important to ask not only *what* creativity is but also *where* it is found; according to the "systems model *"...creativity can be observed only in the interrelations of a system made up of three main parts: ...the domain, ...the field ...and the individual"* The domain consists of *"a set of symbolic rules and procedures"* and the field *"...includes all the individuals who act as gatekeepers to the domain"* [15]. Creativity typically occurs when the symbols from a particular domain are used to generate a new idea which is in turn adopted within the domain by the existing field, or occasionally, new domains can be formed altogether. The systems approach expands common regard of creativity as an endeavor associated with a single person and considers external factors as a catalyst for creative achievement: *"...whatever individual mental process is involved in creativity, it must be one that takes place in a context of previous cultural and social achievements, and is inseparable from them"* [16]. An interesting example of creative output lying beyond the individual sphere is Renaissance Florence, where the timely co-existence of appropriate socio-political conditions enabled the civic encouragement of intensive artistic creation during a short period of time (1400–1425) [15]. Accepting the likelihood of creative production as a result of collective interactions among agents and processes, we speculate herewith on the advantages of using multiple designers for curating inter-connected design workflows.

2.2 From Computational to Augmented Creativity

As a result of recent technological developments in AI a new kind of creative output has been observed: the new field of Computational Creativity (CC) *"...is an emerging branch of artificial intelligence (AI) that studies and exploits the potential of computers to be more than feature-rich tools, and to act as autonomous creators and co-creators in their own right...The CC field seeks to establish a symbiotic relationship between ...scientific and engineering endeavors, wherein the artifacts that are produced also serve as empirical tests of the adequacy of scientific theories of creativity"* [11].

It is important to note that computational creativity should not be equated with AI, as the former incorporates facets of the latter. According to research studies, intelligence is "a necessary-but-not-sufficient condition of creative ability, creative activity and creative achievement" [17]. While a certain intelligence threshold (120IQ) is required for creative thinking, other conditions are also necessary. Nevertheless, the correlation between creativity and intelligence is not exponential; incremental increase in intelligence beyond this threshold does not necessarily favor creative achievement [15]. As creativity is inclusive of intelligence, our mention of "architectural intelligence" in this paper will henceforth refer to design creativity, because we believe it encapsulates the requisite properties for successful design decisions.

Our discussion will argue that creativity is intertwined and related to design agency and authorship within AI systems; collaboration between human and artificial agents depends on human intellect, AI-produced solutions, and their relationship and modes of interaction. This can increase our creative design capacity by leveraging AI's ability to explore a large range of possible successful solutions, establishing a kind of augmented architectural intelligence within a robust and flexible framework. There are several approaches identified by neuroscientists to assess creativity, focusing on person; product; process; press(ure)/place [12]. Our discussion will focus on the "process" approach to establish qualitative benchmarks for increasing creative thinking in design, as we think evaluating the logistics of the overall method is more beneficial than a discrete consideration of outcomes. The significance of crafting "hybrid" processes which leverage on human intuition and machine (AI) has been highlighted by several experts, like Gary Kasparov. 'Kasparov's law' indicates that a *"...weak human + machine + better process was superior to a strong computer alone and, more remarkably, superior to a strong human + machine + inferior process*" [18].

3 State-of-the-Art AI Technologies and Background Literature

In this section, a background literature is offered to situate the research within current research on AI and architecture. In order to identify related work, and our research methods, state-of-the-art concepts and approaches that are most relevant, and employed in our work need to be explained. Those methods include DL methods, Generative Adversarial Networks (GANs), and types of GANs (Pix2Pix: supervised learning; CycleGAN unsupervised learning), explained in the following subsections.

3.1 State-of-The-Art: Deep Learning

Ian Goodfellow et al. discuss the potential of machines to learn and develop a hierarchy of interrelated concepts. In this type of framework, humans curate the formal knowledge needed for the machine [19]. The challenge for an AI lies in acquiring knowledge on its own, extracting and interpolating patterns from datasets. Specifically, it is difficult for the machine to select particular features for extraction, to successfully perform the determined task. This indicates a need for applying Machine Learning (ML) towards representation learning versus representation mapping. In a DL approach, AI can adapt to new tasks and discover features, performing complex tasks not possible for humans [20]. We predict that Computational Creativity can greatly impact the ecology of architectural design.

3.1.1 Generative Adversarial Networks (GANs)

Developed by Goodfellow et al.'s, GANs are defined by two deep neural networks (DNN) which compete against each other: a discriminator network (D) and a generator network (G) [20]. The two networks engage in an adversarial learning process of generating synthetic samples that are incrementally more realistic (Generator) and a process of distinguishing synthetic samples from real samples (Discriminator). While the generator network tries to predict features given a certain category and learn how to improve its output to resemble reality, the discriminator network tries to perform a correct classification (*real* or *fake*), given an instance of data. The objective is to minimize the distance between the two sample distributions.

3.1.2 Unsupervised GAN (CycleGAN)

CycleGAN models represent an innovative approach to image-to-image translation where learning occurs in a process of translating an image from a source domain (X) to a target domain (Y), without image pairing. The objective is to learn how to map (X) to (Y) coupled with an inverse mapping of (Y) to (X), performed by introducing a cycle consistency loss to enforce consistency. In other words, the DL training is targeted towards achieving successful image translation within the cycle, where (X) is translated to (Y) and (Y) should be translated back to (X) accurately [21].

3.1.3 Supervised GAN (Pix2Pix)

Pix2Pix models are a conditional sub-category of GANs for synthesizing images in a supervised logic where pairing is necessary to reconstruct images after learning from a paired dataset. Pix2Pix networks use a loss function to train their mapping from an input image to an output image. This makes it successful at synthesizing images from labeled datasets such as maps, reconstructing parts from edge maps, in addition to other tasks. The method showed success when used in wide applications, and became popular for artists in particular [22].

3.2 Deep Learning Application in Architecture

The research on AI contributes beyond the advancement of machinic or robotic intelligence, additionally providing descriptive and analytical understanding of how human intelligence operates. According to Hiroshi Ishiguro *"the robot is a kind of mirror that reflects humanity and by creating intelligent robots we can open up new opportunities to contemplate what it means to be human"* [23]. AI has been modeled on the human brain's neural networks, and can perform tasks which include image recognition, classification, clustering, etc.. In recent ML developments, Google researchers observed that if an AI can recognize an image, that meant it had understood the structure and representation of that image. As a result, engineer Alex Mordvintsev reversed the direction (flow) of the image recognition network, resulting in a network (DeepDream) which can generate images [9]. In "Machine Hallucinations" (2019) Refik Anadol, combined AI and media art, utilizing 100 million images of New York City for training a StyleGAN. These networks produced an AI-generated future vision of the city [25]. This strategy showed a good example of directing the data generation process to create a creative artistic expression [25, 26]. Early architectural work generated by AI includes a design process developed by Japanese architect Makoto Watanabe. His process applied AI for "Induction Design" in the project" Tsukuba Express/Kashiwanoha-Campus Station" designed in 2004–05. The objective was to retrieve unpredicted "magic-like" aesthetics without pre-determined conception of "what good is". The workflow involved exchanges between the designer and the machine, starting with sketches and allowing the program to infer the design intent and propose other sketches; evaluation occurred repeatedly until satisfactory design results were reached [24]. More recently, Matias del Campo argued for generating an AI sensibility that goes beyond formal aesthetics [27]. He questioned the interpretation of AI as *Style* and the multi-dimensional interpretations it entails, arguing for the use of ML to examine stylistic features [28]. In the work of Güvenç Özel, *"Interdisciplinary AI"*, ML was employed to classify and evolve stylistic approaches borrowed from artistic references. The human's role becomes limited to input dataset curation and selection of ML-produced iterations. His underlying argument is that AI is certainly capable of design creativity when calibrated and customized to do so, opening possibilities for external aesthetics [29]. The work of Immanuel Koh used an AI model to extract spatial patterns for generating new configurations with associated semantics [30]. Another example of applying GANs, in particular Pix2Pix, to optimize the process of generating a diverse set of architectural floor-plans can be seen in the work of Stanislas Chaillou [31]. His research showed AI's potential in automating floor-plan design while still producing high quality design options, in a creative process where AI is sensitive to site conditions, building footprint, spatial configuration, and furnishing.

It is important to note that the scope of the investigation in most examples from current AI research in architecture, is often limited to representational and artistic exploration. Representation has been traditionally identified as a generally challenging and important aspect of this research field even in earlier AI models [10]. Apart from representation, in our exploration of AI's potential, we seek a process-driven approach where AI is employed to tackle multiple architectural systems (Fig. 1).

Motivation for this research stems from a need to investigate and redefine interactive modes between AI and human agents in workflows using multiple interconnected neural networks. A primary objective is to identify strategies to expand the possible solution search space. This capacity of DL models has been demonstrated during applications of AI to solve games like Go. The AlphaGo algorithm used a system composed of a Policy Network, a value Network, and a Monte-Carlo tree search to expand and refine the selective search of potentially successful scenarios [2].

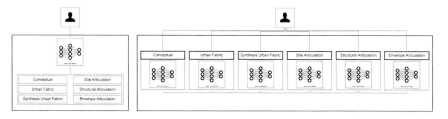

Fig. 1 A sequential design approach vs a parallel design approach

4 Methods

The scope of our research is to investigate the application of AI strategies through a holistic approach to design processes. Adopting John Gero's schema of 'design prototypes' [32], a new framework to architectural design was pursued, employing interconnected DL models and strategies for 2D and 3D. To demonstrate the framework and test its functionalities, a prototype was formulated through a series of experiments at a discrete architectural and design intention level. Our proposed methodology hinges on "systems theory" consideration of architectural design, which involves a *"... shift from the view of architecture as a material object to architecture as a system comprised of and working with a series of interrelated systems"* [33].

Architectural design thinking is by no means linear. The designer negotiates several parameters to find a solution accounting for all criteria in terms of certain hierarchical preferences. Rather than thinking of AI as a "closed" loop of "input–output" in relation to the design process, the authors propose a framework (Deep-Chaining) consisting of (1) a single vs a multi-designer approach; (2) a series of complementary DNNs to examine the potential of logical continuity in AI-driven workflows for a new mode of generative computational design.

4.1 Single Versus Multi-designer Approach: Agency and Feedback Loops

This approach is expected to challenge yet augment the designer's agency, to work in an interactive mode of collaboration with the machine. Computational creativity is exploited and coupled with humans' creativity, in a back-and-forth exchange of design input and feedback, progressively, towards finding new design solutions, aesthetics and higher performance.

When adopting a single-designer unidirectional strategy, the designer determines the architectural task, which is the driving force of the project. Thus, the initial tasks in the design process affect subsequent ones, while the opposite is not true (e.g. Urban Fabric → Street Layout → Building Footprint). The strategy is inherently linear, allowing design intentions and sensibilities to be only forward-propagated in the process. This strategy's unidirectional nature limits the emergence of unexpected solutions and back-propagation, thereby preventing self-organization of designers' intentions.

A multi-designer bi-directional strategy is therefore regarded as more appropriate for creating an open "input–output" loop (the meta level). Within each layer (at the infra level): this strategy manifests as (a) a process where *human and AI agents* interact through self-organization and (b) a process where *multiple human designers* formulate their design intent through self-organization. Therefore, the strategy enables three distinct modes of interaction: designer–designer, designer–AI, and AI–AI (chained networks). It also allows for bi-directional propagation

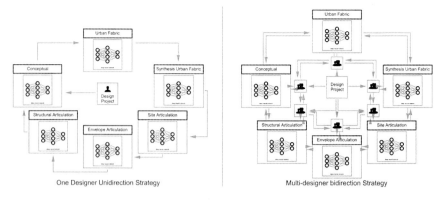

Fig. 2 One designer unidirectional and Multi-designer bidirectional strategy

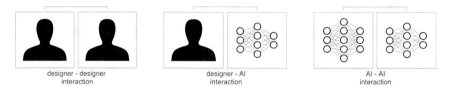

Fig. 3 Modes of interaction between agents (Human–Human; Human–AI; AI–AI)

of designers' intentions and sensibilities, therefore becomes inherently nonlinear, governed by local rules of interactions (Figs. 2 and 3).

4.2 Complementary DNN: Case-Study

An application of the prototype was carried out to demonstrate the developed framework, test its strategies and evaluate its findings. It was structured in the format of a long workshop offered to 24 architects. Participants were clustered in 6 groups (4 people each). Group 1–3 and Group 4–6 focused on urban and architectural scales, respectively. The workshop aimed to reconsider the Architectural Design Cycle, proposing nested generative AI design processes. Implementing an interactive designer–AI feedback loop, a number of complementary DNNs was employed in collaboration with the groups, to examine the potential of a logical continuity in AI-driven workflows for architecture. The workshop structure consisted of a range of DL models deployed to tackle a variety of urban tasks involving *learning from natural patterns*, *urban fabric encoding*, *urban fabric synthesizing*, and architectural tasks of *site articulation*, *tectonic articulation*, and *envelope articulation* (Fig. 4). Each group implemented strategies ranging from fully-chained to hybrid. The chained network combinations can interrogate the potential of supervised versus unsupervised models. The resulting strategies were (1) Fully-Chained Supervised; (2) Fully-Chained Hybrid (Supervised and Unsupervised); and (3) Partially-Chained Hybrid (Supervised and Unsupervised).

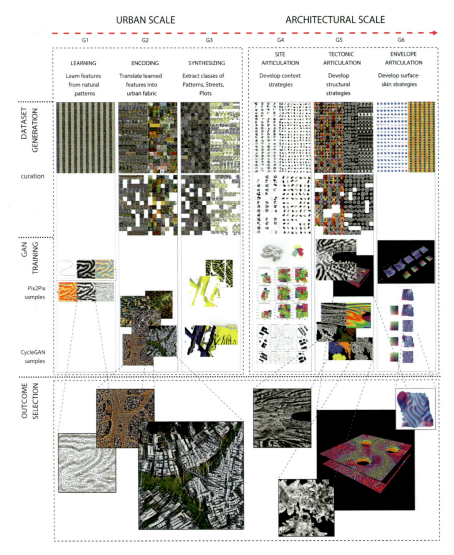

Fig. 4 Implementation of various NNs for discrete tasks, divided into two sections: urban and architectural. The urban scale was divided into three processes: learning, encoding, and synthesizing; the architectural level was divided into site, tectonic, and envelope articulation

4.2.1 Strategy 1: Fully-Chained Supervised

Group 1 used a fully-chained strategy, using supervised Pix2Pix models for image-to-image translation. The work followed a sequential logic to chain all the models (Fig. 5), where an "intuitive curve" can lead to a corresponding urban settlement scheme. The workflow was targeted to assist designers in planning urban settlements which were informed by environmental conditions (i.e. solar radiation) and topographical features. The training set was produced from simple initial sketches—representative of landscape formation intuitions—which mapped a Gray-Scott Reaction–Diffusion pattern to a synthetic topography, which was then analyzed for cumulative solar radiation. Based on local cumulative radiation and topographical features, a settlement was generated. The designers curated the dataset for training the following DL models:

1. **Sketch-to-Pattern**. The first network was trained with a paired dataset of intuitive curve (sketch) and Gray-Scott Pattern, to enable pattern prediction from a sketch.
2. **Pattern-to-Displacement**. The second network used the Gray-Scott pattern prediction from the first network to generate a displacement mesh.
3. **Displacement-to-Radiation Analysis**. The third network used the displacement mesh to generate Solar Radiation analysis predictions.
4. **Radiation Analysis-to-Settlement Strategy (Plots)**. The fourth network used the Solar Radiation Analysis and contour gradients to deploy the settlements into the landscape.
5. **Settlement Strategy (Plots)-to-Settlement Strategy (Roads)**. The fifth network used settlement plots to generate road predictions.

The logic of fully chained supervised networks (Fig. 6) is to establish a continuity for impacting i.e. a pattern or system in systematic manner, by certain contextual factors which allow design scenarios to emerge. Once a change occurs in the beginning of the chain, all the corresponding conditions adapt and update automatically to reorganize the evolving system. It is a hierarchical workflow, where each part (DL model) depends on the previous model. The objective of this strategy is to guide how information impacts early design conceptualization. Once the model is trained, it can simulate an interaction with a human designer, within the learned range of design intentions. This strategy can be applied retrospectively to early design stages as a means to update design solutions. The clarity of this process allows the automation of a given task, which in turn can enable human agents/designers to interact with, and explore the design space in a more open-ended fashion. The economy of

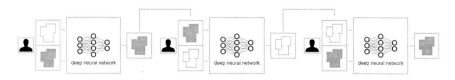

Fig. 5 Diagram of strategy 1, fully-chained supervised DL models

Latent Design Spaces: Interconnected Deep Learning Models ...

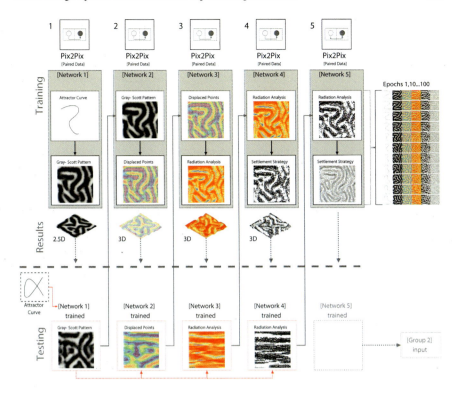

Fig. 6 Strategy 1, Fully Chained Supervised, using sequential logic in connecting four Pix2Pix Networks

time and effort incurred through certain automations can increase the probability of risk-taking in the design ideation, leading to unconventional, possibly more creative decisions in far-reaching areas of the design solution space.

4.2.2 Strategy 2: Fully-Chained Hybrid (Supervised and Unsupervised)

In a fully chained hybrid strategy, (Group 4) unsupervised CycleGAN models were combined with Pix2Pix supervised models. The workflow targeted a design proposition that was responsive to its context (surrounding built environment) and was tailored to a process of multiple fully-chained AI models (four CycleGANs and one Pix2Pix) and parametric translation (Fig. 7). The fully-chained models started with Alpha-plot images and finished with a simple abstract mass model. In between, the models were trained to generate the following sequence of outcomes: (1) Alpha Plots to Plot Outlines; (2) Plot Outlines to Subdivided Plots; (3) Subdivided Plots to Plots with Buildings; (4) Plots with Buildings to One Plot/One Building; (5) One Plot/One Building to Floor Plan Generation; (6) Floor Plan Generation to Colored Floor Plan and Colored Floor Plan to Mass.

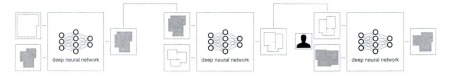

Fig. 7 Diagram of strategy 2, fully-chained hybrid DL models

This strategy suggested an approach where design evolved from physical conditions, surrounding built structures, and responded to those conditions in a less constrained manner, compared to the fully chained supervised strategy. The results demonstrated a higher degree of freedom in design exploration, leading to new or unexpected results in terms of building footprint relative to plot outlines. In general, this strategy facilitates exploration of design ideas that can be vague, latent, and cannot be easily described i.e. when design intentions are qualitative and undefined, opening the chance for reaching solutions outside the original dataset (i.e. through accidental discovery). Additionally, the chaining with supervised models at some step offers designers the benefit of partial automation which can prompt higher risk-taking in design thinking, as already mentioned (Sect. 4.2.1). This approach can be applied as a suggestive method for possible design scenarios in early phases of design conceptualization (Fig. 8).

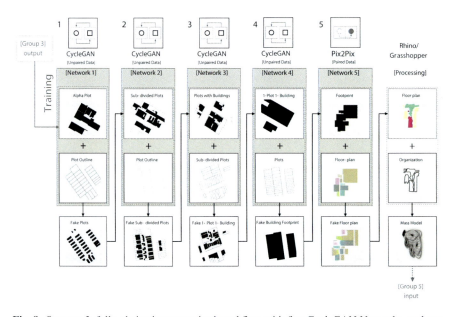

Fig. 8 Strategy 2, fully-chained unsupervised workflow with five CycleGAN Networks, and one Pix2Pix model in the process

4.3 Strategy 3: Partially-Chained Hybrid (Supervised and Unsupervised)

This strategy incorporated a partially chained approach with a combination of supervised and unsupervised networks (Group 5, 6) (Fig. 9). The partial chaining allowed a different workflow; parallel DL training allowed simultaneous experimentation with supervised and unsupervised networks, in a hybrid mode (Fig. 10).

Group 5's workflow comprised four Pix2Pix networks and two CycleGAN networks to produce a spatial configuration of three-dimensional topologies with morphological qualities inspired by natural systems. The process began with a three-dimensional coral model and tried to map differentiated coral patterns on a spatial model, according to surface curvature and color distribution. The experiments included a transition from color swatch to corresponding coral pattern, and subsequently, a transition from color silhouette to coral pattern. Group 6's workflow (Fig. 11) illustrates another example of this strategy, using two DL models, one supervised and one unsupervised. Results from the latter were used as input for the former in testing mode, and then processed in Rhino/Grasshopper for translation into 3D geometry.

This strategy ultimately encapsulates the advantages of the aforementioned strategies, by combining both supervised and unsupervised models, with the flexibility of chained and parallel operation. As a result, designers can capitalize on selected automated tasks, leaving them free to explore in more high-risk search spaces, without precluding the possibility for accidental discovery available through the flexible problem definition with unsupervised networks.

It is important to note that the other groups followed slightly different strategies; Group 2 used unsupervised DL models (CycleGANs) in a parallel, disconnected logic. Group 3 followed a partially-chained unsupervised strategy. All groups' networks were intended to be inter-connected, so the output of the preceding group became the input for the following one, when running the networks inference mode (a neural network infers things about new data it's presented with, based on its training). The workshop resulted in a less linear and more complex workflow, where Group 3 used output from Group 1 and Group 4 used output from Group 5. In addition, early workshop planning specified two or three DL models per group, while some groups ended up training up-to six DL models. This demonstrates the need for experimentation with multiple AI networks to calibrate the workflow. Finally, the case-study reinforced the effectiveness of multi-designer bi-directional approach, discussed in Sect. 4.1.

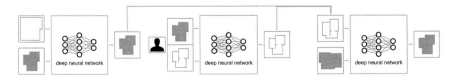

Fig. 9 Diagram of strategy 3, partially-chained, hybrid DL models

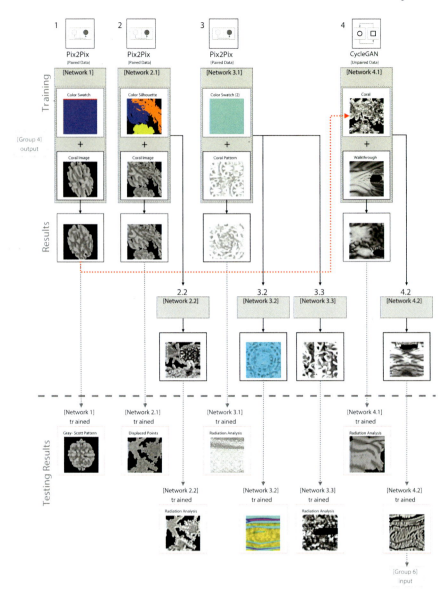

Fig. 10 A similar approach to Strategy 2, partially-chained hybrid workflow with four Pix2Pix and two CycleGAN

Latent Design Spaces: Interconnected Deep Learning Models ...

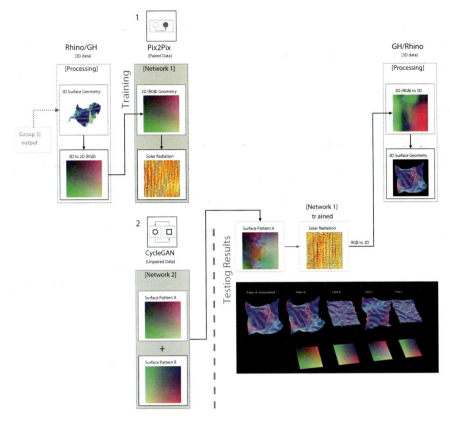

Fig. 11 Partially chained Pix2Pix and CycleGAN process

5 Results Discussion

Our discussion of the case study outcomes focuses on the potential of design strategies to establish a useful *human-machine* agency which assists architectural creativity. A specific examination of the interaction between human designers and ANNs (*human–human, human–machine, machine–machine*) is used to make this assessment. We identify a couple of important things to situate this discussion.

5.1 Design Search Spaces

The design space is an ever-expanding or contracting area of possible solutions. Every decision results in an expanded or constrained search space. Depending on the design objectives, certain domain-specific AI models can facilitate narrowing down the designer's options, expediting and facilitating design exploration [2]. Within our

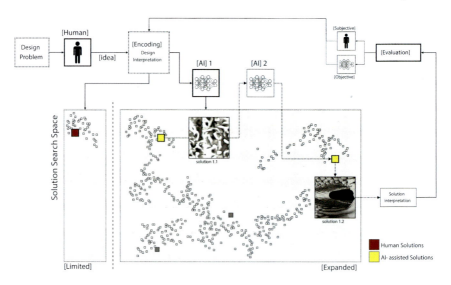

Fig. 12 Introducing AI models for ideation expands the design space of possible solutions. Results can be plugged into further AI models, thereby progressively re-directing the designer to alternative areas in the latent space of solutions

comparison of human and machine cognition, it is worth noting that humans might not tend to explore a certain range of solutions (either because of bias, habit or other unintentional factors), while the DL model can implicitly and explicitly help human designers expand their solution range. For example, the use of AI (DL models) as an assistive tool for selected task automation can encourage the human designer to assume a higher degree of 'risk', considering options which might otherwise not be explored. Furthermore, in spite of the advantage of human cognition with respect to immediate recognition and synthetic ability of reading data [34], human agents are not able to examine a design space at the same resolution and breadth as a DL model, processing, for example, domain translation (Fig. 12).

5.2 *Frame of Reference for Measuring Design Creativity*

We have considered focusing our assessment either, on the agents involved in the workflow (human agent; artificial/computational agent) with respect to their distinct contribution (~person. agent), or towards the overall process by way of its outcomes. In the former case, it may be difficult—if at all necessary—to make such a distinction between agents, because both human and AI agents are meant to assist each other. As a result, we prefer to focus on the process, and identify the agents' optimal function within it (order of use for each agent; way of interaction between human–machine; selection of single versus multiple human agents to coordinate the A.I. process).

By extension, the evaluation of outcomes is facilitated by the nature of the proposed workflow, which examines the distinct parts of the process; this allows regular assessment of the design criteria at various nodes of the workflow. The evaluation of any creative process which includes AI-assisted steps, requires human input, according to experts: The use of AI can help target the second category of creativity, per Boden: *"Exploratory creativity is the type best suited to AI."* (This does not exclude AI's potential to contribute more profoundly in the creative process) *"There are countless examples...However, even exploratory AI depends crucially on human judgement. For someone must recognize—and clearly state—the stylistic rules concerned"* [35]. Boden acknowledges the necessity for collaboration between human-machine in the final part of a design process, the evaluation, and potential re-calibration. Although the computational power of neural networks is useful for heuristic search, human input is still required for evaluation, as available objective metrics like "SSIM" or "Perceptual Similarity Index" used by AI to assess "successful" outcomes may not be coincident with the subjective evaluation humans perform (and which can include other qualitative factors beyond the network algorithm's performance metrics). For instance, visual transfer can be approached through both style and domain-transfer networks; assessing the results of the former may be harder from a human standpoint, as the qualities are transferred visually, not by semantic association, so any due preference would remain very subjective. Assessing the overall process seems to be a more inclusive way to evaluate the proposed design framework and relies on the understanding of relationships among design agents.

5.3 Supervision of the Process by a Singular Versus Multiple Designers

Creative achievement can be situated and discussed as a kind of output which takes place within a specific domain, or sub-domain. Nevertheless, bridging across more than one domain is also important because it presupposes creative thinking. Popular perception of creativity usually reflects one of two situations: either significant contribution to one domain, or contributions to multiple domains [36]. Although the second instance is considered less feasible, it is important because it can direct our attention to the benefits of specialization and polymathy as they relate to our earlier comparison of one-directional and bi-directional design strategies for this project (Sect. 4.1).

Following the design strategies discussed earlier, the expected outcome was a sequential approach between groups, where each group used two or more networks, testing their trained models with input from the previous group, while providing their output for testing the trained model of the following group. However, the implemented workflow of each group produced more than two–three networks, and sometimes exceeded six networks. In practice, some parts of the workflow proved less sequential than expected, and required the use of parallel (simultaneous) training.

The presence of multiple human designer-agents established a bi-directional workflow structure which encouraged the self-organization of design intent through its back-propagation to earlier stages and networks. At the same time, different agents presuppose different qualitative backgrounds which, combined, may lead to original insights for updating the task parameters among the various networks. In some sense, the combinational presence of multiple domain-specific agents (multi-designer strategy) could allow for a "simulation" of *domain-general* understanding which is present in polymaths, but is quite rare to find.

6 Conclusions and Future Work

The methodology described in this paper tackles the application of AI in a process-based workflow. Early DL applications targeted experimental results by focusing on representation; our work utilizes ANNs to handle the complexity and multiplicity of the architectural design process, avoiding a reductionist handling of design decisions by stand-alone ANNs, through the proposed framework ("Deep Chaining"). This type of design approach increases the chance for creative design thinking because it focuses on iterative collaboration, where semantic relationships among the explored design features are not fixed, but can adapt to influence from other features. Furthermore, using a chained process, the designer(s) is able to exploit AI's capabilities to tackle multiple layers of design simultaneously. The idea of creative production through collective synergy has long been proposed by scientists, as Professor Csikszentmihalyi reminds us of Dr. Jonas Salk, who believed interactions among individuals from different domains could help new ideas emerge, which would otherwise have not: *"I find that that kind of creativity is very interesting and very exciting—when this is done interactively between two sets of minds. I can see this done in the form of a collective mind, by a group of individuals whose minds are open and creative and are able to bring forth even more interesting and more complex results....this is in fact part of the process of evolution, and ideas that emerge in this way are equivalent to genes that emerge in the course of time. I see that ideas are to metabiological evolution what genes are to biological evolution "* [15].

Our approach favors clarity of architectural intentions, thereby acting as *objective* benchmarks for evaluating the AI-generated outcomes, unlike otherwise heuristically pursued design exploration by the authors [37]. The method facilitated the breakdown of the architectural design process into systems and tasks with specific layers, allowing more control over local decisions and result evaluation, in contrast to employing an overall AI model for a complex multi-layered task like design.

The "Deep-Chaining" method augments current Human and Artificial Intelligence, by incorporating three modes of interaction: *human–machine, machine–machine* and *human–human*. A case study tested the proposed framework, using a structure of variable architectural systems (i.e. *concept, tectonics, envelope*), applied to both architectural and urban scales, within a workflow of multiple designers collaborating with a series of DL models. Three primary strategies of connected networks were explored, including *fully-chained supervised, fully-chained unsupervised,* and *partially-chained* hybrid models. The case study findings demonstrate the effectiveness of the framework to tackle the two scales at multiple phases, leading to a new design process.

The research limitations involve the complexity of enacting, and more importantly, assessing the creativity of Human-AI collaboration, (*process*) as well as the assessment of AI-produced work (*product*) within this still-experimental research phase of AI in architecture. One problematic issue involves the assessment of the "perfect" or successful mode of interaction of human and artificial agency in generative strategies by identifying the interactions among the involved agents (*Human–machine; Human–Human; Machine–Machine*). Another issue is the complexity of coordinating multi-agents, which can be more challenging compared to a single-designer.

Recent work by the authors has elaborated on this framework development through the integration of more types of GANs (i.e. StyleGAN; StyleGAN nada) which are not addressed here, with other ANNs which are based on natural language processing (NLP) like DALL-E, VQGAN + CLIP and Diffusion Models [38]. Current and future work by the authors may address minimizing the top-down supervision of the chaining; allowing multiple agents to self-organize beyond their local environments and further testing the adoption of further language-based AI models in light of recent advancements in diffusion models (i.e. Midjourney, DALL-E 2).

Acknowledgements This work proposes a new research framework, introducing the notion of "deep chaining", connecting multiple artificial neural networks within an architectural design workflow, and originates on earlier work by the authors, which enabled testing of an experimental design workflow. This was first implemented in 2020, during the *Digital Futures 2020* online conference and workshops event. An online workshop, offered by Daniel Bolojan, Emmanouil Vermisso and Shermeen Yousif, allowed the implementation of a design workflow, advancing a number of strategies from the workshop's teaching structure. The authors would like to thank the participants in the "Creative AI Ecologies" workshop taught during "Digital Futures 2020": Maider Llaguno-Munitxa, Khaled Nahas, Manoj Deshpande, Anuj Modi, Weiqi Xie, Vlad Bucsoiu, José Roberto Arguelles Rodriguez, Danny Osorio Gaviria, Yunling Xie, Parvin Farahzadeh, Daniel Escobar, Ian Fennimore, Shaoting Zeng, Andrei Padure, Heba Eiz, Kaihong Gao, Behnaz Farahi, Pooya Aledavood, Sarath Raj Sridhar, Frank Quek, Ahmed Hassab.

References

1. Alexander, C.: Systems generating systems. Archit. Des. **38**, 605–610 (1968)
2. Hassabis, D.: Creativity and AI. The Rothschild Foundation Lecture: The Royal Academy of Arts. (2018). https://www.youtube.com/watch?v=d-bvsJWmqlc
3. Rossi, F.: Building trust in artificial intelligence. J. Int. Aff. **72**(1), 127–134 (2018)
4. Schwab, K.: The fourth industrial revolution. Currency (2017)
5. Susskind, R.E., Susskind, D.: The future of the professions: How technology will transform the work of human experts. Oxford University Press, USA (2015)
6. Bolojan, D.: Creative AI: Augmenting design potency. Archit. Des. **92**(3), 22–27 (2022). https://doi.org/10.1002/ad.2809

7. Stocking, A.W.: Generative design is changing the face of architecture. Build. Des. (2009)
8. Chen, J., Stouffs, R.: From exploration to interpretation-adopting deep representation learning models to latent space interpretation of architectural design alternatives. (2021)
9. Leach, N.: The AI design revolution: Architecture in the age of artificial intelligence. Bloom. Vis. Arts (2021)
10. Gero, J.S.: Ten problems for AI in design. (1991)
11. Veale, T.C., Amílcar, F., Pérez, Rafael Pérez y.: Systematizing creativity: A computational view. In: Veale, T.C., Amílcar, F. (eds.) pp. 1–19. Springer Nature, Cham, Switzerland (2019)
12. Abraham, A.: The neuroscience of creativity. Cambridge University Press (2018)
13. Chollet, F.: On the measure of intelligence. arXiv® (2019)
14. Boden, M.A.: The creative mind: Myths and mechanisms. Psychology Press (2004)
15. Csikszentmihalyi, M.: Creativity: The psychology of discovery and invention. Harper Perennial, New York (2013)
16. Csikszentmihalyi, M., Wolfe, R.: New conceptions and research approaches to creativity: Implications of a systems perspective for creativity in education. The systems model of creativity, pp. 161–84. Springer (2014)
17. Karwowski, M.D., Jan, Gralewski, J., Jauk, E., Jankowska, D.M., Gajda, A., Chruszczewski, M.H., Benedek, M.: Is creativity without intelligence possible? A necessary condition analysis. Intelligence **57**, 105–17 (2016)
18. Kasparov, G.: Deep thinking: where machine intelligence ends and human creativity begins. Revista Empresa y Humanismo. **23**(2), 139–143 (2020)
19. Goodfellow, I., Pouget-Abadie, J., Mirza, M., Xu, B., Warde-Farley, D., Ozair, S., et al.: Generative adversarial nets. Adv. Neural Inf. Process. Syst., 2672–80 (2014)
20. Goodfellow, I., Bengio, Y., Courville, A., Bengio, Y.: Deep learning. MIT press Cambridge (2016)
21. Zhu, J.-Y., Park, T., Isola, P., Efros, A,A.: Unpaired image-to-image translation using cycle-consistent adversarial networks. CoRR. abs/1703.10593 (2017)
22. Isola, P., Zhu, J.-Y., Zhou, T., Efros, A.A.: Image-to-image translation with conditional adversarial networks. IEEE Conference on Computer Vision and Pattern Recognition (CVPR)2017, pp. 5967–76 (2017)
23. Ishiguro, H.: Hiroshi Ishiguro: Are robots a reflection of ourselves? Hiroshi Ishiguro (in conversation with Maholo Uchida). (Accessed 2019)
24. Watanabe, M.: Algorithmic design/induction design, three kinds of flow/three stations. https://www.makoto-architect.com/kashiwanohaCSt.html (2004). Accessed September 29th, 2020 2020
25. Forbes, A.: Creative AI: From expressive mimicry to critical inquiry. Artnodes. **26**, 1–10 (2020)
26. Mirza, M., Osindero, S.: Conditional generative adversarial nets. arXiv preprint arXiv:14111784 (2014)
27. del Campo, M., Manninger, S., Sanche, M., Wang, L.: The church of AI-An examination of architecture in a posthuman design ecology. (2019)
28. del Campo, M., Manninger, S., Carlson, A.: A question of style. In: Yuan, P.F., Xie, M., Leach, N., Yao, J., Wang, X. (eds.) Architectural Intelligence: Selected Papers from the 1st International Conference on Computational Design and Robotic Fabrication (CDRF 2019), pp. 171–188. Springer Singapore, Singapore (2020)
29. Özel, G.: Interdisciplinary AI: A machine learning system for streamlining external aesthetic and cultural influences in architecture. In: Yuan, P.F., Xie, M., Leach, N., Yao, J., Wang, X. (eds.) Architectural Intelligence: Selected Papers from the 1st International Conference on Computational Design and Robotic Fabrication (CDRF 2019), pp. 103–116. Springer Singapore, Singapore (2020)
30. Koh, I.: The augmented museum—a machinic experience with deep learning. In: Holzer W.N., D., Globa, A., Koh, I. (eds.) RE: Anthropocene, Design in the Age of Humans—Proceedings of the 25th CAADRIA Conference Chulalongkorn University, Bangkok, Thailand (2020)
31. Chaillou, S.: AI+ architecture: Towards a new approach. Harvard University. (2019)

32. Gero, J.S.: Design prototypes: A knowledge representation schema for design. AI Mag. **11**(4), 26 (1990)
33. Menges, A., Ahlquist, S.: Computational design thinking: Computation design thinking. John Wiley & Sons (2011)
34. Zhang R.I., Phillip, Efros, Alexei A., Shechtman, Eli, Wang, Oliver: The unreasonable effectiveness of deep features as a perceptual metric. (2018)
35. Boden, M.: Artificial intelligence: A very short introduction. Oxford University Press (2018)
36. Garcia-Vega, C.W.: Vincent. polymathy: The resurrection of renaissance man and the renaissance brain. In: Jung, R.E.V., Oshin, (eds.) The Cambridge Handbook of the Neuroscience of Creativity, p. 528–39. Cambridge University Press, Cambridge (2018)
37. Bolojan, D., Vermisso, E.: Deep Learning as heuristic approach for architectural concept generation. ICCC2020. pp. 98–105
38. Bolojan, D., Vermisso, E., Yousif, S.: Is language all we need? a query into architectural semantics using a multimodal generative workflow. In: Jeroen van Ameijde, N.G., Kyung Hoon Hyun, Dan Luo, Urvi Sheth (eds.) POST-CARBON—Proceedings of the 27th CAADRIA Conference. Sydney, pp. 353–62. (2022)

From Technology to Strategy: Robotic Fabrication and Human Robot Collaboration for Increasing AEC Capacities

Dagmar Reinhardt and M. Hank Haeusler

Abstract This position paper unpacks the relationship between intangible pre- and post-production and tangible production processes under an Industry 4.0 framework for architecture and design to mitigate the Architecture Engineering Construction (AEC) sectors' contribution to climate change and investigate potentials for SDG 9 (industry, innovation and infrastructure). As Industry 4.0 is describing a business model or strategy foremost that utilises and incorporates technology via a cyber-physical system, we investigate how robotic technologies and human robot collaboration can enable methods, frameworks, and systems for the AEC sector; and what opportunities and challenges outside the tangible production floor can be considered to tie in architecture and construction. By reviewing state-of-the-art tangible production processes, robotic fabrication, and robotic interfaces, we aim to outline potential research domains in intangible pre-and post-production towards Next Gen Architectural Manufacturing. We conclude with objectives for reducing architecture's resources appetite using computation and modern manufacturing strategies and a strategic framework to enable this in the AEC sector. This investigation, its proposed hypothesis, methodology, implications, significance, and evaluation are presented in this chapter.

Keywords Cyberphysical systems · Robotic fabrication · Human robot collaboration · Data-driven design strategies

United Nations' Sustainable Development Goals 9. Build resilient infrastructure · promote sustainable industrialization and foster innovation

---

D. Reinhardt (✉)
ADP School of Architecture, Design and Planning, The University of Sydney, Sydney, Australia
e-mail: dagmar.reinhardt@sydney.edu.au

M. H. Haeusler
ADA/School of Built Environment/Computational Design, UNSW, Sydney, Australia
e-mail: m.haeusler@unsw.edu.au

1 Introduction

The building sector contributes significantly to the current climate crisis on a global scale by using the largest amount of natural resources (~60%), producing excessive emissions (~50%) and waste production (~50%)—all of which are related and interlinked [63]. A discussion of Industry 4.0 in a context of the Architecture Engineering Construction AEC industries must thus have at its centre a reduction of use of resources, a reduction of emission, and a reduction of waste. Consequently, in this position paper, we discuss and outline a strategic approach for Industry 4.0 geared towards enhancing collaboration amongst various departments to increase efficiency and productivity [4] and so assist in improving the AEC industries carbon footprint. We align with two critical comments. As Adams notes, 'the technology of Industry 4.0, while important, is less important than the business model that utilises and incorporates' [3]. Moreover, as MIT economists argue, 'digital is not about technology but strategy' [69]. Section 2 opens the discussion with an overview of second machine age general purpose technologies and cyberphysical systems and continues towards reconsidering production values via intangible pre- and post-production processes. Here we see an important role and contribution of the architectural business sector. Section 3 reviews a case for current AEC with research and knowledge in robotic fabrication as tangible production floor. Section 4 overlays the AEC sectors version of intangible pre- and post-production processes; *synthesis* (creative process), *management* (business process), and *analytics* (data process) and outlines pathways for an integrated, cross-disciplinary framework as strategy to address the building sector's climate problem. Section 5 concludes with overarching objectives for an industry 4.0 framework in AEC sector and potential SDG9 contribution.

2 Smile Curve for AEC Industries, a Development Space

Referred to as a descriptor for developments and advancement of information technology in the German economy in 2011, the term 'Industry 4.0' was rapidly adopted by the Architecture Engineering and Construction industry (AEC). The continued high interest in this concept—of industry, research and academia—is evidenced by recent web discussions, with a keywords search ('Architecture' AND 'Industry 4.0') yielding in Scopus 25,903 document results: and with an exponential rise in the years 2014–2020 (access data 5. July 2022). While this provides by no means a qualitative insight, it clearly presents the growing interest that exists in architecture and construction for the concepts, methods, tools and adoption for an Industry 4.0 framework.

2.1 Industry 4.0 and General Purpose Technologies

The interrelation of technologies that facilitate the emergence of the 'Smart Factory' is a base concept that is highly valuable for architecture design practice and adoption into the construction industries. Design principles that can inform the design to construction workflow and thus enable implementation to Industry 4.0 scenarios include *interoperability* (ability of systems connection); *virtualization* (data-based models, digital twin); *decentralisation* (ability for local decision making); *real-time capability* (data collection, analysis and evaluation); and *modularity* (flexible adaptation through modules) [35].

The fact that new approaches in architecture have become available results from new general-purpose technologies such as digitised and social data analytics, sensors, machine learning, or robotics which allow the automation of cognitive tasks and offer human and software-driven machine substitutes [12]. Similar to electricity or the combustion engine that rendered labour and machines complementary in the First Machine Age, these general purpose technologies are identifiable as single generic system or equipment; recognizable over a lifetime; have scope for improvement, and will be used and enable uses with spill over effects [45]. Significantly, IT driven changes in manufacturing systems are expected to affect product- to service-orientation even in traditional industries (Lasi 2021).

Key advanced technologies associated with Industry 4.0 are manifold, ranging from the Internet of Things (IoT)/Internet of Services (IoS), Cloud Computing, Big Data, Smart Factory, 3D-Printing, Mobile Computing and Radio-Frequency Identification (RFID), the Cyber-Physical Systems (CPS) or Embedded systems, Augmented Reality (AR)/Virtual Reality (VR)/Mixed Reality (MR) and the Human–Computer-Interaction (HCI) [43]. Their adoption brings benefits for design-to-make production processes alongside digital technologies within industrialised construction, which are much needed for certainty of cost, schedule, and scope in the AEC industries [51].

As Fig. 1 shows, methods that become thus available include processes and strategies that enable digitisation and integration of work and construction processes at different stages, where this work alongside Building Information Modelling (BIM) and manufacturing concepts such as Product Lifecycle-Management (PLM) and Modularisation. Out of the range of these general-purpose technologies, as this chapter argues, robotics holds a particular significance, as this enables computational data to being seamlessly integrated with work processes and thus bridging between digital/virtual realms and the physical/real. At the core, robotics opens different strategies in terms of how to approach data and labour. Beyond management (data capture, simulation analysis), robotic applications as part of Industry 4.0 enable connectivity and interoperability between human workforces, data, material and machines, in the domains of robotic fabrication and human–robot interaction (HRI) or collaboration (HRC), as will be further discussed in Sect. 3.

Ross et al. [69] propose in 'Designed for Digital—How to architect your business for sustained success' that the true impact of the digital stems not primarily

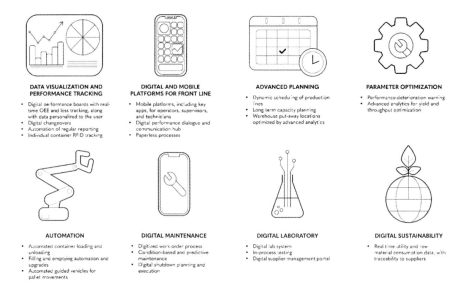

Fig. 1 Available technologies and methods for a context of architecture and construction (after McKinsey report) [53]

from application as a technology but considering these as a strategy. This is important when reviewing existing challenges in the AEC industry [15], which include field-level barriers for strategic innovation; fragmentation as a barrier for collective action; limited understanding for business models and use of digital transformation; investments not driven by strategies; lack of orchestrated or common approaches; and lack of knowledge and skills for digital transformations. Consequently, we need to move towards better discernibility for all phases of plan to production to support companies' productivity increase and add value for design, production and services.

2.2 Increased Productivity Through Cyber-Physical Systems

Internationally, the AEC industry is one of the least digitised and least efficient industries [18, 54, 55]. A growing list of performance and productivity problems are directly linked to the sector's failure to embrace advanced technology [26, 52, 72]. The current divide between advanced manufacturing's move towards Industry 4.0 and architecture's stagnation on a status quo suppresses opportunities to improve cost competitiveness and value differentiation.

In this context, cyber-physical systems (systems linked to computation) present an alternative pathway for architecture and construction; by providing an increased potential for data capture and integration [54]. Cyber-physical systems can further adopt a concept of *digital twin* (a highly complex virtual model that is the exact counterpart of a physical condition, object or entity, process or service). Benefits arise from data being continually updated and mapped against it, which can then be simulated, analysed and evaluated to trial different scenarios and enable decision-making. Such accessibility of future scenarios allows investigating systems performance—and consequently being able to operate, maintain and repair systems with no physical proximity and affordance. Importantly, by bridging mechanisms for communication, control and sensing via sensors, cyber physical systems further enable collaboration between design and manufacture, and interoperability through open-source libraries and hardware. The integration of cyber-physical systems into construction workflows via data processing techniques and inter-device communication allows fabricators, manufacturers and constructors to overcome process fragmentation and directly link physical production processes to computational processes [76]. Cyber-physical systems for coordination play an increasing role in construction and are adopted for surveys, task planning and networking control systems. The introduction of sensors informs on required ad-hoc changes, and enables direct, responsive, intelligent and interconnected workflows through continuous online monitoring on the basis of data acquisition. In opening for diagnostic protocols and adjustments, this allows for overcoming stereotypical, standardised or modularised building methods and construction processes. Yet despite the adoption of Industry 4.0 technologies and ongoing research and development, there remains a considerable gap between research, industry and practice collaborating for manufacturing and construction [17]. There is a strong focus on production activities, yet there is limited exploration on how Industry 4.0 principles could be applied across different phases. Hence, we ask: In which way can opportunities and challenges outside the tangible production floor and beyond CAD/CAM and robotics digital manufacturing technologies enable an Industry 4.0 framework for the AEC sector?

2.3 Smile! Lifting the Pre and Postproduction process for AEC

Implementation of Industry 4.0 solutions empowers manufacturing companies by enhancing collaboration: effectively providing relevant information for people on a real-time basis [4]. We argue that a close look into the distinct and successive phases in manufacturing holds the key to opportunities in the AEC industry in a context of Industry 4.0. Linking three core phases is essential, the pre-production phase with R&D, design and logistics; the production activities with the 'actual' production; and the post-production phase with distribution, sales and service. Since failing in one would sabotage and hinder success in the overall production process, all must be considered for Industry 4.0 as changes affect the entire supply chain, not only the tangible production activities. However, improvements in productivity become more accessible by coupling tangible production activities to include in-tangible pre- and post-production phases. As the so-called 'smile curve' in Fig. 2 illustrates, value can

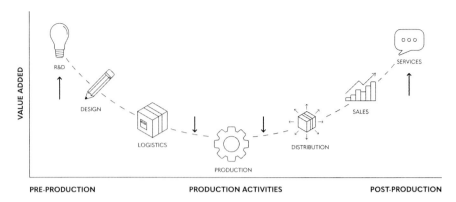

Fig. 2 Value added in a 'smile curve' for manufacturing context, embedding phases of pre-production (R&D, design), production (logistics, production and distribution) and post-production (sales and services)

be added across the different stages of bringing a product on to the market in an IT-related manufacturing industry (Industry Insights).

Daniel Chuter, CEO of the Innovative Manufacturing Cooperative Research Centre (IMCRC), refers to value increase as 'moving up the smile curve' [46], where production (including logistics and distribution) can be largely enhanced by introducing focus (resources, investments, knowledge) to pre-production (R&D, design) and post-production (sales and services). In a context of AEC industries, ignoring phases outside of core production creates a bottleneck for manufacturing industries; and a similar bottleneck exists in the form of architecture and building practices that model and manage design and construction data, used to establish, and later maintain building stock, infrastructures and services. For example, architects provide phase-based information for builder/manufacturer through design and documentation. They are thus external and as a result usually unavailable for partnering in pre- and post-production with advanced architectural manufacturing and construction. Architecture's digital practices, systems and platforms can significantly support the AEC industries (with advanced architectural practices on the basis of advanced computational modelling and scripting and fabrication knowledge, development of Artificial Intelligence, Machine Learning, and Big Data approaches to digital twins) as core technologically intersecting domains and phases via methods and processes, and so connecting parts and sectors of the building industry which are central to improving efficiency and competitiveness and increasing innovation and ensuring direct links to manufacturing [6, 53, 52]. A collaborative workflow through taking cyber-physical systems in full advantage, and early integration of R&D with a focus on modelling within the machining/manufacturing framework can raise the value of architecture and construction equally—a Next Gen architecture manufacturing approach.

Figure 3 illustrates a framework for three systemic lenses: *synthesis, management* and *analysis*, interrelated through new cybersystem technologies, digital twin and digital business models that allow for interoperability. Each of these lenses holds specific potential to align capacities and knowledge in architecture with manufacturing processes. *Synthesis (creative process)* accommodates pre-production (R&D/Design) through advanced computational scripting, parametric modelling (PM) and machine learning (ML) enhances understanding, optimisation and automation of complex, repetitive tasks and 'workflows' in practice. As a result, the creation of more efficient, reliable and machine-readable manufacturing instructions would enable manufacturers to complete new product designs and achieve operational productivity gains for small scale production [13, 60, 18, 33]. Furthermore, this phase is core for integration of industry competencies with architectural design and planning. *Management (business process)* addresses post-production (sale) and targets commercial advantages and risks of 'business as usual' models in architectural business. Changes in a combination of consumer spending patterns, economical, ecological, and external political pressures can support the AB sector to reconsider business models towards new digital 'XaaS' (Anything as a Service) models, thus redirecting towards design (synthesis) and innovative manufacturing [15, 18, 27, 70] Colins et al. (2016).

Analytics (data process) incorporates post-production (services) for simulation, analysis and evaluation of existing and new data (from CAD to BIM to PM) across the architectural service industry [43]. AI, Machine Learning, Parametric Modelling and Big Data can be adopted to establish digital twins of buildings as extracts from architectural data to use for services. Equally used for describe-for-production, these digital twins can be employed to maintain and repair their physical counterparts and increase operational efficiency [7].

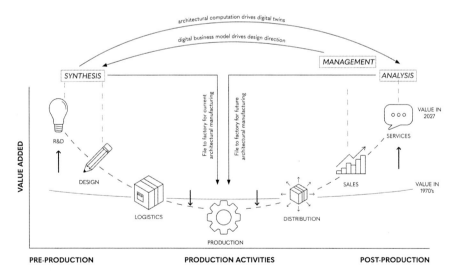

Fig. 3 Co-opting the Australian Government's Modern Manufacturing Strategy's (AGMMS) 'smile curve' [30]

Importantly, knowledge for processes and agency through technology can be increased between research, architectural practice, and industry, by closely reviewing potential intersections for architecture along the manufacturing curve. These systemic lenses inform the 'smile curve' and add greater value from architectural services both in pre-and post-production including ongoing maintenance, specifically for architecture to construction. Consequently, we argue for two ways to increase the potential for architecture and construction to work better together within the technologies of Industry 4.0. Firstly, by enhancing and enabling the core phase of production (ie human labour, machines, data workflow, for example robotics and AR as is discussed in Sect. 3). Secondly, by delivering a targeted approach to intangible and tangible production activities such as robotic fabrication, collaborative robotics and interfaces for architecture.

3 Industrial Robotic Fabrication and Human Robot Collaboration

The AEC industries continue to be slow in integration of six axis industrial robotic arms and defining robotic tasks due to perceived barriers including safety, costs and skill applications [62]. Yet six axis industrial robotic arms, robotic fabrication technologies, sensor systems and haptic interfaces can change the way in which architectural design practice, manufacturing and construction are conducted, through robotic fabrication methods and technology; mobile and onsite robotics; and human–robot collaboration, as is discussed in the following.

3.1 Robotic Fabrication for Architecture Development of Material Applications and Construction Methods

Digital fabrication tools (CAD, CAM) have continuously risen in popularity for manufacturing and fabrication, with articulated arm robots that are reliable and flexible; can effortlessly execute an unlimited variety of non-repetitive tasks, and which have become increasingly affordable, accessible, and usable. Initially adopted for high precision, autonomous workflows and independent locomotion in the automotive industry, robotic applications are now considered a catalyst technology that leverages mass customisation to a more elaborate and even architectural scale. Current robotic system providers include ABB, UR, KUKA, Boston Dynamics, Fanuc, with a large range of differentiated robot specifications and types. These industrial robots provide the ideal combination of human and machine labour, connect to a wide bandwidth of general-purpose technologies of Industry 4.0 (AI, ML, Data, AR/VR) and consequently can support the AEC with a potential for further developing/customising a wide variety of existing building and construction materials. It can

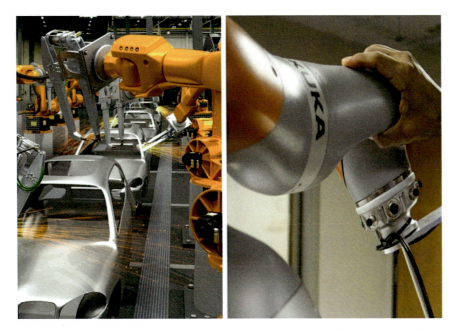

Fig. 4 Development of robots from automotive to architecture: task and workspace restrictions (left) towards intuitive haptic interfaces (right)

already be observed that construction automation technology, STCR approaches, service robot systems, and other microsystems technology are merging with the built environment, becoming inherent elements of buildings, or building components [8] (Fig. 4).

Global research into robotic applications has been developed between multiple partnerships (variably with industry, research, academia, software and robot developers, or architectural practices). Designing for robot production (architecture) and operating robot setups (fabrication) has become accessible through multiple programming languages such as C/C++, Python, Java, C#/.NET, or directly intersect with architectural modelling and scripting data (on the basis of McNeel Rhinoceros and GH plugin), Robot Operating System (ROS); robot programming coupled with motion simulation such as KUKA|prc; or KUKA|crc: Cloud Remote Control.

In the last decades, research on robotic fabrication with six-axis industrial robotic arms has been thoroughly investigated with new potentials for processes, systems, materials, construction methods [31, 16, 74, 78]. This laid the foundations for an enormous spectrum of potential applications for bespoke and customizable fabrication processes and robotic control protocols and has been widely disseminated through bi-annual proceedings [10, 52, 1] and across journals (Springer Construction Robotics; Automation in Construction; Robots in Construction). A non-exhaustive overview for research robotic applications between 2012 and 2018 shows:

- Robotic brick laying [31, 78, 21]

- Robotic modular timber assembly [31];
- Robotic wood processing [64]
- Robotic assembly (Gandia et al., 2018; Snooks Jahn, 2016)
- Robotic subtractive cutting/milling (Clifford McGee, 2011; Clifford et al., 2014; Feringa McGee, 2014)
- Incremental sheet forming (Kalo Newsum, 2014; Nicholas et al. 2016; Ficca 2017);
- Robotic bending (Culver et al., 2016; Tamre et al. 2013);
- Robotic 3D printing (Johns et al. 2012; Oxman et al. 2017; Hyperbody/Bier/Mostafavi, 2015; Dubor 2016; Branch technology, 2015; Feringa, 2017; Huang et al., 2018; Alothman et al. 2018; Battaglia et al., 2018; Gaudilliere et al., 2018)
- Robotic (carbon fibre) weaving (Yablonina, 2016; Doerstelmann et al. 2016; Witt, 2016; Reinhardt et al., 2018).

As a consequence, research on robotic fabrication protocols, systematic applications of a variety of end-effectors that can inform standard construction processes (assembly, positioning, punching, drilling, cutting, sawing, fixing, plastering) and non-standard productions (3D printing, wire cutting, weaving/threading) together with an understanding for workspace scenarios, workflows and operative has been widely disseminated. However, several challenges remain, including (a) knowledge and skill transfer from research to industry applications for methods and techniques of robotic protocols; (b) upscale to industrial building processes and building site; and (c) increased human–machine or human–robot interactions with intuitive feedback and interfaces.

3.2 Onsite Robotics-Large Scale Construction

While the construction industry has not kept up with manufacturing in adopting robotics despite the promise of improvements in quality and enterprise performance and a shortened time-to-market for products [11, 58] this is partially due to underlying conditions that differ strongly in the two sectors: manufacturing uses closed work settings, construction is produced in multi variant, unstable and uncertain environments. Standard construction sites pose challenges for operating industrial robots due to complexity as a consequence of unstructured environments, and thus limiting transfer and direct adaptations to this sector. Other typical characteristics of construction industries impede the adoption of robotics such as a high volume of manual operations, inconsistent deviations and variability of the construction site over long periods, large and heavy building structures, and in general limitations resulting from outdoor operations [50]. Other barriers exist due to scalability—robotic arms and thus workspace and range are restricted to the systems in which they are mounted, for example on a crane, track rail or gantry system, and so robotic reach and work scale is determined by the platform. However, significant achievements to date include gantry based modular components (Sequential Roof, Gramazio Kohler), integrated robotized construction site [28], or mobile robotic construction units [21].

An exert overview of industrial robotic applications with large-scale construction and buildings includes: HadrianX's FastBrick (2018, Australia) demonstrated in a commercial application for a mobile robotic system for construction of block structures from a 3D CAD model in unstructured environments with construction of a residential unit in three days. Odico Construction Robotics (2021, Denmark) developed Factory-on-the-Fly as a platform technology for mobile on-site robotic construction, driven by Sculptor® operating system for constructing with a robotic cell through intuitive interface programming (iPad). DFAB House by NCCR/ETH researchers (since 2016-, Switzerland); a 1:1 demonstrator for implementation of novel robot driven construction methodologies, including In Situ Fabricator (an autonomous on-site construction robot), Mesh Mould (a formwork-free, robotic process for steel-reinforced concrete structures), Smart Slab (integrated ceiling slabs fabricated with 3D-printed formwork); and Spatial Timber Assemblies (robotically fabricated timber structure). SQ4D Inc. Autonomous Robotic Construction System (ARCS) (USA); Robotic 3D printed residential house with use of sustainable materials (mold and fire resistant) and assumed 70% reduction of cost and labour. MX3D/Laarman MetalXL (Netherlands [18] combines a standard industrial robot and power source for 3D metal printing, showcased with a 3D printed steel bridge.

On a smaller scale, robotic technologies that aim to assimilate conventional construction approaches for building construction represent a larger percentage of the developments. These often focus on singular construction activities, tasks or components, such as protocols developed for a single, formatted and modularised material (bricks and blocks), for fluid-controlled material deposition (concrete and clay printing), or for customised protocols (steel-welding), preparation for masonry walls (marking, fixing), plaster deposition or tile laying. Principle knowledge exists in research for robotic fabrication, yet this does not extend to actual knowledge for AEC industries, nor does it connect to performance criteria or values (use, location, cost, durability, performance, materials, construction method), or to business models (integration, service, maintenance). Only recently research has moved from closed robotic protocols to knowledge-based systems [23], which originate with workers' embodied expertise. More research is needed to streamline construction workflows, such as innovative use for robotically processed construction materials, adopt improved robotics interfaces and hardware, to achieve sufficiently versatile, on-site, and human-interactive robotic systems.

3.3 *Cyber-Physical Pathway for Robotic Fabrication*

Commercial building processes are commonly conducted by multidisciplinary teams (architects, engineers, consultants, and on-site/off-site contractors) and consequently the way in which construction information can be successfully communicated to builders or contractors is highly relevant. Data feedback on issues of material affordances, labour and production protocols, or unpredictability and uncertainty across construction sites through cyber-physical systems thus represents a huge potential for connecting human workers, robots, machines, materials and control devices (tablets, computers) for collaborative workflows [75]. Importantly, while this can enable architects to create, test and build in a virtual environment and so support evidence-based collaboration and inclusive decision-making, this also enables architects to connect to data controlling design to fabrication processes inclusive of resource data; enhancing existing building information models with operating data; and overview of lifecycle demands comparative analysis.

Whereas the full control of the computational or building information model, fabrication method, and assembly can be affordable, a highly specialized workflow and customised robot setups in industrial or commercial projects can present challenges related to economics and time. In this context, the adoption of robotic fabrication, and recent developments for human–robot collaboration hold significant potential to change in construction industries. In the last five years, increased efforts have been made to connect robots to the bandwidth of cyberphysical systems with the aim to move beyond static systems, closed operations and linear protocols (Fig. 5).

Coupling digital monitoring, sensor feedback and haptic interfaces with physical robotic manufacturing methods enables robots to sense, analyse and respond to changes in movement, tasks, material resources and thus gives access to new possibilities for production. This includes the potential for startups and entrepreneurship, whereby industrial robotics allows new manufacturing technologies to gradually evolve from initial tests and pilot studies to industrial processes—a new generation of construction [22]. Here, the raw production capacity of industrial robotics brings 'design and build' approaches to construction into view. Robotic startups revisit the

Fig. 5 Cooperation to Collaboration: human support of robotic fabrication process for robotic carbon fibre winding (Reinhardt et al, 2017, left), versus data capture for fabrication techniques and knowledge of motion, force and material resistance (Reinhardt et al, 2016, right).

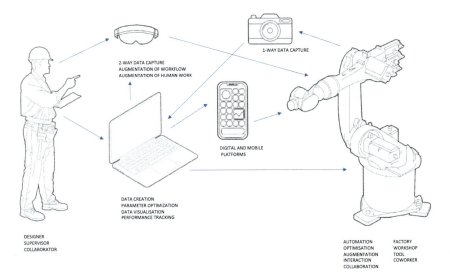

Fig. 6 Network for Human–Robot Cooperation across cyberphysical systems, digital twinning and intuitive interfaces in onsite conditions

idea of an architect-builder through computational design knowledge coupled with the means of production thus opening new career paths and in fact new professions for the AEC industries (Fig. 6).

3.4 Human–Robot Collaborations

The fourth industrial revolution changes the role of humans in operations systems, and so integrating human work contributes towards a successful digital transformation [58]. Whereas industrial robotic robotic arms were previously confined to factory settings with strict safety controls [25] and regulated in standard manufacturing environments by the ISO 15066 safety standard [66] with explicit limitations to robotic work environments to prevent accidents and injuries to human operators, this has opened with the concept of Human–Robot Collaboration (HRC), introduced by Colgate and Peshkin [14]. Collaborative relationships allow human(s) and robot(s) as one of the most important modes, where humans and robots have intersections in space and time domains through the shared work/tasks. This translates to shared working environments and shared working time for human–robot collaborations, with shared non-fenced zones and direct physical interaction. Approaches of human-centred and creative methods, user interfaces and machine learning have started to be developed in recent years, so that more direct access and control over processing of data to machinery—and with that, more direct interaction between human–robot processes—become available [29], with systems and methods for better understanding the human as active agent in the workflow with robotics.

Instead of fully autonomous or automated robot setups, collaborative robots (CoBots) in partnership with humans are the future of construction work: robots can perform tasks that are repetitive, dangerous, harmful, monotonous, or even physically impossible for a human worker while the operator would manage the more skilled work that required more finesse and experience. Instead of using industrial robots as 'human' substitutes, robots can be used intuitively and actively in an immediate interaction between design and motion. The current (r)evolution in human–robot interaction shifts the procedural/prescriptive programming of robots (typically via 'recipes' written in industrial robot languages) towards declarative/if–then scenarios and criteria based robotic protocols.

3.5 *Lifting the AEC Smile Curve Through Robotic Interfaces*

Human Robot Collaboration (HRC) recenters building and manufacturing processes from the product towards a human or user-centric process. Emergent interactive strategies explore sharing of tasks and actions where humans and robots have intersections in space and time and here, the concept of coexistence manifests as a shared working environment and shared working time. Importantly, this means that the smile curve can be significantly lifted through direct integration of production knowledge at the early stages R&D and Design, as a direct and iterative loop for logistics and production. To achieve this, the integration of data, but more importantly interactions or collaborations between human and machine, digital twinning in the form of data visualization of workflows, positions, and reconsiderations of robot workspace and movement protocols in relation to human co-workers is crucial. This means that robot-specific software that couples particular robot programming with robot products, such as aforementioned RobotStudio, ABB robots or KUKA|prc software needs to be expanded to directly and intuitively interact with the user—and thus bringing trained/skilled/embodied knowledge systems from the factory floor/construction site to the architect's office. While these distinct robot languages commonly require a specialist to operate the software and create robot instructions, this is about to change. Recent research explores Java programming via KUKA LBR-iiwa, using seven axes with integrated force-torque sensors to safely safe interact and further enabling entirely new applications that use hand-guiding and utilize the force-sensors to compensate for high tolerances on building sites, like manual assembly tasks [9]. Other alternative methods of interacting with robots that reach beyond previously required complex bus systems or industrial data interfaces include vision-based safety systems for human–robot collaboration [34]; haptic programming approaches where a collaborative robot physically manipulated/taught a movement based on feedback loops between the robot controller and its associated force sensors [19], vision systems [49], application of alternative robot programming through tablet [60] and augmenting robot processes through AR via Microsoft Hololens [42], fologram; [37]. The collaboration for the same goal [5] and research into classification of collaboration levels [40] changes the directive of human–robot collaboration from subservient strategies or a confined placement. Distinguishing between the different levels is important because of the requisite safety issues and for the purpose of designing and evaluating the worker aspects of human–robot interaction (e.g., acceptance and workflow), but moreover, for facilitating best practices to implement future human–robot collaboration on the factory floor [2]. The organisation of industrial robots in a human–robot team approach [78] frees interaction space considerably, where human and robot have agency, task sets and are single, team or multiple constellations (for example a singular robot, a robot-human team, or a multiple robot-multi operator setup). This includes turn-taking, handover, or multi-party and situated interaction [73]. Current developments for defined relationship frameworks include reference models for human–robot interaction [20, 18], and systematically define interactive intention between human and robot [75],with models of 'leader–follower' (human to robot), or status descriptions for both agents including 'active' (leader status), 'inactive' (rest), 'supportive' (following prompts/external control) or 'adaptive' (changing roles). Moreover, this requires a different form of robotic interaction and entails a move from robot to control a process segment, towards development of a communication language with task content, specified interactive nodes, and task process transition between agents (and here is where haptic interfaces are extremely useful in instructing the robot through discrete visual systems adaptation with special symbols and prompts).

4 Discussion

In the following, we discuss the potential of robotic fabrication and human–robot collaboration for new strategic prompts in Industry 4.0, core objectives for the AEC sector, and the Big Picture Thinking for better integration for SDG 9 (industry, innovation and infrastructure) and Climate Change.

4.1 Moving Robotic Technologies Forward

Robotic fabrication, as has been discussed, has solvers for applications in advancements in the construction industries. The increasing need for agile production equipment can be addressed by collaborative robots in dynamic environments, working alongside human workers, and while this require reconfiguration and agile control methods, plug and produce frameworks are currently developed with exchange of hardware modules coupled with agent-based system extending the robot operating system [69]. Consequently, multiple pathways for production chains could be orchestrated, and applications that are highly customised to serve the special user needs, solve specific issues and tasks can be optimised through multi-agent reinforcement learning [24]. What is required are frameworks for integration, collaboration between construction trade and architecture practices to enhance synthesis (creative process) and production processes. A focus on the construction of hardware (sensing technology, end-effector design, etc.) and software (programming AI algorithm) needs to implement tasks and balance exploration and innovation for human–robot collaborations, and methodology and system configuration, to achieve a coordinated development of AEC.

4.2 Developing Core Objectives for Industry 4.0 Frameworks

In addition, core objectives can be outlined for the AEC sector, given the technological advancements available through robotic technologies, cyberphysical systems and digital twinning for tangible production. Firstly, training, upskilling and transfer of process knowledge between architects and manufacturers needs to be increased, to make them fit to deliver complex, high value-add architectural manufacturing. The integration of both business operations between intangible and tangible production will be extremely valuable and further fuelled by population growth and increased demands on the industry. Secondly, contributing to digitalise architecture and engineering firms can increase productivity and potential speed of project delivery, by creating manufacturing-specific design tools and frameworks. The World Economic Forum estimates that full-scale digitalisation has the potential to generate 12–20% in annual cost savings in the construction industry. Thirdly, establishing the methodological foundations for a profound rethinking of the design process in the AEC sector. This chapter has presented a new paradigm, adopting an integrative and cross-sectoral approach encompassing digital business models, computer science, architecture, and engineering for architectural manufacturing. It consequently works toward a future industry where organisations, products, and services are arranged around specific projects or problems rather than distinct disciplines. Fourth, research is required into how to remove the bottleneck between design as an intangible pre-production process with tangible production activities, where cross-lateral training between architects and manufacturers will make construction and production cost effective, feasible and innovative. Lastly, we aim to contribute to accelerate digital transformation in AEC businesses by developing industry and organisational interventions. The digital transformation investigation will include business models (e.g. platform models), future scenarios, ecosystem collaboration (e.g. open innovation) and processes that will enhance organisational efficiency, agility, growth, and profitability.

4.3 Investing in SDG 9 to Counteract Climate Change

Industry 4.0 and circular economy knowledge can radically transform waste management [51], reduce resource consumption in a manufacturing context [45], or contribute towards achieving the Sustainable Development Goals. Understanding opportunities in the tangible and intangible pre- and postproduction in the architecture business sector and enhancing collaboration between different stakeholders is an immediate and important means of systematically connecting technologies available through Industry 4.0–with the larger scope of moving construction towards better resource management, circular economies, and increased building performance. Computational architecture practice coupled with advanced manufacturing and robotic fabrication strategies can unlock opportunities for AEC, when instrumentalised not merely as a technological pathway, but as strategies that can inform and change the way in which we operate. If the AEC sector plans to unpack via computational methods, strategies, and tools, quantity and type of resources in buildings and by applying modern manufacturing strategies—to build less, with less and with new materials and material systems—to mitigate resource demands of the building sector—then we need to develop a strategic framework first for upskilling all parties involved. To this extent the development of integrated, cross-disciplinary, innovative training frameworks from technology to strategy will address the building sector's request for advancement and at the same time provide pathways for answering the current climate problem.

5 Conclusion

This position paper has explored the utilisation of technologies for Industry 4.0 towards advancement of AEC industries, with a focus of applications of cyber-physical systems, digital twins and architecture robotics. We have discussed state-of-the-art tangible production processes in the domains of robotic fabrication and manufacturing, and ways in which these systems impact on challenges and opportunities outside the tangible production floor, with contributions in the form of a framework that integrates pre-and post-production phases and so uplift the manufacturing/construction smile curve, adding value for the AEC sector. We have outlined potentials for increased knowledge integration between architecture practice and the construction sector and defined objectives and potentials for the digital as major change agent in the AEC sector's role and impact in a climate change context.

Increasingly buildings will have a digital twin, a virtual model designed to accurately reflect a physical object. Building file to factory capabilities within architecture will help manufacturers to viably engage in design-led production via file to fabrication, so a pathway to develop sector-specific IP and training for AI-driven specialised architectural manufacturing out of digital twins will be an important aspect of architecture in a context of Industry 4.0. Data on building performance under changing environmental conditions will enable a deeper understanding for individual buildings but more importantly of larger building groups and their interferences, enabling better overview of complex data for building collectives and urban scapes that can respond for subtle and extreme changes, such as increased heat, floods and bushfires of the past years. In that scenario, digital twins could not only pass data back to manufacturing, but robot fabrication could be continued into robotic maintenance, human–robot collaboration embedded in buildings, and extend to different robotic ecologies—including industrial arms, drones, robot swarms and augmented support through interactive and haptic interfaces.

Acknowledgements The authors acknowledge parts of Sect. 2 are aligned with the research and thinking of the Australian Research Council (ARC) Industry Transformation Training Centre for Next-Gen Architectural Manufacturing' (IC2220100030). Findings published under Sect. 2 are the result of research towards the grant application and are the result of the team effort of the applicants.

Funding. Sections (Sect. 2) of this research was funded by the Australian Research Council (ARC) Industrial Transformation Training Centre for Next-Gen Architectural Manufacturing (IC2220100030).

References

1. https://doi.org/10.1007/978-3-319-26378-6_10
2. Aaltonen, I., Sali, T., Marstio, I.: Refining levels of collaboration to support the design and evaluation of human-robot interaction in the manufacturing industry, Procedia CIRP 72 (2018), https://doi.org/10.1016/j.procir.2018.03.214, pp 93–98
3. Adams, J. A.: Critical considerations for human-robot interface development. In Proceedings of 2002 AAAI Fall Symposium (pp. 1–8) (2002), https://www.aaai.org/Papers/Symposia/Fall/2002/FS-02-03/FS02-03-001.pdf
4. Arcot, R.: Cyberphysical systems—the core of Industry 4.0'. Online article, access date: July 7, 2022, https://blog.isa.org/cyber-physical-systems-the-core-of-industry-4.0
5. Bauer, A., Wollherr, D., Buss, M.: Human–robot collaboration: a survey. International J. Human. Robot. 5(01), 47–66 (2008)
6. Benachio, G. L., Carmo Duarte Freitas, M., Tavares, SF. (2020). Circular economy in the construction industry: A systematic literature review. J. Clean. Prod. 260, 121046 (2020). ISSN 0959-6526, https://doi.org/10.1016/j.jclepro.2020.121046
7. Bernstein, P (2018) Architecture | Design | Data, Birkhäuser, Basel
8. Bock, T.: The future of construction automation: Technological disruption and the upcoming ubiquity of robotics. Autom. Constr. **59**, 113–121 (2015). https://doi.org/10.1016/j.autcon.2015.07.022
9. Braumann, J., Stumm, S., Brell-Cokcan, S. (2016). Towards New Robotic Design Tools: Using Collaborative Robots within the Creative Industry. ACADIA //2016: POSTHUMAN FRONTIERS: Data, Designers, and Cognitive Machines [Proceedings of the 36th Annual Conference of the Association for Computer Aided Design in Architecture (ACADIA) ISBN 978–0–692–77095–5] Ann Arbor 27–29 October, 2016, https://doi.org/10.52842/conf.acadia.2016.164, pp. 164–173
10. Brell-Çokcan, S., Braumann, J. (eds.): Robotic Fabrication in Architecture, Art and Design 2012, Springer International Publishing Switzerland (2012). ISBN: 978–3–319-04662-4
11. Brettel, M., Friederichsen, N., Keller, M., and Rosenberg, M. (2014). How virtualization, decentralization and network building change the manufacturing landscape: An Industry 4.0 Perspective. International Journal of Mechanical, Industrial Science and Engineering, 8(1), 37–44
12. Bryonjolfsson, E., McAfee, A. (2014). The Second Machine Age: Work, Progress, and Prosperity in a Time of Brilliant Technologies (WW Norton)
13. Chapman, R (2005) Inadequate Interoperability. ISARC. Ferrara (Italy)
14. Colgate, J.E., Wannasuphoprasit, W., Peshkin, M.A.: Cobots: Robots for Collaboration with Human Operators. Proceedings of the International Mechanical Engineering Congress and Exhibition, Atlanta, GA, DSC-Vol. **58**, 433–439 (1996)

15. Criado-Pérez, C., Shinkle, G., Hoellerer, M., Sharma, A., Collins, C., Gardner, N., Haeussler, MH., Pan, S. (2022). Digital Transformation in the Australian AEC Industry: Prevailing Issues and Prospective Leadership Thinking, Journal of Construction Engineering and Management, Volume 148 Issue 1 - January 2022 p.1–12. DOI: https://doi.org/10.1061/(ASCE)CO.1943-7862.0002214
16. Daas, M., Witt, AJ.: Towards a Robotic Architecture. Actar D (2018)
17. Deltex Clarity (2020) Architecture & Engineering Industry Report (info.deltek.com)
18. Deutsch, R. (2019) Superusers: Design Technology Specialists and the Future of Practice. Routledge
19. Devadass, P, Stumm, S., Brell-Cokcan, S (2019). Adaptive Haptically Informed Assembly with Mobile Robots in Unstructured Environments, Pages 469–476 (2019 Proceedings of the 36th ISARC, Banff, Canada, ISBN 978–952–69524–0–6, ISSN 2413–5844)
20. Djuric, A, Urbanic, R., Rickli, J. (2016), A Framework for Collaborative Robot (CoBot) Integration in Advanced Manufacturing Systems, SAE International Journal of Materials and Manufacturing, Vol. 9, No. 2 (May 2016), pp. 457–464
21. Dörfler, K., Sandy, T.M., Giftthaler, M., Gramazio, F., Kohler, M.: Mobile robotic brickwork. In: Reinhardt, D., Burry, J., Saunders, R. (eds.) (2016) Robotic Fabrication in Architecture, Art and Design 2016. Springer International Publishing, Switzerland (2016). https://doi.org/10.1007/978-3-319-26378-6_10.
22. Feringa, J. (2014). Entrepreneurship in Architectural Robotics: The Simultaneity of Craft, Economics and Design. Special Issue: Made by Robots: Challenging Architecture at a Larger Scale, Volume84, Issue3,https://doi.org/10.1002/ad.1755, pp 60–65
23. Flores, A., Bauer, P., Reinhart, G.: Concept of a learning knowledge-based system for programming industrial robots, Procedia CIRP, Volume 79, 2019. ISSN **626–631**, 2212–8271 (2019). https://doi.org/10.1016/j.procir.2019.02.076
24. Fologram, https://fologram.com/
25. Fryman, J., Matthias, B. (2012). Safety of industrial robots: From conventional to collaborative applications. In ROBOTIK 2012; 7th German Conference on Robotics (pp. 1–5). VDE. https://www.researchgate.net/publication/269411126_Safety_of_Industrial_Robots_From_Conventional_to_Collaborative_Applications
26. Gallaher, M P., Chapman, R. (2004) Cost analysis of inadequate interoperability in the US capital facilities industry. https://nvlpubs.nist.gov/nistpubs/gcr/2004/NIST.GCR.04-867.pdf
27. Gardner, N.: New Divisions of Digital Labour in Architecture. Fem. Rev. **123**, 56–75 (2019)
28. Gharbia, M., Chang-Richards, A., Lu, Y., Zhong, R., Li, H.: Robotic technologies for on-site building construction: A systematic review, Journal of Building Engineering, Volume 32, 2020. ISSN **101584**, 2352–7102 (2020). https://doi.org/10.1016/j.jobe.2020.101584
29. Gillies, M.: Understanding the role of interactive machine learning in movement interaction design. ACM Transactions on Computer Human Interaction (TOCHI), 26(1) 1–34, (2019)
30. Government's Modern Manufacturing Strategy MMS (Industry.gov.au 2022)
31. Gramazio, F., Willmann, J., Kohler, M.: The Robotic Touch: How Robots Change Architecture. Park Books (2015)
32. Hadrian X, FastBrick, https://www.fbr.com.au/view/hadrian-x, access date 07 July, 2022.
33. Haeusler, M., Gardner, N., Zavoleas, Y.: Computational Design—From Promise to Practice, Avedition, Ludwigsburg (2019)
34. Halme, R.J., Lanz, M.: Review of vision-based safety systems for human-robot collaboration. Procedia CIRP **72**, 2018 (2018). https://doi.org/10.1016/j.procir.2018.03.043,pp111-116
35. Hermann, M., Pentek, T. and Otto, B. (2015). Design Principles for Industry 4.0 Scenarios, TU Dortmund, DOI: https://doi.org/10.13140/RG.2.2.29269.22248, acces date: 13/05/2022
36. Industry Insights, Manufacturing and the smile Curve, https://publications.industry.gov.au/publications/industryinsightsjune2018/documents/IndustryInsights_2_2018_Chapter3_ONLINE.pdf
37. Jahn, G., Newnham, C., Beanland, M. (2018). Making in Mixed Reality. Holographic design, fabrication, assembly and analysis of woven steel structures. ACADIA // 2018: Recalibration. On imprecision and infidelity. [Proceedings of the 38th Annual Conference of the Association

for Computer Aided Design in Architecture (ACADIA) ISBN 978–0–692–17729–7] Mexico City, Mexico 18–20 October, 2018, pp. 88–97 https://doi.org/10.52842/conf.acadia.2018.088
38. Jahn, G., Newnham, C., van den Berg, N. (2022). Augmented Reality for Construction From Steam-Bent Timber. Jeroen van Ameijde, Nicole Gardner, Kyung Hoon Hyun, Dan Luo, Urvi Sheth (eds.), POST-CARBON - Proceedings of the 27th CAADRIA Conference, Sydney, 9–15 April 2022, pp. 191–200 http://papers.cumincad.org/cgi-bin/works/Show?caadria2022_296
39. KUKA|prc, https://www.robotsinarchitecture.org/kuka-prc
40. Kolbeinsson, A., Lagerstedt., E, Lindblom, J. (2019). Foundation for a classification of collaboration levels for human-robot cooperation in manufacturing, Production & Manufacturing Research, 7:1, 448–471, DOI: https://doi.org/10.1080/21693277.2019.1645628
41. Kyjanek, O., Al Bahar, B., Vasey, L., Wannemacher, B., Menges, A. (2019). Implementation of an Augmented Reality AR Workflow for Human Robot Collaboration in Timber Prefabrication. Proceedings of the 36th ISARC, Banff, Canada, ISBN 978–952–69524–0–6, ISSN 2413–5844), pp 1223–1230
42. Lasi, H., Fettke, P., Feld, T., Hoffmann, M. (2014). Industry 4.0. Business & Information Systems Engineering:Vol. 6: Iss. 4, 239–242. https://aisel.aisnet.org/bise/vol6/iss4/5
43. Lee, J.H., Ostwald, M.J.: Grammatical and syntactical approaches in architecture. IGI Global (2020)
44. Lipsey, R.G., Carlaw, K.I., Bekar, C.T.: Economic Transformations: General Purpose Technologies and Long-Term Economic Growth. Oxford University Press, Oxford (2005)
45. Liu, H., Wang, L.: Human motion prediction for human-robot collaboration. J. Manuf. Syst. **44**(2017), 287–294 (2017). https://doi.org/10.1016/j.jmsy.2017.04.009
46. Lonergan, D. (2019), Breaking the mould: How Industry 4.0 is modernising the way manufacturers do business. online article, Manufacturer's Monthly, access date: July 7, 2022, https://www.manmonthly.com.au/features/breaking-mould-industry-4-0-modernising-way-manufacturers-business/
47. Lublasser, E., Hildebrand, L., Vollpracht, A., Brell-Cokcan, S.: Robot assisted deconstruction of multi-layered façade constructions on the example of external thermal insulation composite systems. Construction Robotics **1**(1), 39–47 (2017)
48. MX3D, Laarman: 3D printed Steel Bridge, https://mx3d.com/services/metalxl/, access date 07 July, 2022.
49. Malik, A., Bilberg, A.: Developing A Reference Model For Human-Robot Interaction. International Journal On Interactive Design And Manufacturing (Ijidem) **13**(4), 1541–1547 (2019)
50. Marks, A., Muse, A., Pothier, D., Sahwney, A. (2020). Future of work in construction, Autodesk white paper, https://www.rics.org/globalassets/rics-website/media/knowledge/20200522_autodesk_whitepaperconstruction_final.pdf, accessed July 7th, 2022
51. Mavropoulos, A., Nilsen, A. (2020). Industry 4.0 and Circular Economy: Towards a Wasteless Future or a Wasteful Planet?. ISW Wiley
52. McGee, W., Ponce de Leon, M. (eds.): Robotic Fabrication in Architecture, Art and Design 2014. Springer International Publishing, Switzerland. ISBN: 978–3–319–04662–4
53. McKinsey Global Institute. Barbosa, F., Woetzel, J., Mischke, J., Ribeirinho, M., Sridhar, M., Parsons, M., Bertram, N., Brown, S.: Reinventing construction through a productivity revolution, February 27, 2017 (2017)| Report, https://www.mckinsey.com/business-functions/operations/our-insights/reinventing-construction-through-a-productivity-revolution, access date: May 13, 2022
54. McKinsey Global Institute, Ribeirinho, M., Mischke, J., Strube, G., Sjödin, E., Blanco, JL., Palter, R., Biörck, J., Rockhill, D., Andersson, T.: The next normal in construction: How disruption is reshaping the world's largest ecosystem. McKinsey Global Institute, June 4, 2020 | Report. https://www.mckinsey.com/business-functions/operations/our-insights/the-next-normal-in-construction-how-disruption-is-reshaping-the-worlds-largest-ecosystem, access date: May 13, 2022, access date: May 13, 2022
55. McKinsey Global Institute, Gregolinska, E., Khanam, R., Lefort, F., Parthasarathy, P. (2022). Capturing the true value of Industry 4.0, McKinsey Global

56. NCCR, ETH: DFAB House: http://dfabhouse.ch/, access date 07 July, 2022.
57. Odico, Formwork Robotics: Factory-on-the Fly, https://www.odico.dk/technologies#factoryon-the-fly, access date 07 July, 2022.
58. Oesterreich, T.D., Teuteberg, F.: Understanding the implications of digitisation and automation in the context of Industry 4.0: A triangulation approach and elements of a research agenda for the construction industry. Comput. Ind. **83**, 121–139 (2016)
59. Pan, G., Pan, S.L., Lim, C.-Y.: Examining how firms leverage IT to achieve firm productivity. Info. and Man. **52**, 401–412 (2015)
60. Pedersen, J., Neythalath, N., Hesslink, J., Søndergaard, A., Reinhardt, D.: Augmented drawn construction symbols: A method for ad hoc robotic fabrication, International Journal of Architectural Computing 2020, 18(3), 254–269 (2020). DOI: https://doi.org/10.1177/147807712 094316.
61. ROS robot operating system, https://www.ros.org/
62. Reinhardt, D., Haeusler, H. , London, K., Loke, L., Feng, Y., Barata, E., Firth, C., Dunn, K., Khean, N., Fabbri, A., Wozniak-O'Connor, D., Masuda, R.: CoBuilt 4.0 - Investigating the Potential of Collaborative Robotics for Subject Matter Experts, IJAC International Journal of Architectural Computing (2020) https://doi.org/10.1177/1478077120948742
63. Van Rijmenam, M.: The Organisation of Tomorrow. Routledge (2019)
64. Robeller, C., Weinand, Y.: Fabrication-Aware Design of Timber Folded Plate Shells with Double Through Tenon Joints (2016). https://doi.org/10.1007/978-3-319-26378-6_12.
65. RobotStudio, https://new.abb.com/products/robotics/robotstudio
66. Rosenstrauch, M., Kruger, J.: Safe human-robot-collaboration introduction and experiment using iso/ts 15066, International conference on control, automation and robotics (ICCAR) IEEE, pp 740–744 (2017)
67. Ross et al Designed for Digital - How to architect your business for sustained success (2019) https://mitpress.mit.edu/books/designed-digital
68. SQ4D Inc, Autonomous Robotic Construction System (ARCS): Robotic 3D printed house, https://www.therobotreport.com/robot-helps-3d-print-a-home-for-less-than-6000-in-materials/, access date 07 July, 2022.
69. Schou, C., Madsen, O.: A plug and produce framework for industrial collaborative robots. International Journal of Advanced Robotic Systems, 14(4) (2017) https://doi.org/10.1177/172 9881417717472
70. Shinkle, G.A., Gooding, L.H., Smith, M.L.: Transforming strategy into success. Productivity Press, New York (2004)
71. Skantze, G.: Turn-taking in Conversational Systems and Human-Robot Interaction: A Review. Elsevier (2020). https://doi.org/10.1016/j.csl.2020.101178
72. Sobek, W.: Non Nobis - Ueber das Bauen in der Zukunft, Band 1: Ausgehen muss man von dem, was ist. Avedition, Ludwigsburg (2022)
73. Susskind, R., Susskind, D.: The Future of the Professions. Press, Oxford Uni (2016)
74. Vasey, L., Menges, A.: Potentials of cyber-physical systems in architecture and construction, Routledge (2020)
75. Wang, X.V., Kemény, Z., Váncza, J., Wang, L.: Human–robot collaborative assembly in cyber-physical production: Classification framework and implementation. CIRP Ann. **66**(1), 5–8 (2017)
76. Wang, B., Zhou, H., Yang, G., et al.: Human Digital Twin (HDT) Driven Human-Cyber-Physical Systems: Key Technologies and Applications. Chin. J. Mech. Eng. **35**, 11 (2022). https://doi.org/10.1186/s10033-022-00680-w
77. Willmann, J., Block, P., Hutter, M., Byrne, K., Schork, T.: Robotic Fabrication in Architecture, Art and Design 2018. Springer International Publishing Switzerland. (2019)
78. Yuan, P., Xie, M., Leach, N., Yao, J., Wang, X. (eds.): Architectural Intelligence–Selected Papers from the 1st International Conference on Computational Design and Robotic Fabrication (CDRF 2019). Springer (2020). https://doi.org/10.1007/978-981-15-6568-7.pdf. Access date: Jul 7, 2022.

Overview on Urban Climate and Microclimate Modeling Tools and Their Role to Achieve the Sustainable Development Goals

Matteo Trane◉, Matteo Giovanardi◉, Anja Pejovic◉, and Riccardo Pollo◉

Abstract The role of the fourth Industrial Revolution enabling technologies is pivotal if the paradigms of data-driven, performance-based, and optimized design have to become standard practice. Urban climate and microclimate models are increasingly likely to support the design for adaptation, resilience, and mitigation of the heat island in cities. In this context, the objective of this chapter is to emphasize the role of urban climate and microclimate modeling tools to achieve the Sustainable Development Goals at the local level. To this, firstly the authors screened the Agenda 2030 official Targets and Indicators and the European Handbook for Voluntary Local Reviews' indicators, highlighting how they deal with environmental quality, urban climate and microclimate issues. Interlinkages and possible trade-offs were identified among goals and targets, too. Secondly, a robust overview on the main software for climate modeling is provided. Tools were clustered, according to the domain of application, into scale, statistical, numerical, and dispersion/air quality models. Thus, the authors focused on numerical models, identified as proper tools for architects and planners to support urban and micro-urban scale design. A final matrix compares the most used numerical models at a glance, highlighting main features, fields of applications, environmental parameters simulated, and interoperability options.

Keywords Climate models · Computational fluid dynamics · Urban microclimate · Agenda 2030 · Sustainable development goals · ENVI-met

---

M. Trane (✉) · A. Pejovic · R. Pollo
Interuniversity Department of Regional and Urban Studies and Planning, Politecnico Di Torino, 10125 Turin, Italy
e-mail: matteo.trane@polito.it

A. Pejovic
e-mail: anja.pejovic@polito.it

R. Pollo
e-mail: riccardo.pollo@polito.it

M. Giovanardi
Department of Architecture and Design, Politecnico Di Torino, 10125 Turin, Italy
e-mail: matteo.giovanardi@polito.it

United Nations' Sustainable Development Goals 11. Make cities and human settlements inclusive, safe, resilient and sustainable · 13. Take urgent action to combat climate change and its impacts · 3. Ensure healthy lives and promote well-being for all at all ages

1 Introduction

Open public space is crucial when talking about quality of life within urban environments. It reflects the identity of a city in an aesthetic, ecological, and functional sense, thus providing environmental, social, economic, and health benefits and fostering the sense of community [58]. Moreover, redesigning the space between buildings has the great potential to both adapt cities and mitigate the causes of climate change, which, in turn, is the most promising strategy to cope with ongoing climate crises [11]. The role of open public space is also promising in attenuating the effects of the Urban Heat Island (UHI), as designing by combining strategies to cope with both climate change adaptation and mitigation has immediate effects on microclimate, thus quality of life and health [52]. Specifically, the role of greenification—i.e., the use of green infrastructure (green areas, roofs, and façades and vertical vegetation)— benefits the local climate conditions and has a direct impact on human thermal stress and mortality [29]. Indeed, heat waves accounted for 68% of natural hazard-related deaths in Europe in 1980–2017 and many climate models still project a global rise in climate hazards [59].

In such a scenario, the quality of public space is definitely a key element to convey sustainable development, and the Agenda 2030 put specific emphasis on access to safe open public space for all, good air quality, and environmental comfort. As a consequence, modeling the behavior of urban elements—both natural and artificial—is essential for decision and policy making, data-driven urban planning and climate projection-based scenario assessment. The rapid spread of enabling technologies, with the consequent increase in computing capabilities, redefines design practice in favor of computational approaches, which require interoperability, modeling, 'simulability' and connectivity [55] to finally boost the 'generative process' of an optimized design [14]. Specifically, the use of numerical models and tools may support domain experts (e.g., urban planners, architects, landscape designers, environmental engineers) and non-technical professionals (e.g., decision-makers, public and civic stakeholders) [29] in the early stage of a performance-based design process (*ex-ante*), thus allowing better monitoring of the effects of design and policies (*ex-post*). Nowadays, many software and solutions exist, enabling urban climate and microclimate simulations with different spatial resolution and grid size (from less than one meter to hundreds of kilometers) for different purposes (from urban streets to regional modeling).

In the perspective of an increasing full 4.0 awareness for architects and planners, i.e., shifting from a computer-based design to a computational one [3], the purpose of this chapter is to provide a detailed overview on the most used tools for urban climate and microclimate modeling, highlighting their role to ease the transition towards a sustainable development. Moreover, this chapter will focus on how these tools may support the achievement of the United Nations' Sustainable Development Goals (SDGs) at the local level. Indeed, it should be taken into consideration that 65% of SDG targets could not be achieved without a decisive commitment of cities towards the implementation of local sustainability agendas [40], possibly aligned with the UN Goals. In this context, computational design plays a crucial role in shaping cities' transition towards carbon neutrality, combining several dimensions: mitigating the UHI, reducing the resource consumption, promoting the user's comfort, and boosting the resilience to climate change. Indeed, planning for the adaptation claims for a combination of data and scenario assessment, that is, the results of knowledge exchange and contamination among disciplines, to eventually provide policy makers with consistent and sound 'environmental synthesis', grounding on predictive modeling and supported by research [4]. Thus, the role of the 4th Industrial Revolution enabling technologies towards the achievement of the SDGs is fundamental if the paradigms of data-driven, performance-based, and optimized design process must be universally pursued.

This paper is structured as it follows. After the introduction, Sect. 2 describes the methodology adopted. In Sect. 3, the authors map the recurrence of the urban climate and microclimate-related issues in the Agenda 2030, giving emphasis to possible interlinkages and trade-offs among them and their utility for progress measurement. In Sect. 4, the authors clustered urban climate and microclimate tools according to their features and scale of application. Section 5 specifically focuses on numerical models, providing an overview on the ones most frequently used and quoted by the literature. In Sect. 6, the results are discussed by a matrix summing up the main features of the software and providing some insights on how they may further support the Agenda 2030. Finally, in the Conclusions, the authors open up to the possibility of both integrating data from climate simulations into GIS environments and providing more and more local climate-specific boundary conditions, based on extensive sensor networks, time-series and climate projection data for future scenarios simulations and assessment.

2 Materials and Methods

The first objective of this chapter is to put into correlation climate analysis with the Agenda 2030 Sustainable Development Goals. Specifically, the authors highlight how models and tools for climate and microclimate simulation could assist to reach the SDGs (Sect. 3). To this, the authors screened the 17 goals, 179 targets and 239 indicators of the Agenda 2030. Correlations, both positive and negative, are made evident by means of several Sankey diagrams. First, it has been analyzed how Agenda 2030 deals with urban climate and microclimate issues, finding interlinkages with seven goals and eighteen targets. Second, the authors emphasized how urban climate and microclimate models may specifically help measuring progress towards some of the selected indicators, coming from both Agenda 2030 and the *European Handbook for Voluntary Local Reviews* (VLRs) [49]. Third, interlinkages and possible trade-offs both among the selected targets and between them and other Agenda 2030 targets are presented. To this, the European Commission's Joint Research Centre (JRC) *SDG Interlinkages* tool from the *KnowSDG Platform* was used [7]. This tool helps visualizing the cumulated interlinkages from a set of publications, to quickly see and understand for which interlinkages there is strong agreement in the literature. It should be noted that, at the time of finalizing the outputs and figures for this chapter (May 2022), the first edition of the *European Handbook* was available and a previous version of the tool, with less analyzed papers on the SDG interlinkages, was consulted.

In the second part of this chapter (Sects. 4 and 5) a robust overview on the software available for urban climate and microclimate modeling is provided. First, by screening the scientific literature, the authors clustered them into scale, statistical, numerical and dispersion and air quality models (based on [15] and according to the scale of applications (meso- and micro-scale). Second, by screening manuals, websites, and literature providing information on the main commercial solutions available, the analysis focused on numerical models only, identified as the main tools for architects and urban planners to assess the effects of local-scale design options. Third, in the Discussions (Sect. 6) a matrix offers a complete figure to compare the most used numerical models, highlighting their main features, scale of applications, and environmental parameters that can be simulated.

3 Urban Microclimate in Light of the Agenda 2030

3.1 *An Overview: Urban Microclimate, SDGs, Targets*

The 17 Sustainable Development Goals (SDGs) were adopted by the United Nations in September 2015 as part of the 2030 Agenda for Sustainable Development [57]. They replaced the previous Millenium Development Goals by defining a path towards sustainability fitting to all countries. The concept of genuine quality of life for all, to be ensured in respect of the planetary boundaries [50], stays behind the Agenda 2030's vision. Although not legally binding, SDGs have been setting up a normative framework by which cities, as major actors towards climate neutrality, may boost sustainability. However, many challenges remain in fully localizing the SDGs, mostly due to lacking data, the need to establish local priorities in light of policy regulations and a complex framework of highly interlinked targets, and capacity building [54].

Although the term "microclimate" is not explicitly mentioned within Agenda 2030, at deeper look connections among the topic and the SDGs exist and are strong. As recalled by [58], the potential of quality public spaces to contribute to sustainability involve five SDGs (3, 5, 8, 11, 13) and eight targets (3.4, 3.6, 3.9, 5.2, 8.8, 11.7, 13.1, 13.2). In a broader perspective, the authors argue that efforts towards

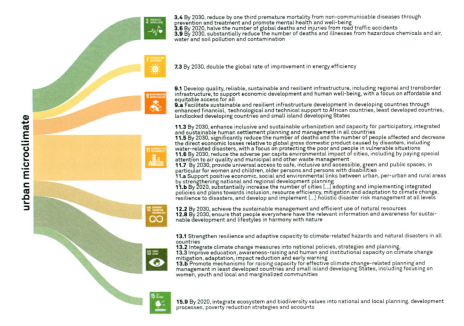

Fig. 1 Interlinkages between mitigating Urban Heat Island and SDG targets

urban microclimate mitigation by public space redesigning may imply benefits in achieving SDG 7.3 ("By 2030, double the global rate of improvement in energy efficiency"), as lowering outdoor temperatures turns into reducing energy demand for buildings [44]. Besides, it may still have positive effects on SDG 9.1, 9.a, 11.3, 11.5, 11.6, 11.a, 11.b, 12.2, 12.8, 13.3, 13.b, 15.9 (Fig. 1).

3.2 Urban Microclimate: Models and Tools for SDGs' Indicators

In the context of the Agenda 2030, climate and microclimate modeling tools may assist to measure progress towards the achievement of SDGs, fulfill the lack of intra-urban scala data by simulating, and analyze the effects of policies and design *ex-ante* and *ex-post* [53]. Specifically, among the target interlinkages identified, climate models may provide a strong support to indicators 3.9.1 ("Mortality rate attributed to household and ambient air pollution") and 11.6.2 ("Annual mean levels of fine particulate matter (e.g. PM2.5 and PM10) in cities (population weighted)"), when it comes to models able to assess air pollution concentration. However, major contribution of climate modeling approaches is linked to SDG 13 indicators, which can benefit the employment of modeling approaches when it comes to assess national policies facing disaster risk mitigation, local strategies for disaster risk reduction strategies, adoption of national adaptation plans, and the communication of risks arising from climate hazards (respectively, indicators 13.1.2, 13.1.3, 13.2.1, 13.3.1). Finally climate models are still central in contributing to the measure of target 11.7.1 ("Average share of the built-up area of cities that is open space for public use for all, by sex, age and persons with disabilities"), 11.a.1 ("Number of countries that have national urban policies or regional development plans that (a) respond to population dynamics, (b) ensure balanced territorial development; and (c) increase local fiscal space") and both 11.b.1 and 11.b.2 ("Number of countries that adopt and implement national disaster risk reduction strategies in line with the Sendai Framework for Disaster Risk Reduction 2015–2030", "Proportion of local governments that adopt and implement local disaster risk reduction strategies in line with national disaster risk reduction strategies") (Fig. 2).

SDG localization is highly challenging: scholars, practitioners, policy makers and practitioners are committed in measuring progress at urban and local level by framing and implementing indicators needing high data granularity. The *European Handbook for SDG Voluntary Local Reviews* (VLRs) [49] provides a "fundamental instrument to monitor progress and sustain the transformative and inclusive action of local actors towards the achievement of the Sustainable Development Goals" in the specific European Union (EU) context. The structure of the *Handbook* (first edition—2020) is lighter than the 2030 Agenda framework and tailored to the EU cities needs: it comes with 72 indicators, each of which is related to one or more SDGs targets. The use of climate models may provide effective support in assessing indicators 49 and 53 (if a pollution concentration module is embedded in the software), 48, 58, 61 (Fig. 2).

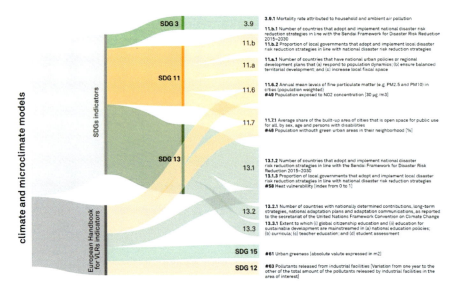

Fig. 2 Climate and microclimate models in support of SDGs indicators measurement

3.3 Mapping SDG Interlinkages and Trade-Offs

The analysis of the potential contribution of climate and microclimate models in achieving the SDGs confirmed that major benefits could be registered in progressing SDG 11 and 13. However, boosting sustainable development in all dimensions (social, economic, environment) implies a systemic approach, assessing interlinkages and possible trade-offs among targets to eventually minimize them [54]. Indeed, according to scholars the general lack of progress in effectively reaching the SDGs [19] is also due to poor understanding and addressing of interactions between goals and targets. As stated by [9], "the challenge lies in identifying and countering inherent conflicts (trade-offs), while harnessing and building on potential synergies (co-benefits) between the 169 targets", to finally unveil the "invisible" nature of Agenda 2030, i.e., enhancing interlinkages among targets to achieve them faster and with minimized drawbacks.

Coming back to highlighted targets, according to the *Interlinkages Visualization tool* (version: May 2022) 11.6, 11.7, and 11.a manifest eight positive interlinkages; among them, strong support to target 3.9, 13.2 and 12 could be provided (Fig. 3). No trade-offs are detected among SDG 11 selected targets and other targets. Moreover, fifty-two positive interactions exist among targets 13.1, 13.2, 13.3 and other targets. Target 13.1 is one with the greatest number of interlinkages (twenty-six), especially with SDG 9 and 11. On the contrary, target 13.2 is the one with major possible trade-offs, especially with goal 14 ("Life below water"). However, no trade-offs are detected among SDG 13 targets and the ones related to urban climate and microclimate (from other SDGs). To conclude, although trade-offs exist and SDG 13 implementation should be managed "with care", no specific drawbacks exist among the selected targets, implying that the targets to which climate modeling may provide support could be achieved independently and eventually providing benefits to other Agenda 2030 targets.

4 Climate Analysis Models: An Overview

4.1 Clustering Climate Models

Scale models. Scale models have mainly been used for the simulation of urban flow, turbulence and dispersion phenomena in wind tunnels and water flumes or outdoors over type-arrays of building-like obstacles. They have been using a range of experimental facilities, maquettes and techniques for measurement of urban canyons that have deeply contributed to the understanding of the urban atmosphere [15]. The development of Computational Fluid Dynamics (CFD) models was also possible thanks to high resolution datasets produced in laboratories. Scale studies were able to demonstrate that small-scale features such as roof shape or tree placement can significantly impact street canyon ventilation rates (Kastner-Klein and Plat 1999).

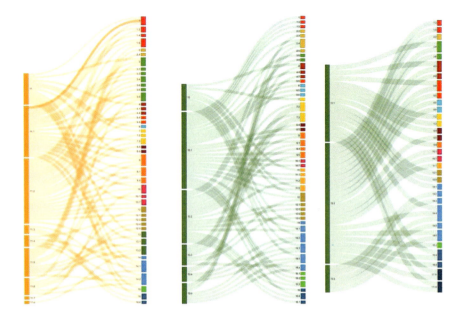

Fig. 3 In orange, the interlinkages among SDG 11 targets and the Agenda 2030. In green, interlinkages (left) and trade-offs (right) among SDG 13 targets and the Agenda 2030 (*Source* JRC's Know SDG Platform—Interlinkages visualization, https://knowsdgs.jrc.ec.europa.eu/)

Only a few studies used city-scale models, demonstrating that building height has a strong influence on the Urban Canopy Layer (UCL) [15]. One of the limitations of scale models is the size, because of the typical size of wind tunnels allowing only a small range of model scales to be evaluated.

Statistical models. Statistical models estimate the effects of cities on climate. Among their main advantages, they usually produce realistic results, require low computational features and input parameters. The disadvantages of certain statistical models imply long observation periods, lack of physical basis and data from many different locations [15]. Their main field of application is the study of the UHI and most of them are simple linear (multiple) regressions of temperature differences between the center of the city and the surrounding area (ΔT), where temperature is calculated as function of several meteorological variables (e.g., wind speed, rural lapse rate, etc.) [22]. More advanced statistical models use spectral analysis, eigenvectors or neural networks to predict risks related to UHI [46]. Statistical relations have been also used for assessing the normalized evolution of UHI, allowing fast forecasts of city temperatures and helping predict future climate conditions [15].

Numerical models. Numerical models allow 3D simulation of urban thermal environments to assess the interactions among (vertical and horizontal) surfaces, solar radiation and wind. They mostly deal with the governing equations of the flow (e.g. Navier–Stokes equations). The use of numerical tools is becoming practice at meso- and local scale to assess multiple design alternatives, informing urban planners, decision makers and designers [29]. They are able to provide information about any evaluated parameter in all points of the computational domain and conduct comparative analysis based on different scenarios [6]. Numerical methods are used to predict impacts of different microclimates and design options on the urban environment. The most used numerical simulation approaches are Energy Balance Model (EBM) and CFD [51].

Dispersion and air quality models. Dispersion and air quality models are used for evaluation of air quality and prediction of environmental phenomena that could finally affect people's health. These applications vary from evaluation of long-term health effects to short-term emergency response [15]. The complexity of the models depends on the computational speed and storage capacity. CFD models are widely used for urban dispersion simulation today. However, all models should include a meteorological model or parameterization scheme to examine urban phenomena such as wind slowing, turbulence increase, the vertical profile of meteorological variables in the UCL etc. Ideally, the models should also evaluate changes in meteorological conditions between different urban districts [15].

4.2 Spatial Scale of Analysis

Tools for climate and microclimate analysis are usually clusterable according to the scale of application and resolution, too. In this perspective, urban climate-related phenomena were classified as meso-, local, and micro-scale [41], while [29] identified two main domains for climate tools application: macroscale and microscale models. Meso-scale represents a region or city, local scale corresponds to a neighborhood or district area, while micro-scale varies from 10 to 102 m and includes street canyons or single buildings [12].

Meso-scale models. Tools for climate analysis on meso-scale focus on the entire city rather than street-level phenomena [29]. The size of the model is usually a few tens of kilometers, with the grid of up to 10 or 20 km on a city scale or up to 100 or 200 km on a regional scale. The spatial resolution is usually 100 or 200 m [29]. Since 1970s, many organizations developed their own models: mentioning a few, National Center for Atmospheric Research (NCAR) and Pennsylvania State University developed their 5th Generation Meso-scale Model (MM5), Colorado State University developed the Colorado State University Meso-scale Model (CSUMM) and later the Regional Atmospheric Modelling System (RAMS). The research on meso-scale climate is mostly done by remotely sensed data and satellite images to derive Land Surface Temperature (LST), which is used for modeling UHI [29, 29], supported by meteorological data coming from urban stations [56]. Other mesoscale studies use numerical models, such as Weather Research and Forecasting model (WRF) [12], designed for "numerical weather prediction systems for both atmospheric research and operational forecasting needs to analyze impact of climate phenomena on the city" [29].

Micro-scale models. Research on microclimate at intra-urban level is mostly done by Urban Canopy Models (UCM) or CFD models [60]. CFD simulations for neighborhood scale models are mostly based on area range of 200 m to 2 km and they are commonly used for assessing the impacts of mitigation measures on the thermal environment of local-scale urban area, the relationship between the form and layout of buildings or blocks and the local urban thermal environment [35], the layout and planning of ventilation corridors and its impact on the thermal environment and the relationship between the thermal environment and human thermal comfort indices [5]. At this scale, "the measurements and analyses are usually the most suitable to estimate local impacts and benefits in terms of human thermal comfort at the pedestrian level and the impacts of the different planning scenarios" [29]. Mathematical modeling and simulations are mainly coupled with field measurements in order to evaluate the effects of the urban environment on the microclimate and thermal comfort [12]. UCMs are mainly used for examining the energy budget of an urban canopy layer and the airflow derives from the energy budget equations, unlike CFD models [34]. Both types can assess the impact of various parameters such as building orientation, street aspect ratio, surface materials, vegetation, pedestrian comfort and urban ventilation. However, they both have certain weaknesses. CFD models have limited domain size due to immense computational cost, while UCMs have less detailed presentation of airflow around the buildings [34].

5 Focus on Numerical Models: A Comparison

5.1 Energy Balance Models (EBMs)

As recalled by Toparlar et al. [51] numerical models can be further clustered into EBMs and CFDs. EBMs are based on the law of energy conservation and have been used extensively in the past. They take into account energy exchanges between surfaces and ambient air in the UCL and can be used to predict the ambient temperatures and surface temperatures of buildings, pavements and streets [47]. Although they are quick to run and provide accurate results, their major drawback is the absence of air velocity [18]. Indeed, they "separate temperature and velocity fields, such that the assumptions used do not always accurately represent the interaction of velocity and temperature in reality" [47]. According to Imran et al. [18], UCMs are derived from energy balance equations "for a control volume such as two adjacent buildings". Surface and control volumes interact with each other like electric nodes, which results in the matrix of humidity and temperatures. UCMs can be the single-layer canopy model or multi-layer canopy. The difference is that the latter takes into account vertical distribution of the canopy features instead of the average building height [18], providing more detailed analysis but needing more computational resources [60]. The great limitation of the EBMs is the absence of a wind flow field, which causes incomplete representation of atmospheric phenomena. They are not able to determine the latent and sensible fluxes due to the lack of flow pattern information [18].

5.2 Computational Fluid Dynamics Models

Compared to EBMs, CFDs have numerous advantages. The main one is that it can assess a wide range of issues "comprising air speed and movement, air quality and pollution diffusion, wind comfort and thermal comfort as well as the effects of relative humidity and vegetation on indoor and outdoor spaces", as well as being readily available and analyzing complex environments [47]. As such, CFD is the most used numerical simulation method at microscale. CFD models include a variety of numerical models from Reynolds-averaged Navier–Stokes, through Large Eddy Simulation (LES) to Direct Numerical Simulation (DNS) [15]. They consider all principal fluid equations in urban areas simultaneously, solving the equations of temperatures, momentum and conservation of mass [18]. CFD simulations can be used to study urban microclimate, user thermal comfort, and pollutant dispersion at both meso- and micro-scale. At micro-scale CFD simulations are conducted with a resolution from less than one meter up to 100 m. Still they are able to represent more realistic information compared to the EBM [18]. However, CFD models have certain limitations. They have high computational requirements due to their complexity and the wide range of environmental parameters and indexes to be assessed (especially for meso-scale models). High resolution representation of the urban geometry is required, along with the knowledge of the boundary conditions and relevant parameters [51]. Furthermore, CFDs may require expertise in governing equations for output interpretation, while "different scales of turbulence in an urban environment require separate modeling and simplified simulation, which may result in inaccurate outcomes" [47].

5.3 Most Used Tools

ENVI-met. ENVI-met is a CFD 3D microclimate modeling software based on the fundamental laws of fluid dynamics and thermodynamics [48]. It was designed by Professor M. Bruse in 1994 and has been under constant development and scientific expansion ever since. The software uses an urban weather generator to predict the meteorological parameters based on which the most probable weather conditions are recreated [2]. It reconstructs the microclimate dynamics of the urban environment through the interaction between climate variables, vegetation, surfaces and built environment [10]. ENVI-met simulates the atmosphere processes such as air flow, air temperature, humidity, turbulence, radiation fluxes and calculates the indexes and factors of comfort in the urban area—such as Physiological Equivalent Temperature (PET), Predicted Mean Vote (PMV), Universal Thermal Comfort Index (UTCI) and many others [53]. ENVI-met has been used widely in the microclimate research, urban design and thermal comfort studies due to its ability to recreate microclimatic conditions within the UCL [1, 23, 39]. The application of the software has been validated through numerous studies under different climate conditions. ENVI-met also includes plant physiology and soil science. Indeed, the software takes into account transpiration, evaporation and sensible heat flux from vegetation and air, as well as physical parameters of the plants, water and heat exchange from soil and plant water uptake [43]. The software is grid-based and has high temporal and spatial resolution, allowing accurate microclimate analysis up to a street level. The horizontal resolution typically ranges from 1 to 10 m and it is therefore suitable for intra-urban scale assessment. By the latest version, ENVI-met v5.0, Python was integrated into the system, to facilitate the management and visualization of output results via the DataStudio module. Some other new features (such as the Indexed VieW Sphere and the renovated Albero module for vegetation modeling) are meant to make simulations even more accurate on sun radiation and greenery effects. Due to holistic approach and complexity of the calculations, ENVI-met has certain limitations, mostly regarding the computation time with high space resolution and wide simulation domains, and the high temporal resolution of meteorological data needed (30-min timesteps for full forcing mode).

SOLENE Microclimat. SOLENE-microclimat is a simulation tool for modeling at neighborhood scale. It consists of a thermo-radiative model, a CFD model and a thermal building model [17]. SOLENE was developed in the 1990s by CRENAU, *Laboratoire de l'école d'architecture de Nantes*. It was first developed for simulating natural light both outdoors and indoors, taking into account direct solar radiation, diffused sky luminous radiation and inter-reflections [33]. New sub-modules were added as the tool continued to evolve, which allowed it to take into account radiative transfers, conduction and storage in walls and soils, airflow and convective exchanges, evapotranspiration from natural surfaces, energy balance for a building in the simulated area [38]. SOLENE-microclimat is a result of coupling two tools—Code-Saturne which is a CFD open-source code developed by EDF and SOLENE [24]. Code-Saturne allows for calculation of wind speed and airflow distribution within the model. When coupled with SOLENE, a thermo-radiative model, it is able to calculate energy and moisture transportation, based on which it can determine physical characteristics of air and its interaction with urban surfaces [17]. The tool enables parameterization and representation of vegetation, natural soil, green walls and roofs, street humidification, but it is difficult to achieve it on a high level, since it requires a very detailed input and parameters such as leaf area index, water availability and soil/building characteristics [8]. The vegetation layer included in the model takes into account energy transfer by evapotranspiration, radiation and convection [37]. SOLENE-microclimat is able to simulate solar radiation and its impacts on urban surfaces, air flow and its effect, impacts of vegetation and water bodies as well as buildings' energy demand for the whole 3D modeled urban environment [17]. As the outcome of the simulation, outdoor air temperature, humidity, wind velocity and surface temperatures are calculated along with comfort indexes (MRT, PET, UTCI). The advantages of using SOLENE-microclimat are the possibility of representing the whole urban environment [36] and its capability of detail describing, consideration of thermal behavior of buildings, taking into account energy and humidity transfers through green devices and the ability to analyze different greenery types on building energy consumption [42]. The main weaknesses of the tool are the need for coupling with a CFD model for a full assessment and the complexity in terms of program use [8]. There is no user manual, so the guidance is mostly provided by the researchers and engineers already working with it [17] or annual meeting for collective formation [8]. Some modules of SOLENE-microclimat have been validated, such as radiative module, soil, green wall and building. However, due to a wide range of parameters concerning various physical phenomena on different locations, the model has not been validated as a whole [8].

UMEP & SOLWEIG. UMEP (Urban Multi-scale Environmental Predictor) is an integrated tool for urban climatology and climate sensitive planning applications [25] developed in Sweden and Great Britain, written as a plug-in to QGIS which makes it easier to use in integrated urban assessments. UMEP analyses can be performed on various scales, from street canyon to city-scale. It is comprised of four main models: the SOLWEIG model (SOlar and LongWave Environmental Irradiance Geometry model), the SUEWS model (Surface Urban Energy and Water Balance Scheme), the BLUEWS model (Boundary Layer Urban Energy Water Scheme), and the LUCY model (Large scale Urban Consumption of energY model) [20]. UMEP enables users to input atmospheric and surface data from different sources, adapt measured meteorological data for the urban environment, use reanalysis or climate prediction data and compare different climate scenarios and parameters. One of the advantages of UMEP is the ability to analyze different scenarios and problems related to climate-sensitive urban design can be addressed within one tool, on various scales [20]. It is also able to integrate relevant processes and use common data across a range of applications. UMEP tools can provide export data for more complex systems and software. Data from more complex models can be imported into UMEP as well [25]. SOLWEIG is a tool for simulation of spatial variation of 3D radiation flux and Mean Radiant Temperature (MRT) in complex urban environments. The tool uses a digital elevation model of buildings and plants for construction of complex urban structures, which results in relatively high accuracy of the simulation [25]. The model requires geographical information, urban geometry, direct, diffuse and global short-wave radiation, air temperature and relative humidity [28]. The tool works best in clear weather conditions, because it does not consider the cloud amount for diffuse radiation calculation, which results in calculation error. It takes into account combined effects of ground, wall and vegetation on reflected radiation. SOLWEIG focuses on radiation scenarios and therefore calculates MRT as the main comfort indicator [8], but PET and UTCI can be calculated as well [27]. MRT is calculated by modeling shortwave and longwave radiation in six directions and angular factors. It also requires continuous maps of sky view factors to determine MRT [26]. The advantage of using SOLWEIG on a neighborhood scale is GIS representation of buildings and other important modeling elements. One of the limitations of SOLWEIG is that it requires spatial datasets that could be difficult to obtain or integrate [8]. SOLWEIG has been widely evaluated and applied at various urban locations [25]. UMEP, as well as SOLWEIG, is a plug-in to

QGIS, allowing the creation of urban climate maps. UMEP is a free tool that encourages users to participate in their development by submitting comments, reporting issues, getting updates and sharing with other users. UMEP and SOLWEIG also provide detailed manuals and training material [20].

RAYMAN. Rayman is a largely validated [8] micro-scale model developed by Professor A. Matzarakis at the University of Freiburg, Germany, in 2007 [28]. It is used for calculation of radiation fluxes in simple and complex environments [30, 31], with the aim of calculating various thermal indices in order to evaluate the thermal conditions of different climates and regions (Matzarakis et al. 2017). RayMan is able to simulate different thermal indices: PMV, PET, UTCI, Standard Effective Temperature (SET), Perceived Temperature (PT) and MRT [32]. The model requires certain input data regarding air temperature, humidity, wind speed, short and longwave radiation fluxes and activity and clothing data for calculating PET [20]. Aside from the thermal index simulation, the software is capable of calculating and graphically presenting the sun paths for each day of the year, as well as sunshine duration and shadows caused by the surrounding obstacles [30]. All the calculations in RayMan are performed for one point in space and are time-dependent. The software considers only the thermal effects of buildings and vegetation [8]. It takes into account both simple and complex environments with their radiation properties, albedo and emissivity. It is able to calculate radiation fluxes due to the possibility of various input parameters and forms such as topography input, environmental morphology input, free-hand drawing and fish-eye photographic input. The software also calculates the Sky View Factor (SVF) in different ways, based on fish-eye images, free drawing of the horizon limitation, topographic raster or obstacle [30]. The main advantages of RayMan are low computational requirements and clear structure allowing the non-experts in human-biometeorology to use it [31]. The software is able to simulate the microclimate in different urban environments accurately, due to the precise calculation of radiation effects of the complex surface structure [16]. The preparation of input data can be time-consuming [8], but the main limitation regards the absence of the wind model.

Ladybug ecosystem. Ladybug Tools is a group of four environmental plug-ins that are built on top of several simulation engines. It is written in Python, which can be plugged into any geometry engine if the geometry library is translated. It is an open-source interface which unites open source simulation engines. It is connected to the 3D modeling software, which allows the geometry creation, simulation and visualization to be performed within one interface. The tool is mainly known as a Grasshopper plug-in. However, it can be connected to other 3D modeling software, such as Dynamo or Blender. Ladybug Tools consist of: Ladybug, Honeybee, Butterfly and Dragonfly. Ladybug tools are able to assess the thermal conditions and integrate it into the design flow, which makes it a comprehensive instrument for architects and urban planners. They are used for simulating climatic conditions on various scales, from canyon to city scale [32]. Ladybug's components are modular, which makes it capable of evaluating different parameters over various stages of design. Ladybug imports EnergyPlus Weather files (*.epw) which provide weather and location data. Integration with Grasshopper's parametric tools allows for immediate simulation of environmental parameters along with the change in design. This gives designers the possibility to explore the direct relationship between environmental data and the generation of the design [45]. Honeybee performs the analysis of daylight through Radiance, energy models using OpenStudio and envelope heat flow with Therm. Ladybug and Honeybee are capable of simulating the envelope interaction with both indoor and outdoor. They calculate heating and cooling demand with EnergyPlus while evaluating outdoor MRT [13]. Butterfly creates and runs CFD simulations using OpenFOAM, which is capable of running several simulations and turbulence models. Butterfly exports geometry to OpenFOAM and runs different types of airflow design, including simulations of urban wind patterns, indoor buoyancy-driven simulations, thermal comfort, ventilation effectiveness and more. Dragonfly is able to model and estimate large-scale climate phenomena such as UHI, climate change and influence of local climate factors. It is connected to Urban Weather Generator (UWG) and CitySim as well as several datasets which contain publicly available weather data and thermal satellite image datasets. With UWG, Dragonfly is able to estimate hourly air temperature and relative humidity in the UCL [13].

6 Discussion

As for the links between urban climate and microclimate issues and the Agenda 2030, the majority of positive interrelations occurs with targets of SDG 11 (six interlinkages), followed by SDG 13 (four), SDG 3 (three), SDG 9 and SDG 12 (two), finally SDG 7 and SDG 15 (one) (Fig. 2). Shifting to targets and indicators, SDG 11, 3 and 9 may specifically benefit from urban climate modeling. Moreover, the indicators addressing (mostly indirectly) urban climate and microclimate issues are fifteen, coming from Agenda 2030 (ten) and the *European Handbook* (five). However, the authors argue that more robust and direct support can be provided to the indicators by the *European Handbook* (specifically to #58 on "Heat Vulnerability"), which are more urban scale tailored with respect to the ones coming from Agenda 2030 but in turn support the achievement of prioritized targets. Besides, we still argue that no indicator within the *European Handbook* supports the measure of people's exposure to Particulate Matter (PM 2.5–10), although it is considered among the main risk factors for cardiovascular, respiratory, and carcinogenic diseases. Major attention to this issue is put by Agenda 2030 (11.6.2), but the annual mean level of exposure may provide no specific insights on site-specific urban level of exposure, which may significantly vary over a city, thus not providing specific support for design.

A wide range of software for urban climate and microclimate modeling to support the achievement of Agenda 2030 exist. According to the main findings, the most used and cited within the scientific literature are presented in Fig. 4, providing a direct comparison among them. Typology, application scale, environmental parameters that can be analyzed, indices simulated, and accessibility are shown for each software. From a general overview two main aspects clearly emerge: first, all the tools analyzed allows simulations at the micro-scale while only a few of them can be used for local and meso-scale; secondly, only about the half of them (e.g., ENVI-met, RayMan, Ladybug tools, Ansys Fluent) offer a wide range of environmental parameters and thermal comfort indices that can be simulated. For instance, air pollution simulation can be carried out by ENVI-met, Ladybug, Ansys Fluent and OpenFoam. The price affordability is another key factor in choosing software. Although most software developed and provided by academia are free, the most used for advanced analysis require an annual license subscription, while free licenses may come with large constraints. Among software analyzed and for design scenario assessment for local adaptation purposes, it is worth mentioning ENVI-met and Ladybug as the two most promising software for microclimate assessment at intra-urban level, considering the wide range of environmental parameters and indexes assessable, great number of papers validating their use, accuracy of the calculation, field scalability, implementability of functions, interoperability with GIS, CAD, Rhino or SketchUp for modeling, and most user-friendly visualization tools and interfaces.

	Type of model		Model scale			Simulated variables						Simulated thermal comfort index							Simulating air pollution	Native module for modelling	Free Accessibility
	EBM	CFD	Meso	Local	Micro	T_a	T_s	W_s	W_d	Q	Other	PMV	PET	UTCI	SET	PT	MRT	Other			
TEB (Town Energy Balance)	●		●	●	●	●	●	●		●	●			●							●
UWG	●		●	●	●		●				●			●							●
ENVI_MET		●		●	●	●	●	●	●	●	●	●	●	●	●		●		●	●	only with very limited features
RayMan	●				●	●	●			●		●	●	●	●	●	●			●	●
SOLWEIG	●			●	●		●			●			●	●			●				●
UMEP	●		●	●	●		●			●			●	●			●				●
Solene-Microclimat		●		●	●	●	●	●	●	●			●	●			●				●
Ladybug Ecosystem			●	●	●	●	●	●	●	●	●	●	●	●			●				open tools for pay per use software
ANSYS FLUENT		●		●	●		●	●	●		●	●	●	●	●		●		●	●	
TownScope	●				●					●								●			
OpenFOAM®		●	●	●	●			●	●			●							●		●
SkyHelios	●				●			●	●	●			●	●		●	●			●	●

Fig. 4 A comparison among the most used numerical software. Abbreviations: EBM (Energy Balance Model), CFD (Computational Fluid Dynamics), Ta (Air Temperature), Ts (Surface Temperature), Ws (Wind speed), Wd (Wind direction), Q (long and shortwave radiation), PMV (Predicted Mean Vote), PET (Physiological Equivalent Temperature), UTCI (Universal Thermal Climate Index), SET (Standard Effective Temperature), PT (Perceived Temperature), MRT (Mean Radiant Temperature)

7 Conclusions

The development of a full awareness by architects, urban planners, designers and policy makers on Industry 4.0 potential to drive cities in transition towards carbon neutrality is central in supporting environmentally sound policies. Parallelly, the paradigm of enabling technologies may allow for greater quality of life, in a context—the urban one—within which people's health is threatened by several structural stress conditions, to finally "leave no one behind". This chapter highlighted the specific role of urban climate and microclimate modeling tools with respect to open space redesign, as a means to accomplish climate adaptation and mitigation of the UHI effects within cities, the main "theater" of sustainability-related challenges. As mentioned, climate modeling can be a practical tool for measuring and achieving several indicators of the Agenda 2030 SDGs, especially 11, 13 and 3 when on-field measurable data is scarce or not available. To successfully implement the SDGs, interlinkages and trade-offs evaluation should be central. Climate models for simulating and minimizing the effect of projects on the urban thermal comfort and resource consumption may help to achieve multiple targets, but other targets may benefit from the usage of these tools.

Although still mostly requiring a high degree of specialization, numerical modeling in the field of micro-urban climatology is crucial when it comes to *ex-ante* assessment of several design options, affecting the user comfort, the UHI mitigation and, thus, the climate adaptation scenarios. Indeed, the authors highly recommend the use of such tools for enhancing the role of architecture and urban planning in supporting climate-proof urban landscapes design, as well as to consider possible trade-offs when implementing a project by embracing the SDG framework. Besides, further recommendations may regard the production and use of more and more site-specific and temporal fine-grain input meteorological boundary conditions to rely on accurate climate and pollutant distribution simulation [53]. Finally, enhancing the interoperability of these tools with GIS may constitute a tremendous opportunity to perform complex multivariate analysis, by which to turn climatic data into georeferenced climate maps to be coupled with other type of information (e.g., population density, pattern of land use and coverage, epidemiological data etc.). Indeed, boosting an evidence-based approach in the technological environmental design at urban and micro-urban scale needs updated and robust datasets, accessible to the stakeholders. Turning data on microclimate into georeferenced information may definitely constitute the next field of research for intra-urban scale modeling and simulation. Indeed, the role of open data on urban climate and microclimate, as proper GIS databases derived by advanced simulations, may support a system of shared information policy makers and domain experts may rely on, towards performance-based design as a praxis. In this perspective, performing climate simulations based on projection of temperature rise could contribute to the assessment of climate change impacts in the near future too, strengthening the role of performative design as a sound means to face current and future environmental emergencies and achieve the SDGs.

Acknowledgements This chapter is part of the work developed by the Research Unit of Politecnico di Torino as partner of the National Research Project PRIN TECH-START (*key enabling TECHnologies and Smart environmenT in the Age of gReen economy, convergent innovations in the open space/building system for climaTe mitigation*). This chapter is the result of a joint research by the Authors, who are equally responsible for it.

Author contributions: Conceptualization, M.T. (Matteo Trane), R.P. (Riccardo Pollo); investigation coordination: M.T., M.G. (Matteo Giovanardi), R.P.; methodology: M.T., A.P. (Anja Pejovic); literature selection and review process, M.T., A.P. (Anja Pejovic); writing—original draft, M.T., A.P.; writing—review and editing, M.T., M.G., R.P. All authors have read and agreed to the published version of the manuscript.

References

1. Ali-Toudert, F., Mayer, H.: Effects of asymmetry, galleries, overhanging façades and vegetation on thermal comfort in urban street canyons. Sol. Energy **81**, 742–754 (2007)
2. Bande, L., Afshari, A., Al Masri, D., Jha, M., Norford, L., Tsoupos, A., Marpu, P., Pasha, Y., Armstrong, P.: Validation of UWG and Envi-Met Models in an Abu Dhabi District, Based on Site Measurements. Sustainability 11 (2019)
3. Barberio, M., Colella, M.: Architettura 4.0. Fondamenti ed esperienze di ricerca progettuale. Maggioli Editore, Santarcangelo di Romagna, Italy (2020)
4. Battisti, A.: Il progetto come volontà e rappresentazione: dai big data all'apprendimento collettivo. In: Perriccioli, M., Rigillo, M., Russo Ermolli, S. Tucci, F. (eds.), Design in the Digital Age. Technology Nature Culture, pp. 335–339. Maggioli Editore, Santarcangelo di Romagna, Italy (2020).
5. Blocken, B.: Computational fluid dynamics for urban physics: importance, scales, possibilities, limitations and ten tips and tricks towards accurate and reliable simulations. Build. Environ. **91**, 219–245 (2015)
6. Blocken, B., Janssen, W.D., van Hooff., T.: CFD Simulation for Pedestrian Wind Comfort and Wind Safety in Urban Areas: General Decision Framework and Case Study for the Eindhoven University Campus. Environmental Modeling & Software 30, 15–34 (2012)
7. Borchardt S., Barbero-Vignola G., Buscaglia D., Maroni M., Marelli L.: A Sustainable Recovery for the EU: A text mining approach to map the EU Recovery Plan to the Sustainable Development Goals. EUR 30452 EN, Publications Office of the EuropeanUnion, Luxembourg, (2020)
8. Bouzouidja, R., Cannavo, P., Bodénan, P., Gulyás, Á., Kiss, M., Kovács, A., Béchet, B., et al.: How to evaluate nature-based solutions performance for microclimate, water and soil management issues – available tools and methods from Nature4Cities European Project Results. Ecol. Ind. **125**, 107556 (2021)
9. Breu, T., Bergoo, M., Ebneter, L., Pham-Truffert, M., Bieri, S., Messerli, P., Ott, C., Bader, C.: Where to begin? Defining national strategies for implementing the 2030 Agenda: the case of Switzerland. Sustain. Sci. **16**, 183–201 (2021)
10. Bruse, M., Fleer, H.: Simulating surface–plant–air interactions inside urban environments with a three-dimensional numerical model. Environ. Model. Softw. **13**, 373–384 (1998)
11. Denton, F., Wilbanks, T.J., Abeysinghe, A.C., Burton, I., Gao, Q., Lemos, M.C., Masui, T., O'Brien, K.L., Warner, K.L.: Climate-resilient pathways: adaptation, mitigation, and sustainable development. Climate Change 2014: Impacts, Adaptation, and Vulnerability. Part A: Global and Sectoral Aspects. Contribution of Working Group II to the Fifth Assessment Report of the Intergovernmental Panel on Climate Change. Cambridge University Press, Cambridge, United Kingdom and New York, NY, USA, 1101–1131 (2014)

12. Elbondira, T.A., Tokimatsu, K., Asawa, T., Ibrahim, M.G.: Impact of neighborhood spatial characteristics on the microclimate in a hot arid climate – a field based study. Sustain. Cities Soc. 75 (2021)
13. Evola, G., Costanzo, V., Magrì, C., Margani, G., Marletta, L., Naboni, E.: A Novel Comprehensive workflow for modeling outdoor thermal comfort and energy demand in urban canyons: results and critical issues. Energy Build. 216 (2020)
14. Figliola, A.: Envision the construction sector in 2050. Technol. Innov. Vert. TECHNE **17**, 213–221 (2019)
15. Grimmond, C.S.B., Roth, M., Oke, T.R., Au, Y.C., Best, M., Betts, R., Carmichael, G., et al.: Climate and more sustainable cities: climate information for improved planning and management of cities (producers/capabilities perspective). Procedia Environ. Sci. **1**, 247–274 (2010)
16. Gulyás, A., Unger, J., Matzarakis, A.: Assessment of the microclimatic and human comfort conditions in a complex urban environment: modelling and measurements. Build. Environ. **41**(12), 1713–1722 (2006)
17. Imbert, C., Bhattacharjee, S., Tencar, J.: Simulation of urban microclimate with SOLENE-Microclimat—an outdoor comfort case study. In: Proceedings of the Symposium on Simulation for Architecture and Urban Design. Delft (2018)
18. Imran, H.M., Shammas, M.S., Rahman, A., Jacobs, S.J., Ng, A.W.M., Muthukumaran, S.: Causes, modeling and mitigation of urban heat island: a review. Earth Sci. **10**(6), 244–264 (2021)
19. Independent Group of Scientists appointed by the Secretary-General: Global Sustainable development report 2019: the future is now-science for achieving sustainable development. United Nations, New York (2019)
20. Jänicke, B., Milošević, D., Manavvi, S.: Review of user-friendly models to improve the urban micro-climate. Atmosphere 12 (2021)
21. Kastner-Klein, P., Plate, E.J.: Wind-tunnel study of concentration fields in street canyons. Atmos. Environ. **33**, 3973–3979 (1999)
22. Kim, Y.H., Baik, J.J.: Maximum urban heat island intensity in Seoul. J. Appl. Meteorol. **41**(6), 651–659 (2002)
23. Krüger, E.L., Minella, F.O., Rasia, F.: Impact of urban geometry on outdoor thermal comfort and air quality from field measurements in Curitiba, Brazil. Build. Environ. **46**(3), 621–634 (2011)
24. Lauzet, N., Rodler, A., Musy, M., Azam, M.H., Guernouti, s., Mauree, D., Colinart, T.: How building energy models take the local climate into account in an urban context – a review. Renew. Sustain. Energy Rev. 116 (2019)
25. Lindberg, F., Grimmond, C.S.B., Gabey, A., Huang, B., Kent, C.W., Sun, T., Theeuwes, N.E., et al.: Urban multi-scale environmental predictor (UMEP): an integrated tool for city-based climate services. Environ. Model. Softw. **99**, 70–87 (2018)
26. Lindberg, F., Grimmond, C.S.B., Gabey, A., Jarvi, L., Kent, C.W., Krave, N., Sun, T., Wallenberg, N., Ward, H.C.: Urban multi-scale environmental predictor (UMEP) manual. University of Reading UK, University of Gothenburg Sweden, SIMS China (2019)
27. Lindberg, F., Holmer, B., Thorsson, S.: Solweig 1.0—modelling spatial variations of 3D radiant fluxes and mean radiant temperature in complex urban settings. Int. J. Biometeorol. 52 (7), 697–713 (2008)
28. Liu, D., Hu, S., Liu, J.: Contrasting the performance capabilities of urban radiation field between three microclimate simulation tools. Build. Environ. 175 (2020)
29. Lobaccaro, G., De Ridder, K., Acero, J.A., Hooyberghs, H., Lauwaet, D., Maiheu, B., Sharma, R., Govehovitch, B.: Applications of models and tools for mesoscale and microscale thermal analysis in mid-latitude climate regions—a review. Sustainability **13**(22), 12385 (2021)
30. Matzarakis, A., Rutz, F., Mayer, H.: Modelling radiation fluxes in simple and complex environments—application of the RayMan model. Int. J. Biometeorol. **51**(4), 323–334 (2007)
31. Matzarakis, A., Rutz, F., Mayer, H.: Modelling radiation fluxes in simple and complex environments: basics of the RayMan model. Int. J. Biometeorol. **54**(2), 131–139 (2010)

32. Mauree, D., Naboni, E., Coccolo, S., Perera, A.T.D., Nik, V.M., Scartezzini, J.L.: A review of assessment methods for the urban environment and its energy sustainability to guarantee climate adaptation of future cities. Renew. Sustain. Energy Rev. **112**, 733–746 (2019)
33. Miguet, F., Groleau, D.: A daylight simulation tool for urban and architectural spaces. application to transmitted direct and diffuse light through glazing. Build. Environ. 37(8–9), 833–43 (2002)
34. Mirzaei, P.A.: Recent challenges in modeling of urban heat island. Sustain. Cities Soc. **19**, 200–206 (2015)
35. Montazeri, H., Blocken, B., Derome, D., Carmeliet, J., Hensen, J.L.M.: CFD analysis of forced convective heat transfer coefficients at windward building facades: influence of building geometry. J. Wind Eng. Ind. Aerodyn. **146**, 102–116 (2015)
36. Morille, B., Lauzet, N., Musy, M.: Solene-microclimate: a tool to evaluate envelopes efficiency on energy consumption at district scale. Energy Procedia **78**, 1165–1170 (2015)
37. Musy, M., Malys, L., Inard, C.: assessment of direct and indirect impacts of vegetation on building comfort: a comparative study of lawns, green walls and green roofs. Procedia Environ. Sci. **38**, 603–610 (2017)
38. Musy, M., Malys, L., Morille, B., Inard, C.: The use of solene-microclimat model to assess adaptation strategies at the district scale. Urban Climate **14**, 213–223 (2015)
39. Ng, E., Chen, L., Wang, Y., Yuan, C.: A study on the cooling effects of greening in a high-density city: an experience from Hong Kong. Build. Environ. **47**, 256–271 (2012)
40. OECD: A Territorial Approach to the Sustainable Development Goals: Synthesis report. OECD Publishing, Paris, OECD Urban Policy Reviews (2020)
41. Oke, T.R.: Towards better scientific communication in urban climate. Theoret. Appl. Climatol. **84**(1–3), 179–190 (2006)
42. Parsaee, M., Joybari, M.M., Mirzaei, P.A., Haghighat, F.: Urban heat island, urban climate maps and urban development policies and action plans. Environ. Technol. Innov. 14 (2019)
43. Petri, A. C., Wilson, B., Koeser, A.: Planning the urban forest: adding microclimate simulation to the planner's toolkit. Land Use Policy 88 (2019)
44. Pollo, R., Elisa B., Giulia S., Bono, R.: Designing the healthy city: an interdisciplinary approach. Sustain. Mediterr. Constr. 11 (2019)
45. Roudsari, M.S., Pak, M., Smith, A., Gill, G.: Ladybug: a parametric environmental plugin for grasshopper to help designers create an environmentally-conscious design. In: Proceedings of BS2013: 13th Conference of International Building Performance Simulation Association, 3128–35. Chambery (2013)
46. Santamouris, M., Mihalakakou, G., Papanikolaou, N., Asimakopoulos, D.N.: A neural network approach for modeling the heat island phenomenon in urban areas during the summer period. Geophys. Res. Lett. **26**(3), 337–340 (1999)
47. Setaih, K., Hamza, N., Mohammed, M.A., Dudek, S., Townshend, T.: CFD modeling as a tool for assessing outdoor thermal comfort conditions in urban settings in hot arid climates. J. Inf. Technol. Constr. (ITCon) 19 (2014).
48. Sharmin, T., Steemers, K., Matzarakis, A.: Microclimatic modelling in assessing the impact of urban geometry on urban thermal environment. Sustain. Cities Soc. **34**, 293–308 (2017)
49. Siragusa, A., Vizcaino, P., Proietti, P., Lavalle, C.: European Handbook for SDG Voluntary Local Reviews. In: EUR 30067 EN, Publications Office of the European Union, Luxembourg (2020)
50. Steffen, W., Richardson, K., Rockström, J., Cornell, S.E., et al.: Planetary boundaries: Guiding human development on a changing planet. Science **347**, 736 (2015)
51. Toparlar, Y., Blocken, B., Maiheu, B., van Heijst, G.J.F.: A review on the CFD analysis of urban microclimate. Renew. Sustain. Energy Rev. **80**, 1613–1640 (2017)
52. Trane, M., Giovanardi, M., Pollo, R., Martoccia, C.: Microclimate design for micro-urban design. A case study in Granada, Spain. Sustain. Mediterr. Constr. 14, 149–55 (2021)
53. Trane, M., Ricciardi, G., Scalas, M., Ellena, M.: From CFD to GIS: a methodology to implement urban microclimate georeferenced databases. TECHNE Journal of Technology for Architecture and Environment **25**, 124–133(2023)

54. Trane, M., Marelli, L., Siragusa, A., Pollo, R., Lombardi, P.: Progress by Research to Achieve the Sustainable Development Goals in the EU: A Systematic Literature Review. Sustainability **15**(9), 7055 (2023)
55. Tucci, F.: Requirements, approaches, visions in the prospects for development of technological design. In: Lauria, M., Mussinelli, E., and Tucci, F. (eds.), Producing Project. Maggioli Editore, Sant'Arcangelo di Romagna, Italy (2020)
56. United Nations, Department of Economic and Social Affairs, Population Division. World Urbanization Prospects: The 2018 Revision (ST/ESA/SER.A/420). United Nations, New York (2019)
57. United Nations: Transforming our world: The 2030 agenda for sustainable development. Resolution Adopted by the General Assembly A/RES/70/1
58. Vukmirovic, M., Gavrilovic, S., Stojanovic, D.: The improvement of the comfort of public spaces as a local initiative in coping with climate change. Sustainability **11**(23), 6546 (2019)
59. Woetzel, J., Pinner, D., Samandari, H., et al.: Climate risk and response. Physical hazards and socioeconomic impacts. McKinsey Global Institute (2020)
60. Wong, N.H., He, Y., Nguyen, N.S., Raghavan, S.V., Martin, M., Hii, D.J., Yu, Z., Deng, J.: An integrated multiscale urban microclimate model for the urban thermal environment. Urban Climate **35**, 100730 (2021)

Industry 4.0 and Bioregional Development. Opportunities for the Production of a Sustainable Built Environment

Luciana Mastrolonardo and Matteo Clementi

Abstract The paper answers the question about how the technologies characteristic of industry 4.0 can support the bioregional development paradigm, focusing on local building production chains to support the energy retrofit of existing buildings. In particular it investigates design choices capable of activating supply chains that can intercept real and already existing spending flows and activate a workforce capable of evolving the anthropic system in the same direction characteristic of the natural systems. The paper focuses on the description of the features that industry 4.0 could assume in the field of building production to support energy requalification, reorienting urban metabolism of neighborhoods or small settlements towards local territories. It describes the characteristics of technologies adopted in the Industry 4.0 paradigm, in the context of open data, open-source software alongside low-cost microcomputers and sensors, and illustrates their potential in the possible activation of generative local microeconomies. Acceleration, digitization and automation of the construction sector are showing the relevance of having open real-time information to support decision-making processes. In particular the text focuses, on the one hand, on the applicability of such technologies and devices to areas characterized by very limited resources (such as for example the internal and more fragile areas in Italy); on the other hand, on how they enable community of prosumers and local cooperatives to complex productive activities characteristic of the energy communities.

Keywords Circular architecture · Data-driven design strategies · Real environmental performance-based design · Time digital representation and visualization · Enabling community

United Nations' Sustainable Development Goals 7. Ensure access to affordable, reliable, sustainable and modern energy for all · 8. Promote sustained, inclusive

---

L. Mastrolonardo (✉)
Dipartimento di Architettura, Università G.d'Annunzio Chieti, 65126 Pescara, Italy
e-mail: l.mastrolonardo@unich.it

M. Clementi
Dipartimento di Architettura e Studi Urbani, Politecnico di Milano, 20133 Milano, Italy
e-mail: matteo.clementi@polimi.it

and sustainable economic growth, full and productive employment and decent work for all · 9. Build resilient infrastructure, promote inclusive and sustainable industrialization and foster innovation · 12. Ensure sustainable consumption and production patterns · 13. Take urgent action to combat climate change and its impacts

1 Industry 4.0 for the Bioregional Development

Industry 4.0 has been focusing on the use of integrated technologies for improved efficiency and flexibility in the production. Its role in Europe recently emerged and evolved in Industry 5.0 on the social, societal and environmental concept of:

(1) Human-Centric Approach: involving human need in the production process adopting technology usage and the production process to the workers' needs;
(2) Sustainability: technology should be used to optimize resource efficiency, minimize waste, reduce energy consumption and greenhouse emissions and use local renewable sources and develop circular processes.
(3) Resiliency: industrial production should be adaptable in the times of crises, such as geopolitical shifts and natural calamities with resilient strategic value chains, adaptable production and flexible processes [1].

Industry 4.0 and the democratization of machines open up new possibilities and support the construction industry to implement Circular Economy standards and achievement of the S.D.G.s, promoting resource efficiency from project inception to end-of-life. By incorporating circular design requirements and innovation into the technologies, such as design for disassembly, deconstruction, and recycling, properly manage the waste stream, and close the materials loop and by increasing the efficiency in material use and reducing the use of non-renewable natural resources, promoting the use of renewable resources. The terms circular and renewable find an effective synthesis in bioregional development as man-made systems in which energy and matter demand and supply tend to meet on the same territory, increasing the efficiency of use of resources, using mainly locally available renewable ones. The same model finds support in the theories relating to a generative economy [2].

The need for sustainable communities (UN-SDG11) sees the model of bioregional development as the most eco efficient one, since it most closely reflects the characteristic dynamics of evolving living systems. That is, in accordance with the maximum empower principle, towards the maximum use efficiency of solar energy incidents on the local territory [3]. It constitutes a promising reference model that combines environmental sustainability and circular economic development [4–6].

The construction industry is facing the opportunity to benefit from the open availability of digital data and online digital access, to really interconnect physical and digital information. This evolution transforms the business models, enabling transitioning the current reactive to a predictive practice and reducing waste.

The link between Industry 4.0 and construction was primarily based on the potential of digitization and identified four key areas: digital data, automation, connectivity and digital access [7]. They are represented by a confluence of three main trends: industrial production and construction, cyber-physical systems and digital technologies. The increasing importance of concepts such as sustainability, human centric, resiliency and therefore bioregional development imposes tighter constraints based on the intensification and optimization of relations between the built environment and the territory. Open availability of digital data and online digital access can provide the communities inhabiting the territories with tools capable of creating new economies based on the optimal use of territorial metabolism starting from a theoretical basis already consolidated over time.

2 Bioregional System, Built Environment and Upcycling

Industry 4.0 is particularly interesting in its relationship with bioregions, defined through physical and environmental features, including watershed boundaries and soil and terrain characteristics and emphasizes local populations, knowledge, and solutions. Bioregion's environmental components (geography, climate, plant life, animal life, etc.) directly influence ways for human communities to act and interact with each other which are, in turn, optimal for those communities to thrive in their environment. As such, those ways to thrive in their totality—be they economic, cultural, spiritual, or political—will be distinctive in some capacity as being a product of their bioregional environment [8].

In a path towards local self-sustainability, the construction of informed and connected flows of energy and material, through technologies, defines the possibility of a low-tech sector at the service of its territory, perfectly integrated with agricultural resources (through sensors, constant evaluation of performance and robotization of production phases), to disused urban stocks and urban mining, and variables of local energy demand that can be defined through real-time evaluations of energy and sun light flows, as climatic conditions [4].

The Industry 4.0 model insists on system innovation related to robotization as an indispensable instrumental method for renewal of production processes. The integration of machine-to-machine communication and the internet of things (IoT) help to increase automation and improve communication and self-monitoring. Industry 4.0, when combined with the circular economy, represents an industrial paradigm enabling new strategies of natural resource use [7, 9]. In turn, the circular economy is driving more attention in support of innovative life cycle management, communication, and the development of smarter systems reconceptualizing waste as intrinsically valuable. This industrial revolution relies on digital technologies such as wireless connectivity and sensors that connect everything to everyone to gather data to analyze, visualize the entire production system, and provide applicable information to the business system [10].

Built environment is part of city metabolism as a palimpsest of urban stock, a record of human history partly erased and rewritten. The urban structure grows during time, through a succession of reactions and processes that develop from previous stages, a territorial morphology expresses a continuity of development, and a vital and unifying exchange with the environment. Thus, time appears as the great 'master builder' [11] to which human settlements owe their most durable, fittest configuration within a given environment: time becomes a key factor in this evolving system of resources, describing a variable configuration model, related on season, resilient capacity and region based. In Olgyay's work, the term "region" relates to the solar altitude angle, and the regional characters of buildings are defined in terms of heat conservation, thermal demand, orientation, natural lighting and ventilation. Olgyay explained how basic building forms, in traditional architecture, answer to the needs of comfort and protection against the local environmental difficulties; yet there is no prosaic determinism in this approach.

Upcycle principles encourage designers—and anyone interested in this new approach –to see everything on Earth as having potential to be something else. Realize that there is not a garbage can, forget the concept of "trash" and consider that everything has to go somewhere and become useful [12]. The challenges we face are multifarious—population pressure, climate change, energy security, environmental degradation, water and food availability—to name a few. To address them we will have to change our business and production processes, become significantly more energy and resource efficient, and minimize waste. Advanced technologies will accelerate this process of transformation and will play a pivotal role in addressing the immediate need of resource conservation and climate change concerns. Technology supports in utilizing renewable energy resources, developing smart grid systems, making efficient public infrastructure producing environmentally friendly materials and products will be crucial to achieve the overall goal of sustainability.

3 A Method from Big Data to Territorial Metabolism GeoData

3.1 The Informed Open Source Material/Energy Databases

The text presents a method and specific tools aimed at creating open source databases to support the development of integrated material/energy systems for the production of building material and energy from locally available resources. The proposed methodology was developed starting from the open data available on Italian territory, with the intention of allowing local communities to implement this information with self-detected data. The objective is enabling the local community to trigger processes aimed at bioregional development, activating local economies and transferring existing good practices, through an information platform. This objective is achieving through:

1. The preparation of a georeferenced information system starting from open data currently available to all minor municipalities throughout the country.
2. The implementation of geodatabase with data made available to the local community through self-detection activities using low-cost sensors.

The final result would constitute a Territorial Metabolism Geodatabase (TMG) to support the development and application of design strategies. In order to meet the emerging role of Industry 4.0 on Human-Centric Approach, Sustainability and Resiliency, the method aims to map specific information on the same IT medium.

With regard to the Human Centric Approach, the methodology arises from the awareness that in order to trigger new circular and local economies in small settlements, it is necessary to intercept the local expenditure flows starting from the expenditure flows manageable by the public initiative. The provision of this information allows to identify and quantify the money leaving an area and therefore formulate a solution for a sustainable local development processes. Housing, food and transport make up the main spending streams of the settled population [5] (Fig. 1).

The pursuit of environmental sustainability, together with the need to activate circular economies at the local scale, imply matching local demand for energy and material and available resources, in compliance with the regenerative cycles of ecosystems. For example in the sustainable retrofit of existing buildings [13, 14], in order to redesign integrated supply chains of energy and building materials that intercept the largest share of expenditure flows, it is essential to supplement the data relating to the local demand for energy and materials, with the outflows from the system and the renewable resources available and not used (Fig. 2).

The increase in Resiliency is instead closely linked to the possibilities offered by the method in managing the temporal variability of the data. In fact, the information associated with the demand for energy and matter allows us to understand its variability over time and that the same happens for the mapping of flows leaving the system in the form of waste and other renewable resources available locally. For example, this type of data availability allows the easy passage from one supply

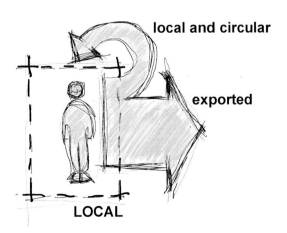

Fig. 1 Human centric approach for circular economies associated with spending flows

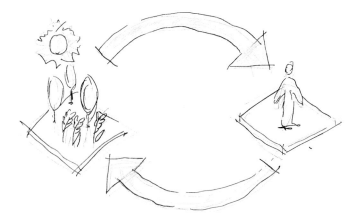

Fig. 2 Sustainability for circular economies

Fig. 3 Resiliency and temporal variability for circular economies

source to another simultaneously available, increasing the resilience level of the supply system. In order to have useful data available to understand the variability of the territorial metabolism in different periods of the year and over several years, it is essential to define appropriate time intervals to describe the complexity of the territorial system and at the same time predict the possible effectiveness of improvement practices. The time intervals considered refer to the monthly, annual and five-year average. The monthly average figure is a useful compromise in order to make such complexity manageable (Fig. 3).

3.2 The Local Territorial Metabolism Data Analytics

The elaboration of thematic maps and the publication on a shared portal make it possible to relate the estimated data to each other [15], and to integrate such assessment with the data actually collected. By integrating new methods of data collection and analysis, stakeholders would be able to take evidence-based decisions on resources allocation and product use and query their processes to meet the local energy necessity. Open source tools, low sensors and microcomputers give the possibility to map the punctual availability of underutilized resources, such as the amount of solar energy on the ground and on buildings envelope, of biomass or other resources such as hydro and wind power (including in the same dataset agricultural wastes, municipal solid waste, and waste heat in the sewage system). In summary, the scope of Big Data Analytics would turn into Local Territorial Metabolism Data Analytics [4, 9] and is closely linked to the local community's ability to produce and self-manage open information oriented towards the generative use of local resources.

The proposed methodology and application tools are divided into the following phases (Fig. 4):

1. LDEM—Mapping the local demand for energy and matter, and therefore the availability of expenditure of the settled community.
2. LR—Mapping local renewable energy potential.
3. EF—(Exported flows) Mapping material flows out of buildings/open spaces.
4. AW—Mapping the potential workforce and skills available locally.
5. DB—Sharing a database of local closing cycles practices.
6. Defining parameters to verify the transferability of best practices; starting from the information mapped and collected in a single open source Geographic Information System.

3.3 Open Tools and Open Data

The current development of the open source software allows operations of big complexity and it gives the possibility of using complex data even to actors who cannot purchase proprietary software such as local administrations, especially minor ones, and designers who are not directly involved in urban planning. These tools represent an important opportunity to process and communicate information to support decisions aimed at designers, planners, and local communities. The open source feature of these tools meant possible to merged some tools into specific platforms and the creation of specific institutions and related websites that report news relating to the current level of development of such tools, such as the Open Geospatial Foundation. Among the open source GIS tools made available, the elaborations of this work used Quantum GIS (QGIS) and GRASS-GIS.

The method starts from the information normally made available to citizens to create support tools for decisions that can be managed from below. Therefore, to ensure the replicability of the procedure to a wide audience the research relies exclusively on open data provided by government and institutions. The open-access data used for the development of the present case study are described in Table 1. In order to be able to create a common geo-referenced database and correlate the information, the data on the geometry of the buildings available or other data relating to smaller areas, have been aggregated by the census block.

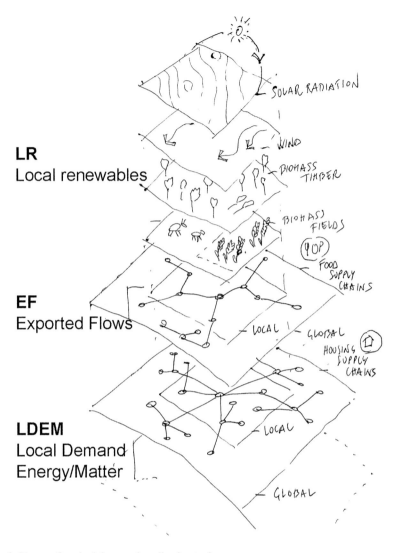

Fig. 4 Phases of methodology and application tools

4 The Case Study

4.1 Local Territory

To test the effectiveness of the methodology, an internal area of the Abruzzo region representative of the Italian minor municipalities was chosen as a case study: San Valentino in Abruzzo Citeriore (Pe). The smaller municipalities of inland areas are

Table 1 Source data table

Data	Data type	Source
Populations, households, industry, services	These datasets include information on the height of the buildings, the number of dwellings, the construction system and period together with population age group and number of inhabitants	Italian National Institute of Statistics [16]
Regional data consumption	Data on household consumption in the Abruzzo region	CRESA [17]
Urban waste flows	Data from the regional urban waste observatory	Regional observatory [18]
Energy	Preliminary assessment of possible energy consumption, divided by different consumption categories in the municipality of San Valentino	SEAP San Valentino [19]
Land use	Productive land available, using the Copernicus georeferenced database which publishes the different land use destinations and the related geometric configuration for the entire European territory	CLC [20]

Table 2 Comparison with average national data

	Total productive land (sqm/inhab)	Agricultural (sqm/inhab)	Woodland (sqm/inhab)	Pasture (sqm/inhab)
Italy	4.362	2.540	1.519	303
Abruzzo	7.682	3.675	2.322	1.685

characterized by a large availability of resources per capita and allow for experimental alternative models of local development. The availability of natural capital per capita is in fact higher than the Italian average. The data in the Table 2 constitute comforting information on the vocation to sustainability of the minor municipalities of the Abruzzo area and make it a candidate for possible experiments oriented towards bioregional development (Fig. 5).

4.2 LDEM Mapping of the Main Spending Flows of the Local Community

The results associated with the progress of the geodatabase are developed in the municipal area of San Valentino with the mapping of the main expenditure flows. With 1922 inhabitants, the energy needs in 2008 was estimated at 31,053.21 MWh/y and probably a small negligible part comes from biomass (in particular wood). Otherwise the municipality depends on petroleum products, natural gas and electricity. Petroleum with 18,735.38 MWh/y is the most substantial part (60%) of the total

Fig. 5 Profiles of the municipality of San Valentino and neighboring ones (in orange) superimposed on the profiles of the census blocks of the municipality (in light blue)

energy of which diesel—among the by-products—reaches the primacy of the overall supply with 12,458.59 MWh/y (40%) followed by natural gas with 8,504.65 MWh/y (28%), gasoline with 6,267.64 (20%) and electricity with 3,715.54 MWh/a (12%). The LPG does not represent an important part at the percentage level. The data below can be integrated with information relating to food costs, which together with housing and transport costs represent the main consumption categories of a community (Table 3).

By associating the unit costs of fuels with the main energy consumptions reported in the SEAP [19], it was possible to quantify the costs related to household utilities and transport and compare them with food costs. In particular, the graph in Fig. 6 compares the consumption categories that are most suitable for an experimental use of local resources and for the activation of possible synergies for housing. The interceptable dynamics that require the least investments are related to the supply chain of cereal products, vegetables and legumes and the supply of fuel for winter heating. However, this assumption is not sufficient and it is necessary on the one hand to map information useful for verifying the effectiveness of these strategies, and on the other hand to understand the local weight of public initiative in promoting strategies oriented towards integrated supply chain planning.

Table 3 Energy use by sectors in San Valentino

Sector	Transport	Residential	Tertiary	Agriculture	Public
Energy use	54%	29%	5%	4%	3%

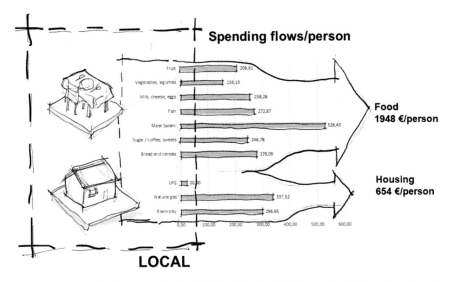

Fig. 6 Main spending flows associated with a San Valentino inhabitant and relating to the categories of consumption, food and home

Once the data on the resident population and on the physical configuration of the urban space have been collected, the geolocation of the information associated with the expenses of the local community can be carried out in two specific ways. A first preliminary assessment associates the data on household expenses with the individual census blocks information, as the number of inhabitants. A second method associated some items of consumption refined with a specific assessment of the energy consumption of buildings, with the different building shapes and technological features. At the current state of research, the assessment of the expenditure flows of the settled community was conducted using the first method. Among the energy costs associated with housing, the main one is related to winter heating followed by electricity consumption: it gives indications on the possible willingness to pay for alternative solutions, or the willingness to face an investment beyond the public incentive systems.

Future developments intend to refine the current development status of the TMG with specific data used for further processing. Data mapped by census block and based on estimates of needs and statistical regional data have to be implemented with the direct contribution of the inhabitants with a participating monitoring. In particular, the data made available to the SEAP [17] are associable to the consumption in a specific year (2008) to each polygon relating to publicly owned buildings. The continuous updating of these spending flows in the TMG itself would be extremely useful.

4.3 LR: Mapping Unused Renewable Energy and Material Potential

According to the various land use destinations that characterize the local area of reference, this phase of the methodology aims to assess and map the type and amount of renewable energy available. Whereas it is not a very densely inhabited area, the research project maps the solar renewable energy potential together with the possible production of energy and materials from renewable sources obtainable from types of territory such as arable land, pasture and wood. In San Valentino the municipal area amounts to 16.39 km^2, the availability of land per person is 8531.56 m^2. The productive area per person is equal to 8.332 m^2 (of which 7.129 for agricultural use and 1.203 forest) [20] (Fig. 7).

The availability of productive land present in the municipal boundaries divided by the number of inhabitants. The potentially usable area for the installation of solar collection devices is a percentage of the built up area, and gives indications on the amount of land made biologically unproductive usable for the collection and use of solar energy. The topographic database (RTDB) allows to enrich the TMG with information on the geometry of buildings. It associates the eaves height to the building's polygons. That information, together with the street level contour lines, allows for generating high-resolution digital elevation models (DEM-1 pixel/50 cm) useful for mapping solar radiation incidents on open spaces and roofs, thanks to tools such as the one proposed by Hofierka [1]. If accompanied by isopleths, these maps can be consulted using common web services such as Google Earth (Fig. 8). The quantities relating to solar radiation, thus included in the TMG, would indicate the precise conditions of solar radiation representative of the monthly average associated with diffuse and direct solar radiation, and to distinguish the portion that affects the roofs from the one related to the open spaces (Figs. 9 and 10). These themes, superimposed on the mapping of publicly owned buildings, are able to identify those census blocks in which the availability of incident solar radiation on roofs or open spaces of public management is greater [21].

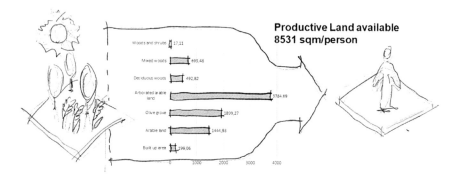

Fig. 7 Surface of production area available per person in the municipal area of San Valentino

Industry 4.0 and Bioregional Development. Opportunities ... 281

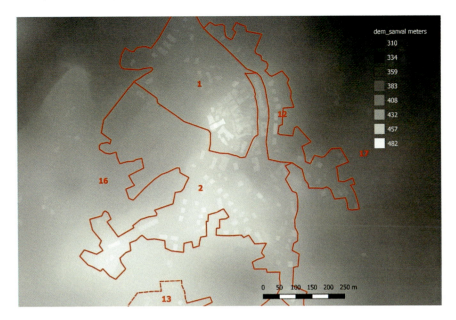

Fig. 8 Digital elevation model of a portion of the urban area of San Valentino (resolution 1 pixel side: 0.5 m) with in red the profiles of the census blocks

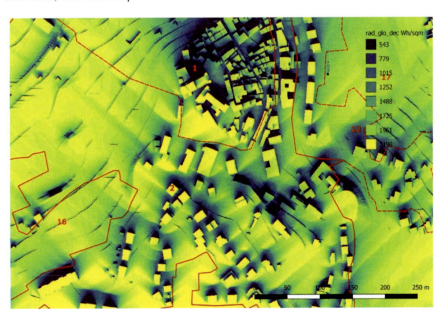

Fig. 9 Map of direct solar radiation representative of clear sky day of December (unit of measurement Wh/sqm * day), in the red profile of the census blocks

Fig. 10 Map of direct solar radiation representative of clear sky day of June (unit of measurement Wh/sqm * day), in the red profile of the census blocks

4.4 EF—(Exported Flows) Mapping Material Flows Out of Buildings/Open Spaces

The mapping of outgoing material flows as solid urban waste was carried out by associating the per capita yearly flows and associate outgoing household waste material flows for the area. By multiplying the quantities of waste generated with the number of residents, the total amount of MSW generated within the area can be obtained.

The current level of resolution of the TMG allows to visualize:

- information on the geometry of the surfaces occupied by different land use destinations;
- maps of the relative solar irradiation representative of the monthly average that take into account the geometry of the buildings;
- the flows of waste emissions by census block.

The use of low-cost sensors and facilitated forms of database self-compilation would make it possible to refine the database by inserting information on future crop planning in advance. In the case of solar irradiation maps, sensors positioned on some roofs allow to compare the data relating to the monthly statistical average with the real irradiation data. Finally, as regards the emission flows of TMG waste, it could include the flows actually conferred by individual users, through the weighing of the tubs (Figs. 11 and 12).

Industry 4.0 and Bioregional Development. Opportunities ...

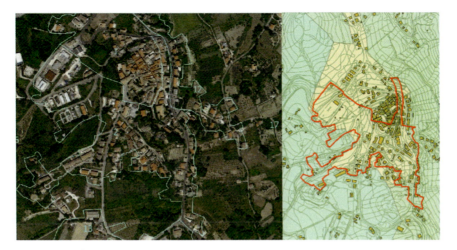

Fig. 11 Geometric configuration of some of the census blocks of the more urbanized area of San Valentino; CB2, in the center, records the presence of publicly owned high-level solar roofs

Fig. 12 Geometric configuration of some of the census blocks of the San Valentino area most suited to plant production. The west boundary of block 16 is covered by publicly owned woodland

4.4.1 AW: Mapping the Potential Workforce and Skills Available Locally

Among the unused local resources, person-hours are a fundamental component. The real possibility of triggering local prosuming processes depends on local availability of human resources. Useful information on the available person-hours are obtainable by crossing age-groups with employment status within the census data (Fig. 13).

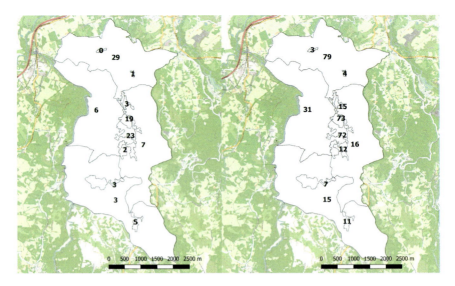

Fig. 13 The map on the left illustrates the number of unemployed by census block, the map on the right illustrates the number of unemployed and housewives by census block

For the present case-study the following categories were selected:

- Total resident population more than 15, unemployed seeking new employment
- Total resident population aged 15 and over, non-labor force
- Total resident population aged 15 and over, house husbands and housewives
- Total resident population aged 15 and over students.

Due to the restrictions associated with the current pandemic emergency, the data relating to the presence of unemployed are changing considerably. It is therefore of fundamental importance to integrate the data already mapped with what will be detected by the current census activity at the end of 2021 [22].

5 Closing Local Loops: Abacus of Best Practices

5.1 Defining Scenarios

The data stored in the TMG constitute an information base aimed at becoming a collaborative platform, updated in real time through sensors, and associated with further quantitative community data. The update is in real time through algorithms shared with the community that intervenes on some categories of data:

Table 4 Good reference practices

Group	Product	Uses	Level of tech	Case study
Energy				
Timber	Pellet or wood chips	Heat, electricity generation	Medium	Trentino forest management
Photovoltaic	Electric energy	Public building or private facilities	Low	Energetic communities
Local energy	Biogas	Heat or electric generation	Low	Local farm
Material				
Cellulose	Recycled cellulose	Insulation, cardboard, construction	Medium to hight	Applied CleanTech
Hemp	Shives (woody bark)	Building materials (blocks, insulation panel)	Low to medium	Edilcanapa in Abruzzo
Data	Public dataset	Collaborative platform	Theoretical	–

- through sensors associated with cultivation, buildings, production activities
- through manual data implementation methods.

Starting from existing good practices, different scenarios are simulated aimed at bringing together local expenditure flows with the potential local production. The identified expenditure flows are related to the resources available locally and respond to local needs by making materials and energy available, focusing on the seasonality of local resources. The table shows some good reference practices which refer to the possible circularity of resources and highlight connections that intercept the potential for local triggering of entrepreneurship or the availability of public goods capable of hosting production at km.0 (Table 4).

5.2 Example Scenarios

This paragraph briefly presents two examples that illustrate the potential use of the data stored in the TMG to verify the transferability of some good practices by developing simplified scenarios. In particular, scenario 1 works on the possible transfer of the good practices comparing them with strategies aimed at local food production (Fig. 14):

- BP1—use of woody biomass from forest maintenance to cover the thermal needs for winter heating of buildings;
- BP2—use of photovoltaic roof systems, connected to the conventional grid, to produce electricity from solar sources.

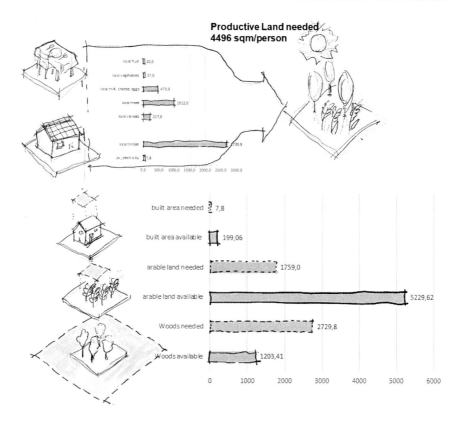

Fig. 14 Productive land necessary to locally cover energy and material needs and data relating to the production area available per person and the area necessary to cover energy and material needs (Scenario 1)

From the comparison between the TMG data and relating to the extension of production area needed and actually available, it emerges that the energy needs relating to winter consumption for heating homes exceeds the actual local availability of woody biomass (considering with the local regenerative cycles). The activation of possible local micro-economies aimed at intercepting the flows of expenditure associated with the energy needs for heating homes will therefore have to develop integrated supply chain projects. They will be oriented on the one hand to cover local energy needs with locally available resources, on the other hand to activate useful strategies to increase the energy efficiency of the homes. The local production of insulating material such as hemp, thanks to the availability of arable land has high potential for success.

Scenario 2 provides for the possibility of growing hemp to be used as an insulating material, in particular the graph in the Fig. 15 shows how a reduction in the thermal transmittance of the building envelope equal to 0.17 W/sqm*K, would correspond to a reduction in consumption energy equal to 2/3 of the initial value. The implementation of activities aimed at the integrated development of the energy supply chain and the insulating material would involve a remodeling of the surfaces of the necessary production area. Optimistically hypothesizing to produce the insulating material in an annual cultivation cycle, the production land balance would see an increase in the arable land component equal to 800 sqm/person, to the advantage of a reduction in the extension of woods useful for producing energy of about 1700 m^2 (Fig. 15).

The comparison between available productive land and necessary land in Scenario 2 shows that the needed surfaces do not exceed the amount of available areas (Table 5).

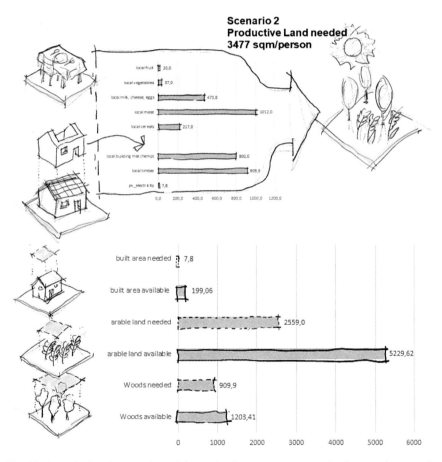

Fig. 15 Scenario 2: Above, surface of the production area necessary using hemp to improve the energy performance of the existing buildings. Below: data relating to the production area available per person and the area necessary to cover the energy and material

Table 5 Available productive land and necessary land in Scenario 2

Practice	Energy/Material	Unit	Productive land	Area (m^2)	Source
BP1 woody biomass	Thermal energy	sqm/kWh*y	Woods	0.763	[1]
BP2 rooftop pv	Electricity	sqm/kWh*y	Rooftop area	0.00714	[16]
BP3 hemp mat	Insulating material	sqm/kg	Arable land	1.33	[18]

6 Conclusions

New emergent technologies and methodologies generate new enabling possibilities for local self-sustainability. Technologies 4.0 is not only an evolution of the sector, but also pleads for its radical transformation based on three concepts: re-industrialization, digitization and sustainability. Three concepts directly bomb a sector that is recognized as a non-industrialized sector (only the production and manufacturing of its materials and products can be considered as factory-made, so under the usual lean management philosophy), and then, industrialized construction is strictly part of the new paradigm [10]. The simplified assessments proposed in the scenarios give a rough idea of how to use the data mapped in the TMG for an informed platform, and exchange information with sensors for giving time response energy and material solution. The coexistence in the same georeferenced database of information on land features and on the population allows to relate data to the expenditure flows of a community with data that can be associated with the production capacities of a territory in compliance with the regenerative cycles of local ecosystems. The georeferencing of this information allows the data to be expressed in general terms or per capita terms, providing clear information on the possible strategies that can be pursued. The TMG itself gives the possibility to include information on the metabolism associated with public properties, i.e. both data related to the demand for energy and matter and information associated with the potential renewable supply. This data availability is fundamental for planning triggering activities of local microeconomies, it will in fact be in the census blocks where public ownership is most present that the start of the first experimental activities will be expected.

In this local systemic design the governance of local production with computerized systems, digital manufacturing, collaborative platforms, self-assembly, robotization of production, Communities 4.0, can provide answers to the economic circularity of the construction sector oriented towards the systemic sustainability of the territories, focusing on humans and social sustainability and enabling local communities in data management to increase their production capacities.

Acknowledgements The contribution has been tackled jointly by the authors and the introduction and the conclusion are to both authors. The 2, 3, 4.1, 4.2 parts are attributed to L. Mastrolonardo, the 4.3, 4.4, 4.5, 5 are attributed to M. Clementi.

References

1. Breque, M., De Nul, L., Petridis, A.: Industry 5.0: Towards a Sustainable, Human-Centric and Resilient European Industry, Publications Office (2021). https://data.europa.eu/doi/10.2777/308407
2. Kelly, M.: Owning Our Future. San Francisco, Berrett-Koehler Publishers (2012)
3. Odum, H.T.: Environmental Accounting. Wiley and Sons, New York, US (1996)
4. Baccini, P., Brunner, P.H.: Metabolism of the Anthroposphere, 2nd edn. MIT Press, Cambridge (2012)
5. Ferrao, P., Fernandez, J.E.: Sustainable Urban Metabolism. MIT Press, Cambridge, US (2013)
6. Odum, H.T., Odum, E.C.: A Prosperous Way Down, Principles and Policies (2001)
7. Battisti, A.: Il progetto come volontà e rappresentazione: dai big data all'apprendimento collettivo. In: Perriccioli, M., Rigillo, M., Russo Ermolli, S. Tucci, F. (eds.) Design in the Digital Age. Technology Nature Culture | Il Progetto nell'Era Digitale. Tecnologia NaturaCultura. Maggioli Editore (2020)
8. Magnaghi, A.: La bioregione urbana nell'approccio territorialista. Contesti. Città, Territori, Progetti (1), 26–51 (2019). https://doi.org/10.13128/contest-10629
9. Loiseau, E., Aissani, L., Le Féon, S., Laurent, F., Cerceau, J., Sala, S., Roux, P.: Territorial Life Cycle Assessment (LCA): what exactly is it about. J. Clean. Prod. **176**, 474–485 (2013). ISSN 0959-6526
10. Bolpagni M. et al. (eds.): Industry 4.0 for the Built Environment, Structural Integrity 20 (2021). https://doi.org/10.1007/978-3-030-82430-3_1
11. Fontana C., Xie S. TECHNE Special Series Vol. 2 (2021)
12. Mastrolonardo L.: Progettazione ambientale a chilometro zero, Maggioli, Sant'Arcangelo di Romagna (2018)
13. Sassanelli, C., Urbinati, A., Rosa, P., et al.: Addressing circular economy through design for X approaches: a systematic literature review, Comput. Ind. **120** (2020)
14. Zampori, L., Dotelli, G., Vernelli, V.: Life cycle assessment of hemp cultivation and use of hemp-based thermal insulator materials in buildings. Environ. Sci. Technol. **47** (2013). ACS Publications
15. Clementi, M.: Progettare l'Autosostenibilità Locale, Edizioni Ambiente (2019)
16. ISTAT: Italian National Institute of Statistics. Basi territoriali e variabili censuarie, censimento della popolazione e delle abitazioni 2011 (2012). www.istat.it. Accessed March 2022
17. CRESA: Centro Regionale di Studi e Ricerche Economico Sociali La spesa per consumi delle famiglie abruzzesi. Edizione (2016)
18. Osservatorio regionale rifiuti Regione Abruzzo 2022 https://www.regione.abruzzo.it/content/osservatorio-regionale-rifiuti. Accessed May 2022
19. Comune di San Valentino in Abruzzo Citeriore (2010): Piano D'Azione per l'Energia Sostenibile, (SEAP) Sustainable Energy Action Plan, The Covenant of Mayors Program
20. CLC Corine Land Cover Database (2012). https://land.copernicus.eu
21. JRC PVGIS Photovoltaic Geographical Information System. https://re.jrc.ec.europa.eu/pvg_tools/en/. Accessed May 2022
22. Regione Abruzzo: Regional Territorial Database, scale 1:5000 (2022) http://opendata.regione.abruzzo.it/catalog//term%3D19. Accessed March 2022

Towards Construction 4.0: Computational Circular Design and Additive Manufacturing for Architecture Through Robotic Fabrication with Sustainable Materials and Open-Source Tools

Philipp Eversmann and Andrea Rossi

Abstract There is a constant increase in demand for new construction worldwide, which is one of the main contributors of worldwide CO_2 emissions. Over the last decades, such increase led to scarcity of raw materials. Although design methods have been developed to increase material efficiency, this has not yet led to a widespread reduction in material consumption. This is due to a variety of factors, mainly related to the inability of conventional fabrication methods to produce the complex shapes that result from such computational methods. Industrial robots, while offering the potential to produce such optimised shapes, often rely on inflexible interfaces and highly complex industry standards and hardware components. In response to this dual sustainability and technology challenge, this article describes a series of research projects for the design and manufacture of architectural components using renewable materials and robotics. These projects are based on novel additive robotic building processes specifically designed for renewable and bio-based building materials, ranging in scale from solid wood elements to continuous wood fibres. We propose methods to optimise the distribution of such materials at their respective scales, as well as manufacturing methods for their production. In this context, the use of novel and automatable joining methods based on form-fit joints, biological welding and bio-based binders paves the way for a sustainable and circular architectural approach. Our research aims to develop intuitive open-source software and hardware approaches for computational design and robotic fabrication, in order to expand the scope of such technologies to a wider audience of designers, construction companies and other stakeholders in architectural design and fabrication.

P. Eversmann (✉) · A. Rossi
Department of Experimental and Digital Design and Construction, University of Kassel, Kassel, Germany
e-mail: eversmann@asl.uni-kassel.de

Keywords Additive manufacturing · Sustainable materials · Circular architecture · Open-source software · Material efficiency · Flexible prototyping · Robotic fabrication · Computational design

United Nations' Sustainable Development Goals 11. Make cities and human settlements inclusive, safe, resilient and sustainable · 12. Ensure sustainable consumption and production patterns · 13. Take urgent action to combat climate change and its impacts

1 Introduction

1.1 Construction Technologies and Sustainability

There is a constant and growing increase in demand for new construction worldwide, which is one of the fundamental drivers of increase of atmospheric CO_2, and hence one of the leading causes of climate change [1, 2]. Over the last decades, such increase and logistic limitations led to scarcity of building materials [3]. This is also partly because massive construction techniques remain the most popular choice of construction [4, 5]. Instead, circular construction techniques, material-saving design methods, and manufacturing processes based on renewable, carbon-storing resources [6] are required.

1.2 Renewable Materials

Since afforestation is not infinitely scalable, additional fast-growing renewable resources are needed. Novel construction techniques with bamboo can be mentioned [7, 8], but unfortunately the material mainly grows well in Asia and south America, leading to a large impact of transportation in a Life-Cycle Assessment for other countries [9]. An alternative, which can be grown in Europe, are willow rods, which can be processed into filaments for additive manufacturing and textile architecture [10]. Natural fibres have good structural properties but are energy intensive to produce, resulting in a positive carbon footprint [11]. The fibre/resin ratio of 30–40% [12] is significantly lower than the material/adhesive ratio in engineered timber products (85–98%) [13]. Recently, construction techniques with mycelium, the root network of fungi, have been developed, i.e. as insulating material, plaster base [14], and mycelium bricks [15, 16]. Even intelligent behavior of mycelium in architecture and buildings is investigated [17, 18]. First companies are developing mycelium-based products, as acoustic elements, and floor panels [21].

1.3 Towards Construction 4.0

Definitions. The term "Industry 4.0" was coined in 2011 as part of a high-tech strategy project of the German government that promoted the computerisation of manufacturing [22]. In this perspective, information and communication technologies such as the Industrial Internet of Things (IIoT) will enable automated, responsive manufacturing in high volumes and with a high diversity of variants. Although there is no single definition of "Construction 4.0" (C4.0), it seems to revolve, similar as Industry 4.0, around a decentralised connection between physical and cyberspace through ubiquitous connectivity. However, C4.0 is not limited to these technologies but draws on a broader spectrum, the most important being Building Information Modeling (BIM), Additive Manufacturing (AM) and Cyber-Physical Production Systems (CPPS) [23]. The great promise of C4.0 lies in the almost complete automation of the entire project life cycle, not only through highly efficient design, fabrication and material use, but also enabling circular construction through digitally traceable reusability of components, material flows and recycling processes [24].

Current state. Digitization in planning has already advanced compared to construction, due to its complexity with many sub-systems, materials and details. Automation of existing processes is often still very difficult and inefficient, being originally designed to be carried out manually. Although little has yet reached construction sites, significant progress has been made in researching novel robotic construction techniques [25, 26]. Due to a strong acceptance and long history of digitization, timber construction is nowadays the most automated building sector [27], with a seamless integration of existing processes as well as new technologies [28, 29]. Automation in concrete construction is generally much lower. Examples are precast building parts and semi-finished components, such as partially precast slabs [30]. The most substantial digital advancements can be found in entirely new construction processes, as 3D printing through extrusion. Besides a series of attempts in the last decades [31, 32], automation in masonry construction is still hardly used, with the exception of prefabricated walls parts [33]. In steel construction, formatting is mostly automated, while assembly and welding are still performed mostly manually. Research is being done on form-fit steel connections [34] and additive manufacturing with steel, and automated assembly and welding for large structures [35, 36].

Key technologies: BIM and Computational Design. Building Information Modeling (BIM) is a concept of associating all kinds of data to 3D Models of buildings [37]. Within C4.0, the major challenges are the successful integration and management of data over the whole conception, planning and life cycle, including the tracing of materials and components in accessible databases to enable digital circular design flows, the integration and merging of Computational Design (CD) with BIM, and the integration of design to fabrication workflows. The term CD is commonly used for design methods associated with scripting, parametric modeling and all kinds of digital form finding methods [38]. Towards a "Computational BIM", two approaches are currently pursued: one is the integration of parametric modeling and simulation in one large software package [39], the other one is creating software bridges, allowing to freely combine multiple smaller packages for their respective strengths [40].

Key technologies: Additive Manufacturing. Subtractive machining processes describe an approximation to the final contours of a workpiece by material removal. Formative manufacturing processes are a volume-constant change in the workpiece geometry [41]. Additive manufacturing processes, on the other hand, describe the successive construction of a workpiece by adding individual elements or layers [42]. In the construction industry, additive manufacturing techniques are currently most advanced in concrete construction with already available solutions on the market [43–46]. In steel construction, first prototypical projects exist, showing the potential for larger-scale applications [47, 48]. For timber construction, research is done on extruding timber particles with different binders [49, 51, 52]. For small-scale applications for design and prototyping polymer-based filaments exist and applications for replicating the textures of wood are available [53, 54]. In the literature, the term "additive manufacturing" or "additive construction" has been broadened to include also assembly processes as in robotic timber construction, referring to the additive nature of the process in contrast to subtractive processes as CNC milling [55–57].

Key technologies: Cyber Physical Production Systems (CPPS). CPPS link real (physical) objects and processes with information-processing (virtual) objects and processes through ubiquitous information networks [58], such as production systems with sensors and feed-back processes, capable to adapt to collected data or provide human interactivity. An example of this is an assembly system we developed for irregular wooden shingles for façades, where geometry scans are sent to a connected computer, which calculates a desired position and sends the data back to the robot for execution [59]. Unlike repetitive processes of past industrialization [60], C4.0 requires highly adaptable workflows. This requires highly accelerated time intervals in which automation systems must be adapted and changed. CPPS promise to achieve these formerly extremely difficult tasks through new sensor technology, kinematic simulation, coding technology, a wide range of industrial components and the ability to create industrial networks with customisable functionality. In the following sections, a distinction is made between the kinetic simulation needed for the conception of CPPS, and the additional control mechanisms of the various components for robotic tools and processes.

Kinetic Simulation. In CPPS, automation needs to be flexible to accommodate for highly variable components. Therefore, kinematic simulation techniques are necessary to plan the robot motions, check for reachability and prevent collisions. Most manufacturers provide their own kinematic simulation software, as ABB RobotStudio [61], KUKA.sim [62], Fanuc Robotguide [63] and Universal Robots URSim [64]. Also, manufacturer independent commercial solutions exist, as Visual Components [65], Gazebo [66], CoppeliaSim [67], HAL Robotics [68] and RoboDK [69]. There are also a range of software solutions developed by research institutes, often based on Rhinos Grasshopper interface, as Robot Components by University of Kassel [70], Compas FAB by ETH Zurich [71], KukaPrc by RobArch [72] and robots by the Bartlett [73]. Research for the development of new robot systems is mainly done using the Robot Operating System (ROS) for programming, path planning and direct robot control [74]. Here, it is also possible to use automated path planning algorithm packages such as Moveit [75].

End-effector design and control. C4.0 requires fast adaption of processes and new robotic tool development. Current industrial automation engineering approaches are very time-consuming and expensive, and few plug-and-play products exist. Therefore, fast, adaptable, configurable prototype-able and cost-effective approaches are needed. This concerns both the communication with the different end-effector components as sensors, moving parts, etc., and the fabrication of the tool. Depending on the task, these components can be differently combined, as i.e., an intelligent industrial camera to detect an object position, and a pneumatic gripper to pick it and place it to its final position. Most components can be controlled by a simple on/off mechanism through digital IOs (Input/Output) [76], as pneumatic valves, simple sensors, or actuation of an electric device. For more complex control, as i.e., controlling the speed, timing etc. of a stepper motor, open-source electronics platforms such as Arduino can be used [77]. These allow for preprogrammed behaviors, which can be further integrated with signals from the robot control system. Also, analog IOs [78], and older protocols, such as serial messaging, can be used for transmitting the actual values of sensor data [79]. Most robot controllers provide I/O boards, to which the tools and sensors can be hardwired. Newer BUS systems as IO-LINK using industrial networks can be digitally configured and require less hard-wiring [80], even via wireless industrial networks [81]. Some systems require more direct tool control and data transmission, which can be achieved through direct connection to the axis computer for real-time adaption, which can be necessary i.e., for force-sensing and adaption [82].

2 Aim/Motivation

The aim of this article is to describe methods to combine these recent advancements in computational design, digital fabrication and robotics with the requirements to reduce material usage, rely on biological and renewable processes, and develop circular design systems. This is achieved by directly linking digital fabrication processes with computational methods for material optimization, as well as adapt them to the unique characteristics of fabrication with renewable, bio-based and living materials. Furthermore, circular strategies for reuse, and for recycling and compositing at end-of-life, are achieved by developing reversible and/or bio-joining systems between elements of a construction system. This is supported and enabled by the development of an adaptive and intuitive robotic programming and actuation framework, relying on open-source software and hardware tools.

The article is organized as follows. In Sect. 3, we present computational design methods for circular and material-efficient components and corresponding robotic fabrication technologies. In Sect. 4, we present a series of case-studies, ranging from discontinuous material placement, as robotic assembly and joining techniques for timber components, to continuous assembly, as veneer winding, additive manufacturing with timber filament, biofabrication with timber reinforced mycelium and compressed bio-bound wood particles. These case-studies are ordered through the scale and resolution of used materials. In Sect. 5, we conclude by highlighting the potentials of the described approach, as well as identifying the further steps to enable a more direct connection between sustainability goals and digital fabrication strategies.

3 Methods

3.1 Computational Design Methods

We apply computational design logics at three different levels: firstly, through a comprehensive circular design approach at the component level and in material selection; secondly, through optimization methods for material distribution within the components; and thirdly, by organizing the information of our design models directly for robot manufacturing.

Circular Design Approach. We define a circular design approach through a holistic integration of the following topics: a computational modular design approach for building components, and a structured choice and characterization of material systems in relationship with their digital processing, assembly and joining technologies, grading and up/recyclability. Instead of highly specific, single purpose building components, modular design systems are developed to allow for elements which could be potentially reused multiple times in different circumstances. While modular same-sized elements often result in uniform designs, computational combinatory logics enable high level of variation in the final assemblies, using only minimal number of component types (see Fig. 1).

Furthermore, often current approaches to modularity rely on a small number of identical components applied in a variety of different conditions within the assembly, often resulting in over-dimensioned elements, and hence high levels of material waste. A more material-efficient approach to modularity, as we propose it, is to assume the outer geometric form and interface areas of components as fixed, but to then develop methods to customize its internal structure to the unique needs of each component, given its specific structural and environmental requirements in its location within a larger assembly. However, when aiming for higher customization within a circular material system, particular attention must be placed on the material assembly and joining techniques, to ensure both reversibility of assembly and recyclability and/or composability of the components within their life cycle. For this reason, we understand design of components as the combinatorial integration between material, binding/joining techniques, fabrication technologies and specific performance requirements (Fig. 2).

Towards Construction 4.0: Computational Circular Design and Additive ...

Fig. 1 A modular design approach combined with combinatorial logics enable a range of different geometries and possible user scenarios as demonstrated in the 3DWoodWind research prototype project (see Sect. 4.2)

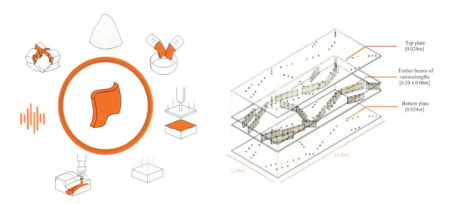

Fig. 2 Left: Circular material process: construction methods using compressed waste-wood particles in combination with bio binders. The material can be shredded and reused multiple times in the same process. Right: Robotic assembly and adhesive-free joining of beams and plates for material-efficient building components, that can be easily recycled

Material Distribution. Current approaches to the use of sustainable materials in construction are often based on solid building components, such as clay, bio-bricks or cross-laminated timber constructions. If the goal is to increase the scope and quantity of renewable resources, it is therefore necessary to develop methods for more efficient use. Computational design tools and digital fabrication technologies offer an alternative, enabling the development of processes for efficient material distribution under different load and support conditions. Furthermore, they also enable to account for the combination of materials of different grades, embedding the unique characteristics of each material as parameters. We propose various computational methods and adapt them to the specific requirements of design with sustainable materials and reversible joining techniques. This includes (see also Fig. 3):

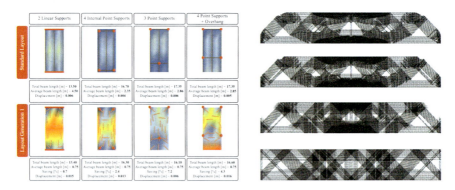

Fig. 3 Left: Distribution of timber beams in a robotically assembled hollow slab component based on stress-lines tracing (see Sect. 4.1). Right: Distribution of willow filaments in a beam manufactured through robotic additive manufacturing based on topology optimisation (see Sect. 4.3)

- Tracing principal stress lines on slab components with the aim of locating and orienting support beams in the areas where higher stresses occur.
- Topology optimization [85, 86], for which we developed a variety of discretization processes of the continuous material distribution obtained through a TO method [87]. Various applications of such method include the placement of beams in slab components, as well as the placement and orientation of timber fibers in our custom AM process.
- Shape optimization, through physical simulation and real-time stress material stress analysis, as we showed in a study on cold bending of glass [88].

These methods offer a flexible matrix of strategies to define an efficient material distribution. To this regard, one of the goals of our research is also to redefine the concept of material efficiency itself, going beyond the exclusive usage of material amount as the dominant metric, and rather attempt to also include material dimensions and grading, as well as the possibilities of reusing existing stocks of waste material from other sources rather than primary ones. Redefining material efficiency in this way allows for a more open and flexible design workflow, able to account for the unique characteristics of architectural construction (see Fig. 4).

Design for Robotic Fabrication. Using the above-mentioned computational methods for material distribution often results in very complex and irregular geometries, making conventional production processes highly inefficient for manufacturing. For this reason, we propose the use of robotic arms as flexible fabrication platforms, able to perform a variety of tasks within a single process. However, this requires designers to account for the unique characteristics and limitations of industrial robot arms already in the design phase, to ensure the buildability of the generated structures. Our approach relies on the definition of any fabrication process as first step in the development of a new design and production workflow. Using computational design tools, it then becomes possible to embed the limitations of the developed fabrication

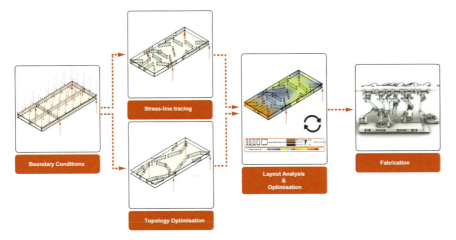

Fig. 4 Overall diagram of a continuous workflow of a hollow-core beam and plate system with form-fit, adhesive-free joining, linking together material layout generation, layout analysis & optimization, material sizing, grading and robotic assembly

process already within the design tools, hence enabling fabricability while generating the design itself. Moreover, having already determined the requirements for fabrication also means that design models can embed all information required for fabrication, which can be regenerated for each design iteration parametrically. One of the limitations to be addressed is the scale of fabrication and transportation equipment. Such limitations come to define the main manufacturing constraints, which in turn determine the maximum allowable size of a building component. Thinking in systems rather than in monolithic structures, these dimensions form the base for the development of individual elements, which can be further combined in modular building components fitting within an overall construction system.

3.2 Robotic Fabrication Methods

The use of robots for additive processes can be largely categorized in:

– Discontinuous Assembly: Pick-and Place processes
– Continuous Assembly: Additive Manufacturing.

For a successful implementation of these methods, an adaptive robotic setup, modular open-source tools, and automated material joining methods are required.

Adaptive robotic setup. In the framework of a DFG major research instrumentation grant, we developed a research facility for Robotic Architectural Production (RAP-Lab), which enables a wide range of architectural, building construction and materials

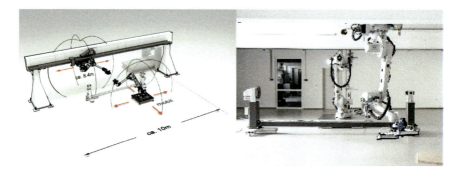

Fig. 5 Robotic Architecture production lab, experimental and digital construction, University of Kassel

science research projects. A suspended ABB IRC 4600 industrial robot on a TMO-2 Güdel gantry was combined with an ABB IRBPL rotational axis, and a mobile, same-sized robot that can be moved and used in cooperative workflows. The system is designed for maximum flexibility through the possibility to run the robots and axis in multiple combinations separately or as a real-time synchronized, 14-axis multi-move system. For tool control, Baluff Profinet IO-Link hubs and digitally configurable pneumatic valves from Festo are integrated on each of the robots, so that only a small amount of additional wiring is required and changes in the control structure can be made via software. The tools themselves are seen as modular functional entities within different control layers. This means that i.e., a tool or tool-system is not integrated in the main controller, but has all the functionality and data collection integrated, so it can be actuated by different robots with simple means. Since the lab is mainly used for research purposes, the entire robot hall was defined as a safety zone. Four mobile three-level enabling safe-switches allow several people to operate the robots in collaborative processes. Even though the facility is unique in its design, adaptability, security and control methods, we used only standard equipment, therefore enabling simple transmission and adaption by industry (Fig. 5).

Modular open-source tool ecosystem. Integrating our various manufacturing processes requires a flexible approach to both robot programming and end effector control. Previously described existing approaches rely either on a centralized approach, where a single tool attempts to address all issues, often resulting in low flexibility, or on the custom combination of separate tools, which needs to be redeveloped for each project, resulting in low transferability of the results. Borrowing the metaphor of the "cathedral vs. bazaar" of Raymond [89], we propose an approach which combines software and hardware elements of a robotic fabrication process through a series of shared interfaces, without imposing a fixed workflow [83]. This is achieved through an approach combining tool components fabrication using low-cost FDM 3d printing and readily available electronic components, with a software interface able to integrate and control such elements in a coherent workflow, and to allow their communication with the robot control code.

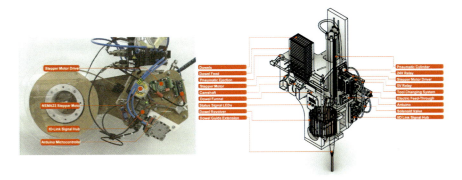

Fig. 6 Modular robotic endeffector examples. Left: Winding tool integrating a stepper driver and its motor, an Arduino microcontroller, and an IO-Link hub interfacing the signals with the robots ProfiNET network. Right: Doweling tool with various components for dowels loading, alignment and press-fitting through a pneumatic piston. A series of motors guide the dowels from the magazine to the insertion position, and sensors track the process

To achieve such goal, we developed two separate open-source tools:

- Robot Components [90], an intuitive robot programming and simulation framework for the Grasshopper visual programming environment, which enables to easily transform CAD geometry in robot instructions for ABB robots.
- Funken [91], a serial communication interface for the Arduino framework, which allows the programming and actuation of electronic components from various computational design and creative coding tools, such as Grasshopper, Python, Processing and NodeJS.

Together, these two tools, combined with our modular and open approach to tool design and manufacturing (see Fig. 6), enable the level of flexibility and adaptability necessary to the unique needs of research in architectural fabrication.

Automated Material Joining. Both for discontinuous and continuous fabrication processes, methods for joining materials are required. The joining technologies and materials also have a large impact on sustainability, since they determine the reusability and recyclability of the components. It is necessary to investigate the specific sustainability requirements according to the base material choice, and to evaluate the performance requirements (structural, building physics, acoustic), to determine the feasibility of natural binding methods. Wherever these requirements allow for it, natural binding methods should be used with bio-based adhesives or even adhesive-free methods as form-fit connections or bio-welding (see Fig. 7). Form-fit connections are created through shaping the material in ways that it can be interlocked in direction of the applied forces. This can be achieved through CNC milling [84], laser-cutting [34] or 3d printing [92]. Also, higher-grade material can be used only for the joining part, as for dowels [93] and other forms of interlocking joining [94]. It is also possible to design the whole component in a way that it topologically interlocks with the neighboring ones [95]. Bio-based adhesives can be based on starch, lignin,

Fig. 7 Left: Robotic willow filament placement process. The use of bio-based binders is possible for non-structural applications. Middle: 3D lattice of maple veneer for integration with mycelium as a bio composite. The joints are realized adhesive-free, through ultra-sonic welding. Right: Robotic press-fitting of beech wood dowels, section cut

tannin, proteins and chitin, among others [96]. We experimented with protein and glutin-based adhesives in additive manufacturing with solid willow filaments (see Sect. 4.3). We have also been investigating starch and protein for binding wood particles (see Sect. 4.5) [97], as well as chitin-based foils to laminate timber filaments. It must be noted that they have not yet been reaching similar performance levels as synthetic ones, particularly concerning structural capacity, humidity and temperature resistance. Material inherent binding agents can also be activated, as lignin in wood, as research on friction welding of timber shows [98]. We recently started investigating ultra-sonic welding of thin veneer filaments with promising results (see Sect. 4.4) [100]. For high-performance components as for structural use, chemical binders might still be necessary. Here, a strategy can be to apply these adhesives, in analogy to 3d printing, only where there are actually necessary in digitally controlled precise amounts (see Sect. 4.2).

4 Case-Studies

Five case studies exemplify the previously described methods at different scales, ranging from the robotic assembly of entire structural slab components to additive manufacturing with wood filaments to investigations on new materials based on wood fibres and biogrowth. The studies are described through the employed materials, computational design approaches, robotic fabrication processes and tooling.

4.1 Assembly and Joining Methods for Modular Timber Components

This project researches computational design and integrated structural joining methods for robotic assembly in timber construction. The focus was on planar surface

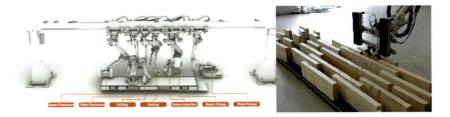

Fig. 8 Left: Diagrammatic perspective view of the robotic fabrication setup showing the individual steps of the assembly process: Pick and place procedure of the beams, laying the top plate without any fixation, nailing the plates for fixation, drilling, injecting dowels. Right: Image of pick and place procedure of the beams

elements, the most common typology of building component used in architecture, as floor and wall elements, often requiring adapting to variable supports and loads. We investigated a novel connection system with robotically placed wood nails and dowels, adapted to force direction. As materials, we used standard solid wood beams of variable dimensions, OSB Wood strand plates, beech wood dowels and nails. We investigated computational optimization approaches (tracing principal stress lines, topology optimization and sizing) in combination with multi material optimization (different timber grades and materials (softwood, beech) (Fig. 3), to optimize for goals as the reduction of the total material volume, reduction of the total costs of the structure, by using lower grade timber where possible and reduction of the global warming potential (GWP) of the used materials [99].

Robotic fabrication was structured as a "pick and place" process (see Fig. 8). Since the joining is effectuated through wood nailing and dowelling, no additional chemical adhesives are required. In terms of automation, we integrated pneumatic two-finger grippers for beam assembly and surface vacuum grippers for handling plate elements. For assembly fixation, we used a manual wood nail-gun, which we automated through pneumatic control. For automated dowel insertion, we developed a custom tool with 3d printed components for material stock, loading through stepper motors and a pneumatic piston for automated dowel insertion (Fig. 6). The dowels were pre-dried for simpler insertion.

By using a layered system of wood plates and short beams, the system is continuously extendable, and it was possible to break away from the reliance on non-reversible gluing processes, as well as to potentially allow the reuse of material from construction demolition, the use of offcuts from production or even to differentiate material grades depending on specific project requirements.

4.2 Robotic Winding of Hollow-Core Timber Elements

Winding technologies are commonly used for the manufacturing of highly resistant lightweight synthetic fiber composites. Our research investigated new potentials for lightweight, material-efficient construction with timber through robotic winding of thin continuous timber strips, providing intact, continuous, and tensile natural fibers. Linear and freeform components with varying cross-sections can be created, as architectural and structural components in construction: furniture and industrial design objects, columns, beams, floor slabs or façade components. The material system is composed of locally sourced hard- and softwood-based veneers strips of only 0.5 mm thickness and certified moisture-curing polyurethane adhesives for structural applications.

We developed both NURBS and mesh-based winding-line generation methods which minimize buckling or twisting of the filaments using locally-geodesic curves [101]. Through computational control of the winding layout, surface geometry and winding sequence, the material can be optimally distributed and customization possibilities as variable patterns with open and closed surfaces and braiding effects can be achieved. The structural filament layout optimization is currently investigated through FE-modeling.

Different robotic setups were used during the development of the project. Initially, the winding process was prototyped on a small robotic arm with a custom-developed rotational axis, controlled via Arduino. For further development, we used a 4 m long horizontal winding axis, which was integrated with the robot controller for precise synchronous movements. The winding tool integrates pneumatic devices for filament cutting, stepper motors for extrusion and an automated adhesive application system (Fig. 6). The proposed robotic framework enabled to switch between the different setups with minimal changes to the design and programming setup, allowing to adapt the production process to the different fabrication equipment available in different stages (Fig. 9).

Fig. 9 The 3DWoodWind project uses strips of thin veneers to and industrial robots to wind structural components as columns and modular ceiling components for full-scale architectural structures. Right: The BBSR Research Prototype was presented at the Digitalbau 2022 in Cologne

4.3 Additive Manufacturing with Solid Timber Filament

In timber construction, currently few large-scale 3D-printing methods exist. Present processes use wood in pulverized form, losing material-inherent structural and mechanical properties. This research proposes a new material, which maintains a complete wood structure with continuous and strong fibers, and which can be fabricated from fast-growing locally harvested plants [102]. We investigated binding and robotic additive manufacturing methods for flat, curved, lamination and hollow layering geometric typologies, and characterized the resulting willow filament and composite material for structural capacity and fabrication constraints [103]. As materials, we used willow filament, developed within the project FLIGNUM at Uni Kassel [10].

To be able to design while keeping material and manufacturing constraints in mind, as the maximum overhang angle, heights, and minimal radii, we created computational design methods that integrate and control resulting geometries for fabricability directly [104]. We also investigated a high-resolution material distribution through topology optimization (Fig. 3) constrained to the dimensions and linearity of the willow filament [23].

The robotic fabrication is characterized as a continuous process of material application, apart from cutting the wood filament when other direction lines are applied. For joining we investigated a range of chemical adhesives, as contact adhesives, uv curing adhesives, and PU hotmelt adhesives. We also investigated natural adhesives as Glutinglue, Kasein, and Chitosan. Automation of the process is performed by stepper motors for extrusion, pneumatic devices for cutting the filament, adhesive application through extrusion, and an automated UV curing LED system (Fig. 10).

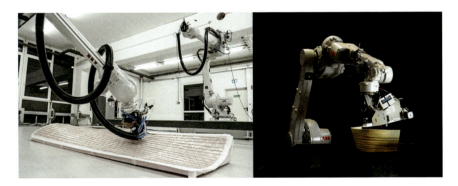

Fig. 10 Additive Manufacturing with a new material: a solid, endless filament made of split willows, an extremely fast-growing plant that can be harvested in Europe. Bio-based adhesives were investigated

4.4 Biofabrication with Timber Reinforced Mycelium

In this research project, a wood-mycelium composite construction method for CO_2-neutral, circular interior systems was investigated. Since mycelium has a low load-bearing capacity, reinforcement and 3D lattices were developed via additive manufacturing techniques from local wood species, which serve as a matrix for biogrowth. This allows to increase the scale of mycelium-based components from the available acoustic panels [21] to a wood-reinforced interior building system, while keeping the 100% bio-based, compostable material qualities. We used Ganoderma lucidum and hemp hurds for mycelial growth and maple veneer for reinforcement. [105].

Design rules for 2d and 3d wood lattice structures were developed both for structural efficiency and fabrication constraints. Initial tests were performed using 2-dimensional flat grids of veneer placed on the outer surfaces of the mycelium components. While this slightly increased bending resistance, it resulted in shear failures within the material. Hence, a novel 3-dimensional grid was developed, where the two flat grids on the outer surface are connected with veneer stripes through the material cross-section, providing the required shear resistance to the final panel [105].

A continuous robotic fabrication process was employed for laying the veneer strips, using extruders, pneumatic cutting devices and robot motion for the placement. For adhesive-free-joining of the veneer layers, we developed an ultrasonic welding method (Fig. 6). Thereby, the wood inherent lignin is performing the material joining [100]. The bio-growth process must be precisely controlled and has to go through the following steps: Substrate inoculation (collection, sterilization of hemp hurds, inoculation with G. lucidium grain spawn, 2 weeks in incubation room with temperature and humidity control); Molding (3D lattice of maple veneer is filled with mycelium substrate, 3–6 days in incubation room), demolding (Fig. 11).

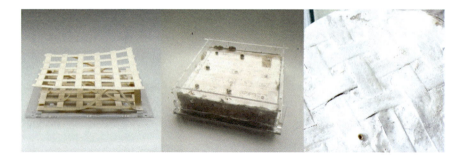

Fig. 11 The HOME project investigated the combination of additively manufactured timber 3d lattices in combination with bio-growth of mycelium, producing a novel reinforced biomaterial, that is fully compostable

Fig. 12 Left: Compacted and milled acoustic elements from used wood particles and starch. Right: Investigations in a 3D-printable, 100% bio-based wood paste

4.5 AM—Compression and Reuse with Wood Particles

The aim of this current research project is to reuse waste wood in a zero-waste process to produce components with a high degree of curvature, for use as lightweight, fireproof and precisely manufactured insulation and acoustic elements, enabling a complete circular material cycle (see Fig. 2). Applications for interior, model and furniture construction as well as concrete formwork are also conceivable. The waste wood is completely recycled and reshaped in the form of wood chips with the addition of sustainable biogenic types of binding agents and with the help of digital manufacturing processes and transferred to new circular construction applications with a requirement for complex geometries. For this we investigated the use of starch, animal and plant-based proteins, and chitosan binders, which are biogenic and result from industrial waste flows.

At the end of the life cycle, the particle-based components and the wood chips from the shaping process can be recycled and reused, creating a closed material cycle. Also, additive manufacturing processes are investigated through the development of an extrudable wood paste (see Fig. 12). The material has a promising compressive strength, enabling a possible substitution for concrete printing in the future. For preliminary investigations, we automated a manual clay extruding device, and added 3d printing parts for material storage and actuation.

Computational design methods are being developed for the surface pattern and acoustic design in conjunction with digital fabrication constraints.

5 Conclusion

This article addresses developments needed in the construction industry to reduce CO_2 emissions and enable strategies to achieve the goal of "more with less" through circular design approaches and lightweight, material-efficient structures using sustainable materials. We provide an assessment and review of current research, automation technologies and practice, to place our research in the context of industry 4.0. Integrative design and manufacturing approaches are needed that simultaneously provide holistic, circular design approaches and fully exploit the advanced geometric capabilities of robotic manufacturing for material efficient components. Digital design methods not only provide the data for their constraints, but also enable a direct and seamless workflow between design and manufacturing. Towards these goals, our research provides computational methods, material development, and robotic automation technologies.

We proposed combinatorial logics in conjunction with modular components that can be functionally expressive and highly optimised to enable component reuse for different application scenarios. Through computational tools, we are able to combine optimisation technologies for material efficiency with the constraints of robotic manufacturing.

This is demonstrated through five case studies in different scales, ranging from the robotic assembly of entire structural slab components to additive manufacturing with wood filaments to investigations on new materials based on wood fibres and biogrowth. The studies were distinguished through the used materials, computational design approaches, robotic fabrication processes and tooling. The difference in size of the employed materials thereby determines the possible resolution of assembling and has an impact on the application scenarios of these technologies. The additive nature of the described systems allows for continuous extensions in multiple directions, allowing for a maximum of design freedom and functional use through geometric variability. These studies show that not only the materials, but also the joining methods have an impact on the sustainable potentials.

We showed that timber beams and plates can be connected through form-fit wood-wood joint methods in automated processes. Therefore, after their service life, no material separation is necessary, making the components easily recyclable. We presented methods for material-efficient, hollow, lightweight components through 3d winding of timber in combination with adhesives. Since certified bio-based adhesives for structural applications are not yet available, we developed a highly precise digital application system, which minimizing adhesive use. An alternative to timber is fast-growing willow plants, that we used to produce an endless solid filament. We investigated novel robotic additive manufacturing processes with bio-based joining, that can be used for semi-structural applications and interior fittings and furniture. Our research on mycelium-wood composites shows that biomaterials that are reusable and fully compostable after their end of life could be used in large-scale applications in architecture. The rethinking wood project starts already a step further in a circular process in taking as base material timber particles from waste wood. The bio-binding process allows also to repeat the process several times from waste of components derived from this method.

Furthermore, tackling the challenge of lowering the access threshold to the adoption of such technologies in construction, our research demonstrates the development and application of open-source tools for the programming of industrial robots, as well as for the actuation and orchestration of the various hardware and software components involved in the production. By providing the tools as open-source packages, as well as developing more accessible and intuitive software and hardware approaches to robotic fabrication, we aim to extend the range of applicability of such technologies to a broader audience of designers, construction firms and other stakeholders in the domains of architectural design and fabrication.

Combined, these proposed approaches aim at providing a model for integrated design and fabrication of sustainable and efficient building components for architecture, providing an effective alternative to conventional design and production processes.

Acknowledgements Research presented in this article was funded by the following grants: Flignum (FNR, #22026418), 3DWoodWind (BBSR, 10.08.18.7-20.24 and 10.08.17.7-21.10), Robotic Timber Assembly (DFG, 436451184), RAP-Lab (DFG, #416914951), Rethinking Wood (BBSR, #10.08.18.7-21.22), Home (BBSR, 10.08.18.7-21.48).

Research was developed in collaboration with Prof. Dr.-Ing. Stefan Böhm (Flignum), Prof. Dr.-Ing. Julian Lienhard (3DWoodWind Research Prototype), Prof. Heike Klussmann (Flignum), Prof. Dr.-Ing. Werner Seim (Robotic Timber Assembly), Prof. Dr.-Ing. Jan Wurm and Prof. Dirk Hebel (Home).

Scientific development and creative contributions to the different projects were done by our chairs team members Anne Liebringshausen, Andreas Göbert, Julian Ochs, Julia Hannu, Eda Özdemir, Nadja Nolte, Kristina Schramm, Hannah Hagedorn, Guido Brinkmann and former team members Zuardin Akbar, Mohammed Dawod, Arjen Deetman, Carl Eppinger, Christoph Schlopschnat and Benedikt Wannemacher.

References

1. Statistisches Bundesamt: New orders in main construction industry (2020). https://www.destatis.de/EN/Themes/Economic-Sectors-Enterprises/Construction/_Graphic/_Interactive/new-orders-main-construction.html. Last Accessed 08 Dec 2020
2. WorldGBC: New report: the building and construction sector can reach net zero carbon emissions by 2050 (2019). https://www.worldgbc.org/news-media/WorldGBC-embodied-carbon-report-published. Last Accessed 08 Dec 2020
3. Shortage of building materials (2022). https://www.ifo.de/en/node/68972. Last Accessed 04 May 2022
4. Building statistics for different construction techniques in Germany (2020). https://www.destatis.de/DE/Themen/Branchen-Unternehmen/Bauen/Publikationen/Downloads-Bautaetigkeit/baugenehmigungen-baustoff-pdf-5311107.pdf?__blob=publicationFile. Last Accessed 03 May 2022
5. Caulfield, J.: A new report predicts significant demand growth for mass timber components. Build. Des. Constr. (2020). https://www.bdcnetwork.com/new-report-predicts-signifcant-demand-growth-mass-timber-components. Last Accessed 08 Dec 2020
6. Churkina, G., Organschi, A., Reyer, C.P.O., et al.: Buildings as a global carbon sink. Nat. Sustain. **3**, 269–276 (2020). https://doi.org/10.1038/s41893-019-0462-4
7. Hebel, D.E., Javadian, A., Heisel, F., Schlesier, K., Griebel, D., Wielopolski, M.: Process-controlled optimization of the tensile strength of bamboo fiber composites for structural applications. Compos. B Eng. **67**, 125–131 (2014). https://doi.org/10.1016/j.compositesb.2014.06.032
8. Hong, C., Li, H., Xiong, Z., Lorenzo, R., Corbi, I., Corbi, O., ... Zhang, H.: Review of connections for engineered bamboo structures. J. Build. Eng. **30**, 101324 (2020). https://doi.org/10.1016/j.jobe.2020.101324

9. Zea Escamilla, E., Habert, G., Correal Daza, J.F., Archilla, H.F., Echeverry Fernández, J.S., Trujillo, D.: Industrial or traditional bamboo construction? Comparative life cycle assessment (LCA) of bamboo-based buildings. Sustainability **10**(9), 3096 (2018). https://doi.org/10.3390/su10093096
10. Silbermann, S., Heise, J., Kohl, D., Böhm, S., Akbar, Z., Eversmann, P., Klussmann, H.: Textile architecture for wood construction. Res. Cult. Architect. **113** (2019). https://doi.org/10.1515/9783035620238-011
11. De Beus, N., Carus, M., Bart, M.: Carbon Footprint and Sustainability of Different Natural Fibres for Biocomposites and Insulation Material. Nova Institute (2019). https://renewable-carbon.eu/publications/product/carbon-footprint-and-sustainability-of-different-natural-fibres-for-biocomposites-and-insulation-material-%e2%88%92-full-version-update-2019/. Last Accessed 29 April 2020
12. Mindermann, P., Gil Pérez, M., Knippers, J., Gresser, G.T.: Investigation of the fabrication suitability, structural performance, and sustainability of natural fibers in coreless filament winding. Materials **15**, 3260 (2022). https://doi.org/10.3390/ma15093260
13. Dataholz. Sustainability evaluation of timber construction products (2021). www.dataholz.eu. Last Accessed 08 Dec 2021
14. Heisel, F., Hebel, D.E., Sobek, W.: Resource-respectful construction—the case of the Urban Mining and Recycling unit (UMAR). In: IOP Conference Series: Earth and Environmental Science, Vol. 225, No. 1, p. 012049. IOP Publishing. https://doi.org//10.1088/1755-1315/225/1/012049
15. The Living, Hi-Fi Tower. http://thelivingnewyork.com. Last Accessed 03 April 2022
16. Moser, F., Trautz, M., Beger, A.L., Löwer, M., Jacobs, G., Hillringhaus, F., … Reimer, J. (2017, September). Fungal mycelium as a building material. In: Proceedings of IASS Annual Symposia, Vol. 2017, No. 1, pp. 1–7. International Association for Shell and Spatial Structures (IASS)
17. Fungal architectures. https://www.fungar.eu. Last Accessed 03 May 2022
18. Adamatzky, A., Ayres, P., Belotti, G., Wösten, H.: Fungal architecture position paper. Int. J. Unconv. Comput. **14** (2019). https://doi.org/10.48550/arXiv.1912.13262
19. Hebel, D.E., Heisel, F.: Cultivated Building Materials: Industrialized Natural Resources for Architecture and Construction, 1st edn. Birkhäuser Verlag GmbH, Berlin, Germany and Basel, Switzerland (2017). https://doi.org/10.1515/9783035608922
20. Heisel, F., Lee, J., Schlesier, K., Rippmann, M., Saedi, N., Javadian, A., Nugroho, A.R., Van Mele, T., Block, P., Hebel, D.E.: Design, cultivation and application of load-bearing mycelium components: the MycoTree at the 2017 Seoul Biennale of architecture and urbanism. Int. J. Sustain. Energy Dev. **6**(1), 296–303 (2019). https://doi.org/10.20533/ijsed.2046.3707.2017.0039
21. Mogu. https://mogu.bio. Last Accessed 5 April 2022
22. Industry 4.0. www.plattform-i40.de. Last Accessed 23 June 2022
23. Begić, H., Galić, M.: A systematic review of construction 4.0 in the context of the BIM 4.0 premise. Buildings **11**, 337 (2021). https://doi.org/10.3390/buildings11080337
24. Construction 4.0. https://www.buildingtransformations.org/articles/construction-4-0. Last Accessed 23 June 2022
25. Manzoor, B., Othman, I., Pomares, J.C.: Digital technologies in the architecture, engineering and construction (AEC) industry—A bibliometric—qualitative literature review of research activities. Int. J. Environ. Res. Public Health **18**, 6135 (2021). https://doi.org/10.3390/ijerph18116135
26. Graser, K., Baur, M., Apolinarska, A.A., Dörfler, K., Hack, N., Jipa, A., … Hall, D.M.: DFAB HOUSE—A comprehensive demonstrator of digital fabrication in architecture. In: Fabricate 2020: Making Resilient Architecture, pp. 130–139 (2020). https://doi.org/10.2307/j.ctv13xpsvw.21
27. Popovic, D., Fast-Berglund, Å., Winroth, M.: Production of customized and standardized single family timber houses—A comparative study on levels of automation. In: 7th Swedish Production Symposium Vol. 1 (2016)

28. Homag Robotic Timber Framing. https://www.homag.com/en/product-detail/robots-in-timber-framing. Last Accessed 5 May 2022
29. Robeller, C., Hahn, B., Mayencourt, P., Weinand, Y.: CNC-fabricated dovetails for joints of prefabricated CLT components. Bauingenieur **89**, 487–490 (2014)
30. Reichenbach, S., Kromoser, B.: State of practice of automation in precast concrete production. J. Build. Eng. **43**, 102527 (2021). https://doi.org/10.1016/j.jobe.2021.102527
31. Altobelli, F., Taylor, H.F., Bernold, L.E.: Prototype robotic masonry system. J. Aerosp. Eng. **6**(1), 19–33 (1993)
32. Bock, T.: Construction automation and robotics. Robot. Autom. Constr. 21–42 (2008). https://doi.org/10.5772/5861
33. Krechting, A.: Prefabrication in the brick industry. In: 13th International Brick and Block Masonry Conference, July. Amsterdam, pp. 4–7 (2004)
34. Lienhard, J., Walz, A.: Digitaler Formschluss–Zahn-Steckverbindungen für komplexe Stahlbauknoten. Stahlbau **87**(7), 673–679 (2018)
35. Kerber, E., Heimig, T., Stumm, S., Oster, L., Brell-Cokcan, S., Reisgen, U.: Towards robotic fabrication in joining of steel. In ISARC. In: Proceedings of the International Symposium on Automation and Robotics in Construction, Vol. 35, pp. 1–9. IAARC Publications (2018). https://doi.org/10.22260/ISARC2018/0062
36. Ariza, I., Rust, R., Gramazio, F., Kohler, M.: Towards adaptive detailing with in-place WAAM connections. In: BE-AM 2020 Symposium and Exhibition, p. 34 (2020)
37. Aish, R.: Building modelling: the key to integrated construction CAD. In: CIB 5th International Symposium on the Use of Computers for Environmental Engineering related to Building, 7–9 July (1986)
38. Faux, I. D., Pratt, M.J.: Computational geometry for design and manufacture (1979)
39. Dynamo. https://dynamobim.org. Last Accessed 23 June 2022
40. Rhino.Inside. https://github.com/mcneel/rhino.inside. Last Accessed 23 June 2022. Burns, M.: Automated fabrication. Improving productivity in manufacturing. Ennex, Los Angeles (1993)
41. DIN e.V.: DIN 8580. Fertigungsverfahren - Begriffe, Einteilung (2020)
42. VDI 3405 Additive Fertigungsverfahren. Grundlagen, Begriffe, Verfahrensbeschreibung
43. Xtree concrete 3d printing. https://xtreee.com. Last Accessed 05 May 2022
44. Buswell, R.A., De Silva, W.L., Jones, S.Z., Dirrenberger, J.: 3D printing using concrete extrusion: a roadmap for research. Cem. Concr. Res. **112**, 37–49 (2018). https://doi.org/10.1016/j.cemconres.2018.05.006
45. Peri concrete 3d printing. https://www.peri.com/en/business-segments/3d-construction-printing.html. Last Accessed 05 May 2022
46. 3d printing house factory. https://www.3d.weber/news/worlds-first-house-3d-printing-factory-opens-in-eindhoven-netherlands. Last accessed 05 May 2022
47. Ding, Y., Dwivedi, R., Kovacevic, R.: Process planning for 8-axis robotized laser-based direct metal deposition system: a case on building revolved part. Robot. Comput.-Integr. Manuf. **44**, 67–76 (2017). https://doi.org/10.1016/j.rcim.2016.08.008
48. Joosten, S.K.: Printing a stainless steel bridge: an exploration of structural properties of stainless steel additive manufactures for civil engineering purposes (2015)
49. Henke, K., Treml, S.: Wood based bulk material in 3D printing processes for applications in construction. Eur. J. Wood Wood Prod. **71**(1), 139–141 (2013). https://doi.org/10.1007/s00107-012-0658-z
50. Lamm, M.E., Wang, L., Kishore, V., Tekinalp, H., Kunc, V., Wang, J., ... Ozcan, S.: Material extrusion additive manufacturing of wood and lignocellulosic filled composites. Polymers **12**(9), 2115 (2020). https://doi.org/10.3390/polym12092115
51. Wimmer, R., Steyrer, B., Woess, J., Koddenberg, T., Mundigler, N.: 3D printing and wood. Pro Ligno **11**(4), 144–149 (2015)
52. Markin, V., Schröfl, C., Blankenstein, P., Mechtcherine, V.: Three-Dimensional (3D)-printed wood-starch composite as support material for 3D concrete printing. ACI Mater. J. **118**(6), 301–310 (2021). https://doi.org/10.14359/51733131

53. Forust 3D printed wood. https://www.forust.com/
54. Stute, F., Mici, J., Chamberlain, L., Lipson, H.: Digital wood: 3D internal color texture mapping. 3D Print. Addit. Manuf. **5**(4), 285–291 (2018). https://doi.org/10.1089/3dp.2018.0078
55. Helm, V., Knauss, M., Kohlhammer, T., Gramazio, F., Kohler, M.: Additive robotic fabrication of complex timber structures. In: Advancing Wood Architecture: A Computational Approach, pp. 29–42 (2016). https://doi.org/10.4324/9781315678825-3
56. Labonnote, N., Rønnquist, A., Manum, B., Rüther, P.: Additive construction: state-of-the-art, challenges and opportunities. Autom. Constr. **72**, 347–366 (2016). https://doi.org/10.1016/j.autcon.2016.08.026
57. ASTM Committee F42 on Additive Manufacturing Technologies, & ASTM Committee F42 on Additive Manufacturing Technologies. Subcommittee F42. 91 on Terminology. Standard terminology for additive manufacturing technologies. ATSM International (2012)
58. VDI Industrie 4.0. www.vdi.de/ueber-uns/presse/publikationen/details/industrie-40-begriffe-terms-and-definitions. Last Accessed 23 June 2022
59. Eversmann, P.: Robotic fabrication techniques for material of unknown geometry. In: Humanizing Digital Reality. Springer, Singapore (2018). https://doi.org/10.1007/978-981-10-6611-5_27
60. Gasparetto, A., Scalera, L.: A brief history of industrial robotics in the 20th century. Adv. Hist. Stud. **8**, 24–35 (2019). https://doi.org/10.4236/ahs.2019.81002
61. ABB RobotStudio. https://new.abb.com/products/robotics/de/robotstudio. Last Accessed 06 May 2022
62. Kuka.sim. https://www.kuka.com/de-de/produkte-leistungen/robotersysteme/software/planung-projektierung-service-sicherheit/kuka_sim. Last Accessed 06 May 2022
63. Fanuc Roboguide. https://www.fanuc.eu/de/en/robots/accessories/roboguide. Last Accessed 06 May 2022
64. UR Sim. https://www.universal-robots.com/download/software-e-series/simulator-non-linux/offline-simulator-e-series-ur-sim-for-non-linux-594/. Last Accessed 06 May 2022
65. Visual Components. https://www.visualcomponents.com/. Last Accessed 06 May 2022
66. Gazebo. https://gazebosim.org/. Last accessed 06 May 2022
67. CoppeliaSim. https://www.coppeliarobotics.com/. Last Accessed 06 May 2022
68. HAL robotics. https://hal-robotics.com/. Last Accessed 06 May 2022
69. RoboDK. https://robodk.com/. Last Accessed 06 May2022
70. Robot Components. https://www.food4rhino.com/en/app/robot-components. Last Accessed 06 May 2022
71. Compas FAB. https://github.com/compas-dev/compas_fab. Last Accessed 06 May 2022
72. Kuka prc. https://www.food4rhino.com/en/app/kukaprc-parametric-robot-control-grasshopper. Last Accessed 06 May 2022
73. Robots. https://www.food4rhino.com/en/app/robots. Last Accessed 06 May 2022
74. ROS – Robot Operating System. https://www.ros.org/. Last Accessed 06 May 2022
75. Moveit. https://moveit.ros.org. Last Accessed 06 May 2022
76. Ueda, M., Iwata, K., Shimizu, T., Sakai, I.: Sensors and systems of an industrial robot. In: Memoirs of the Faculty of Engineering, Vol. 27. Nagoya University (1975)
77. Arduino. https://www.arduino.cc/. Last Accessed 14 May 2022
78. I/O definitions. https://www.electricalclassroom.com/digital-i-o-and-analog-i-o/. Last Accessed 14 May 2022
79. Kiencke, U., Dais, S., Litschel, M.: Automotive serial controller area network. SAE Trans. 823–828 (1986)
80. IO-Link. https://io-link.com. Last Accessed 14 May 2022
81. Frotzscher, A., Wetzker, U., Bauer, M., Rentschler, M., Beyer, M., Elspass, S., Klessig, H.: Requirements and current solutions of wireless communication in industrial automation. In: 2014 IEEE International Conference on Communications Workshops (ICC), June, pp. 67–72. IEEE (2014). https://doi.org/10.1109/ICCW.2014.6881174

82. Stolt, A., Linderoth, M., Robertsson, A., Johansson, R.: Force controlled robotic assembly without a force sensor. In 2012 IEEE International Conference on Robotics and Automation, May, pp. 1538–1543. IEEE (2012). https://doi.org/10.1109/ICRA.2012.6224837
83. Rossi, A., Deetman, A., Stefas, A., Göbert, A., Eppinger, C., Ochs, J., Tessmann, O., Eversmann, P.: An open approach to robotic prototyping for architectural design and construction. In: Gengnagel, C., Baverel, O., Betti, G., Popescu, M., Thomsen, M.R., Wurm, J. (eds) Towards Radical Regeneration. DMS 2022. Springer, Cham. https://doi.org/10.1007/978-3-031-13249-0_9
84. Robeller, C.: Integral mechanical attachment for timber folded plate structures (No. 6564). EPFL (2015). https://doi.org/10.5075/epfl-thesis-6564
85. Søndergaard, A., Amir, O., Eversmann, P., Piskorec, L., Stan, F., Gramazio, F., Kohler, M.: Topology optimization and robotic fabrication of advanced timber space-frame structures. In: Reinhardt, D., Saunders, R., Burry, J. (eds.), Robotic Fabrication in Architecture, Art and Design 2016, pp. 190–203. Springer (2016). https://doi.org/10.1007/978-3-319-26378-6_14
86. Lienhard, J., Eversmann, P.: New hybrids—From textile logics towards tailored material behaviour. Architect. Eng. Des. Manag. **17**(3–4), 169–174 (2021). https://doi.org/10.1080/17452007.2020.1744421
87. Bendsoe, M.P., Sigmund, O.: Topology Optimization: Theory, Methods, and Applications. Springer Science & Business Media (2003). https://doi.org/10.1007/978-3-662-05086-6
88. Eversmann, P., Schling, E., Ihde, A., Louter, C.: Low-cost double curvature: geometrical and structural potentials of rectangular, cold-bent glass construction. In: Proceedings of IASS Annual Symposia, September, Vol. 2016, No. 16, pp. 1–10. International Association for Shell and Spatial Structures (IASS) (2016)
89. Raymond, E.: The cathedral and the bazaar. Knowl. Technol. Policy **12**, 23–49 (1999). https://doi.org/10.1007/s12130-999-1026-0
90. Experimentelles und Digitales Entwerfen und Konstruieren, Universität Kassel. Robot Components. https://robotcomponents.github.io/RobotComponents-Documentation/. Last Accessed 16 June 2022
91. Stefas, A., Rossi, A., Tessmann, O.: Funken—Serial protocol toolkit for interactive prototyping. In: Computing for a better tomorrow—Proceedings of the 36th eCAADe Conference, vol. 2, pp. 177–186, Lodz, Poland (2018). https://doi.org/10.52842/conf.ecaade.2018.2.177
92. Schwicker, M., Nikolov, N.: Development of a fused deposition modeling system to build form-fit joints using an industrial robot. Int. J. Mech. Eng. Robot. Res. **11**(2) (2022). https://doi.org/10.18178/ijmerr.11.2.51-58
93. Guan, Z., Komatsu, K., Jung, K., Kitamori, A.: Structural characteristics of beam-column connections using compressed wood dowels and plates. In: Proceedings of the World Conference on Timber Engineering (WCTE), Trentino (Italy), June (2010)
94. Robeller, C., Von Haaren, N.: Recycleshell: wood-only shell structures made from cross-laminated timber (CLT) production waste. J. Int. Assoc. Shell Spatial Struct. **61**(2), 125–139 (2020). https://doi.org/10.20898/j.iass.2020.204.045
95. Tessmann, O., Rossi, A.: Geometry as interface: parametric and combinatorial topological interlocking assemblies. J. Appl. Mech. **86**(11) (2019). https://doi.org/10.1115/1.4044606
96. Hemmilä, V., Adamopoulos, S., Karlsson, O., Kumar, A.: Development of sustainable bio-adhesives for engineered wood panels–A Review. RSC Adv. **7**(61), 38604–38630 (2017). https://doi.org/10.1039/C7RA06598A
97. Industriell nutzbare, nachhaltige und wiederverwendbare Schalungen zur Realisierung von doppelseitig gekrümmten Betonfertigteilen für energieeffizientes, ressourcenschonendes und klimagerechtes Bauen. https://www.zukunftbau.de/projekte/forschungsfoerderung/1008187-1830. Last Accessed 14 June 2022
98. Stamm, B., Natterer, J., Navi, P.: Joining wood by friction welding. Holz Roh Werkst **63**, 313–320 (2005). https://doi.org/10.1007/s00107-005-0007-6
99. Schramm, K., Eppinger, C., Rossi, A., Braun, M., Brueden, M., Seim, W., Eversmann, P.: Redefining material efficiency—Computational design, optimization and robotic fabrication methods for planar timber slabs. In: Gengnagel, C., Baverel, O., Betti, G., Popescu, M.,

Thomsen, M.R., Wurm, J. (eds) Towards Radical Regeneration. DMS 2022. Springer, Cham. https://doi.org/10.1007/978-3-031-13249-0_41

100. Özdemir, E., Saeidi, N., Javadian, A., Rossi, A., Nolte, N., Ren, S., Dwan, A., Acosta, I., Hebel, D.E., Wurm, J., Eversmann, P.: Wood-Veneer-reinforced mycelium composites for sustainable building components. Biomimetics **7**, 39 (2022). https://doi.org/10.3390/biomimetics7020039
101. Göbert, A., Deetman, A., Rossi, A., et al.: 3DWoodWind: robotic winding processes for material-efficient lightweight veneer components. Constr Robot **6**, 39–55 (2022). https://doi.org/10.1007/s41693-022-00067-2
102. Silbermann, S., Böhm, S., Klussmann, H., Eversmann, P.: Textile tectonics for wood construction. In: Hudert, M., Pfeiffer, S. (eds.) Rethinking Wood: Future Dimensions of Timber Assembly. Birkhäuser. https://doi.org/10.1515/9783035617061
103. Eversmann, P., Ochs, J., Heise, J., Akbar, J., Böhm, J.: Additive timber manufacturing: a novel, wood-based filament and its additive robotic fabrication techniques for large-scale, material-efficient construction. 3D Print. Addit. Manuf. 161–176 (2022). https://doi.org/10.1089/3dp.2020.0356
104. Ochs, J., Akbar, Z., Eversmann, P.: Additive manufacturing with solid wood: Continuous robotic laying of multiple wicker filaments through micro lamination. In: Design Computation Input/Output (2020). https://doi.org/10.47330/dcio.2020.jzan7781
105. Rossi, A., Javadian, A., Acosta, I., Özdemir, E., Nolte, N., Saeidi, N., Dwan, A., Ren, S., Vries, L., Hebel, D., Wurm, J., Eversmann, P.: HOME: wood-mycelium composites for CO_2-Neutral, circular interior construction and fittings. In: Berlin D-A-CH Conference: Built Environment within Planetary Boundaries (SBE Berlin 2022). IOP Publishing (2022). https://doi.org/10.1088/1755-1315/1078/1/012068

RFId for Construction Sector. Technological Innovation in Circular Economy Perspective

Matteo Giovanardi

Abstract The transition towards the Circular Economy (CE) sets new challenges in the construction sector. In addition to reduction of resource consumption and "closing the loop" concept, CE requires the dematerialization of services and products. Building processes and products need to be rethought to ensure sustainable and circular management of the asset. In this context, the progress in the field of Industry 4.0 technologies, such as Internet of Things (IoT) and Radio Frequency Identification (RFId), promises interesting scenarios in fostering circular transition. Indeed, information technologies can assume a critical role in achieve the Sustainable Development Goals 9 and 12. For about fifteen years, RFIds are used by several industries to automate process, optimize cost, and manage asset information through data-driven approach. This paper aims to investigate the feature of RFId technologies and its application in construction sector. In the perspective of promoting CE principles, such technologies can play an enabling role. Thorough the analysis of scientific literature review and experiences in the market, 20 of the most innovative case studies are presented. A clustering analysis of the case studied presented clarifies the most investigated fields and those where research should focus in the future.

Keywords RFId · IoT · Industry 4.0 · Construction industry · Circular economy

United Nations' Sustainable Development Goals 9. Build resilient infrastructure, promote inclusive and sustainable industrialization and foster innovation · 12. Ensure sustainable consumption and production patterns

---

M. Giovanardi (✉)
Dipartimento di Architettura e Design, Politecnico di Torino, 10125 Torino, Italy
e-mail: matteo.giovanardi@polito.it

1 Introduction

The environmental and economic limits of the current linear development model highlight the need for a rapid ecological and circular transition. As required by the European Union [1], moving towards Circular Economy (CE) means renewing products and processes by overcoming models that are no longer sufficient and crystallized in a habitual vision of the present [2]. In this perspective, the progress in the field of Information and Communication Technologies (ICT) can be identified as a driving force for change, as an exogenous phenomenon that points the way for economic and social transformation [3]. Indeed, digitalization and dematerialization of products and processes are considered key factors to support CE and one of the "Six Transformations to Achieve the Sustainable Development Goals" (SDGs) [4]. More precisely, such approach, identified in the pervasive use of Industry 4.0 technology such as Internet of Things (IoT), Big Data, cloud computing, is now essential in fostering fair, responsible, and sustainable innovation (SDG 9: target 9.4, 9.b) to raise resource-use efficiency (SDG 12: target 12.2, 12.5) [5]. The strategic role of data in the ecological and circular transition is confirmed by several scholars [6, 7] and verified by many sectors (e.g. automotive, aerospace, retail, etc.).

Such approach opens extremely interesting scenarios in the construction sector. Still considered among the main sector exerting the strongest pressure on the environment [8] and accounting for almost 9% of European GDP [9], the construction sector plays a crucial role in circular transition, and it has high scope for digitization. In this perspective, the Industry 4.0 technologies represent, on the one hand, the most recent phase of industrial activities digitization, and on the other, a constantly evolving paradigm [10] that stimulates building product innovation through new dematerialized value. Integrated, connected, and collaborative cyber-physical systems, identified on the IoT, can thus facilitate circular transition and restructuring of industries' capital profitability.

An emblematic case concerns Radio Frequency Identification (RFId) technologies. Greater transparency and efficiency in asset management has pushed RFId technology into various sectors such as manufacturing, retails, and logistics. The ability to create, share, and transform data into information along value chains is the key to creating a circular approach using resources in a more efficient way [11]. With a view to stimulating the introduction of circular approaches, this paper aims to clarify the application potential of RFId technologies in the construction sector. A collection of 20 case studies in the last 15 years are presented to map the state-of-the-art. The clustering analysis of such experiences shows the most investigated fields and those where research should focus in the future.

2 Radio Frequency Identification

2.1 RFId Technology

RFId sensors are considered the new paradigm of the IoT [12]. Although the first applications of the technology date back to World War II for "friend or foe" recognition of anti-aircraft [11], recent progress in chip miniaturization and industrial process production have allowed a drastic reduction in price making the technology extremely versatile for many applications [13]. In the last twenty years, interest in this technology has been discontinuous. After a period of great interest between 2004 and 2007, the trend grew again after 2016 driven by an increased focus on IoT technologies. This trend is confirmed by Google Trend, too. Although to be considered from a qualitative point of view, the Fig. 1 showing the interest in Google searches for the terms 'RFId' or 'Radio Frequency Identification' in the last 18 years, compared to 'IoT' [14].

Used in logistics, automatic payment, access control or the identification of components or animals, RFId systems are already mature technologies that have found widespread application in many market sectors. Basically, it is a technology that allows the remote recognition of an object by using radio communication. The system architecture consists of two main elements: a reader with a data processing module and an antenna to generate the electromagnetic field, and a tag, a device placed on the object to be identified, consisting of an antenna, an integrated circuit (IC), and a substrate. Once entered in the radio signal range, the reader queries the tag, reads data, and organizes it in databases and/or shares it over the network. Several types of RFId technologies exists and different classifications can be made according to.

The presence/absence of a battery in the tag. RFId systems can be divided into passive, active, and semi-passive/semi-active. Passive tags are the most popular type, they do not have a battery and receive their power from the RFId reader. Active tags have an on-tag power supply such as a battery, which emits a constant signal containing identification information [15].

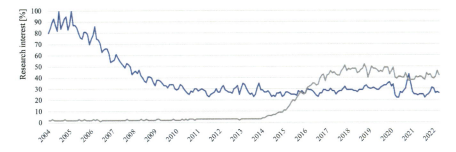

Fig. 1 Search trend on Google [19]: "RFId" (blue) and "IoT" (grey)

Table 1 RFId technology classification

Type	Frequency	Range	Active/Passive	Cost	Main applications
LF	125–134 kHz	1–10 cm	Passive	Very low	Access control, car anti-theft systems, etc.
HF	13.56 MHz	10 cm–1 m	Passive	Low	Labels for retail, safety, etc.
UHF	433 MHz 866–868 MHz (EU) 902–928 MHz (USA)	1–100 m 1–12 m	Active Passive	High	Identification of moving objects, etc.
Mw	2.45–5.8 GHz 3.1–10 GHz (USA)	1–2 m Up to 200 m	Active	High	Alarms, speed gauges, automatic openings, etc.

The frequency of the signal. Low Frequency (LF), High Frequency (HF), Ultra High Frequency (UHF), and Microwave (Mw) are the main types. Generally, LF systems operate in the 125 kHz to 134 kHz range and have a read range of up to 10 cm. HFs operate in the 13.56 MHz range and provide reading distances of 10 cm to 1 m. UHF systems have a frequency range between 433 and 938 MHz depending on the context, offer read ranges up to 2 m, and have faster data transfer rates. Microwave frequency are less common for RFId technologies [15].

The material tag. The chip and antenna are mounted on a substrate, which can be paper, polyethylene terephthalate (PET) or some other type of plastic. The choice of material depends mainly on the type of transponders and the types of actions to which the asset is subjected. Price and durability of the tag depend strongly on the material used [16].

The memory capacity. In current tags, identified as Gen2 RFId tags, the IC contains four types of memory: Reserved Memory, EP Memory, TID Memory, User Memory. The capacity can vary from 8 bits in the case of passive technologies up to 8kbits. Tags with memory can be of the "read only" or "read/write" type when the data can be modified dynamically [17].

The main features can be summarized in Table 1.

2.2 Lesson Learnt by Other Sectors

In recent years, the widespread application of RFId technologies in many product sectors have shown environmental and economic benefits. There are three main experiences that can be stimulating for the construction sector: the auto-identification of assets, geolocation and spatial monitoring of assets, and environmental parameters monitoring. Lessons learnt from other industries can facilitate faster deployment in the construction industry.

Auto-identification. It is one of the main advantages of using RFId systems. Such technology can rapidly raise data collection in supply chains by automatically identifying the assets' features creating new business values by data. The most striking example emerges from the aerospace field [18]. For around 15 years, RFId technologies have been used in the production of airplanes to rationalize production and maintenance time and costs. The Boeing 787 DreamLiner, for example, consists of around 6 million components, provided by around 40 different suppliers. The production cost and maintenance phases require constant monitoring of the status of its components. To improve the control of supply chains, contain costs, and reduce time for inventory, 1.750 RFId tags are used to track aircraft components. Moreover, the digitalization of the inventory process drastically reduces human errors ensuring a greater quality [11]. The civil sector is also moving in this direction, too. Consolis, a world leader in the production of prefabricated tunnel segments and rail sleepers, uses tags embedded in each tunnel segment to identify assets during construction phase, manage information during assets life-cycle, and ensuring greater control and process transparency over the supply chain [19].

Geolocalization monitoring. Real-time geospatial sensing of assets is one of the main scope for RFId technology. Monitoring position of assets in a space could result in greater efficiency in controlling supply chains and governing the timing of a process more accurately. Delivery data, shipment tracking and material stock management are extremely useful actions for logistics and efficient management of complex processes. In many productive companies, doors such as readers monitor the entry and exit of goods and/or people to control matter flows, occupancy, and increase safety in the workplace [20]. In addition to this, position sensors, GPS, accelerometers, and vibration, sensors can be used with RFId systems for performance monitoring of an asset. For some years, structural monitoring is an extreme topic in the academia research, several researchers over the years have tested RFId based solutions. RFId systems have been studied for the structural monitoring of bridges, viaducts, roads and infrastructure works [21]. The ability to track statical position and fluctuations allows operators providing operation and maintenance to intervene in advance, thus reducing the cost of corrective maintenance and the risk of failure. Vizinex, a leading RFId systems company, has developed a passive UHF sensor for automatic maintenance schedule management [22]. In Missouri, the Center for Transportation Infrastructure and Safety has carried out a research project on the implementation of RFId sensors for bridge monitoring [23]. To monitor the corrosion status of a physical element and evaluate its load stress history, a passive RFId sensor was developed.

Environmental parameters monitoring. A further field of application is the use of RFID systems to monitor environmental parameters. Joint use with wireless sensor (WSN) network provides to RFId systems a greater versatility. The typical hardware platform of a WSN node consists of a sensor, a microcontroller, a radio frequency transceiver and a power source. Each node is equipped with a sensor to detect parameters such as temperature, humidity, light, sound, pressure or other physical parameters. Power consumption, chip size and computing power, as well as on-chip memory are very important features to define in an integrated RFId system. A crucial feature of a sensor node is the power source and battery management, especially in WSNs, where the battery can't be replaced [15]. In this context, significant experiences come from pharmaceutical and food chain fields. The cold chain control ensures that products, drugs or food can reach the final consumer in the best condition. Changing in temperature during production and distribution process could lead to alteration of the chemical features of the good with possible damage to the health of the user. In 2016, European MC Donald's and other leaders in the food industry funded the testing of an RFId temperature loggers that could track the food product temperature throughout the supply chain [24]. This ensures high quality in the product sold to the consumer and offers new tools to regulate the relationship with suppliers. A similar example is provided by PostNL's (Belgium). The delivery company has developed together with SenseAnywhere a device for detecting internal conditions in vehicles used for the delivery of pharmaceutical materials [25].

3 Materials and Method

Defining the maturity level and the state-of-the-art of a technology that is difficult to place in a specific context is a tricky issue. Indeed, depending on the application field the maturity level of RFId technology can be drastically change. In recent years, technology development has been taking place more within companies than in academia. This sometimes leads to real difficulties in finding technology-specific information. As confirmed by Costa et al. [15] it emerges that by 2015 the number of patents registered using RFId technology was far higher than the number of scientific papers. This justifies a degree of maturity of the technology that can be widely used in the market. In Europe, more than 16,000 patents have been filed using RFId technology [15]. For these reasons, the collection of case studies used a hybrid approach integrating scientific literature review with experiences from the market.

From the literature review on SCOPUS and ResearchGate databases, of the total 43.000 articles that can be traced through the keywords "RFId" or "Radio Frequency Identification", about 1.283 (about 3%) also integrate the term "construction". An initial analysis of the metadata obtained shows how the distribution of scientific research between 1995 and 2022 (April) follows the trend of Google research show in Fig. 2. The shift of the peak in 2013 is probably due to the research and publication time. Since 2015, the number of papers per year has been constant with around 70 papers per year. Computer Sciences, Automation in Construction and Applied Mechanics and Materials are the three journal papers that include the largest number of articles.

Screening the most innovative cases led to the selection of 20 case studies. Applied research and market application from 2006 to 2022 are presented. The case studies analysis and clustering according to the purpose of the technology provides an overview of the RFId system and highlight new emerging trends for future research.

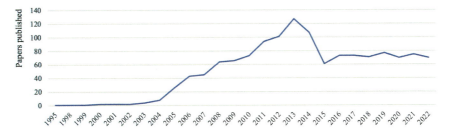

Fig. 2 Number of papers published per year containing "RFId" or "Radio Frequency Identification" AND "Construction" in title, abstract or keywords

4 RFId in Construction Industry

Table 2 shows the case studies with a brief description of the purpose of the technology. Information regarding the year, location, and companies involved helps to clarify the evolutionary process. The case studies are listed in chronological order.

5 Discussion

Form the case studies presented emerges a very wide context. The RFId experiences, as shown in Fig. 3, show a still heterogeneous degree of maturity, proving that it can be considered a recent technology in the construction sector. The market pervasiveness of the cited technology are strongly dependent on national regulatory barriers, the economic value of data provided by technology, and technical issues related to digital and physical integration. Access control for site management or restricted zone in the building represents the most mature and popular in the market. In this case, a lower integration level of tags facilitates the use of cards and keys RFId-based. An emerging and interesting application field is production and supply chain management. Although still used to optimize and control material and component flows within the same company or between a few stakeholders, the creation of a digitized supply chain opens up interesting scenarios for construction sector. A current barrier is the functional and technological unity of the building component, where to install the tag. The promotion of prefabricated systems and dry construction, which facilitates the idea of a building organized by parts that can be replaced over time, may allow a greater application of such tracking systems. The issue becomes more complicated, and the degree of technological maturity and market application declines, when the typical challenges of facility management and computerized asset management arise. Although the advantages from a rational use of resources point of view are obvious, the time dimension, the physical integration, the number and role of stakeholders becomes an obstacle. On the one hand, the technological obsolescence of RFId technology does not (yet) allow its use for components with a long

Table 2 RFId in construction sector case studies

Name, main info	Goal and technology description
RFID-Based Facilities Maintenance. 2006, Frankfurt Airport (DE)	All fire shutters are equipped with RFId tags that store maintenance related information. The technicians identify themselves by scanning their badge and the tag attached to the fire shutter. After performing the checking or the maintenance, the tag is scanned a second time to record updated information. The transponders are designed to be attached to metal. The reading range is only 3 cm and frequency is 13.56 MHz [26]
Intelligent Concrete. 2006, Tilst (DK). Dalton, Aarhus Innovation Lab	An integrated microchips for controlling production and supply chain and optimizing facilities activities for concrete panels. Concrete panel data are shared via internet and organized in specific database. By means of a personal digital assistant with a special reader mounted on the back, the men at the construction site can find all information about the panel immediately (e.g. measurements, weight, serial number, production history, exact mounting instruction and maintenance instructions) [27]
New Meadowlands Stadium. 2010. New jersey (USA). Skanska USA	In 2010, Skanska USA used RFId tag to track over 3.200 pre-cast concrete panels and visualizes it on BIM model for the New Meadowlands Stadium project. Each concrete panel, which weighed around 20 tons, was fitted with a RFId tag to allow it to monitor supply chain information in real-time. The RFId technology allowed to identify and solve problem early in the process, reducing in this way the construction period by 10 days and saving US$ 1 million [28]
Service-Oriented Integrated Information Framework. 2011, Seoul (KR). Sungkyunkwan Univ., Doalltech Co	This research aims to develop a seamlessly integrated information management framework that can share logistics information to project stakeholders. To provide "just in time" delivery for construction sector, the research group have developed an integrated framework to digitalize component and building material flows. The pilot test showed that it can improve time efficiency by about 32% compared to the traditional supply chain management. The result of this research is expected to be utilized effectively as a basic framework to manage information in RFId/WSN based construction supply chain management environments [29]
Door Control System (DCS). 2012, Bielefeld (DE). Schüco	The DCS control system provides access control using RFId technology. A passive RFId card tag for operators is the digital key to access a specific room. Integrated into the door, a RFId reader recognizes the operators and enables them to pass through. Such application is interesting for security and access control to private areas such as hospital, bank, offices etc. [30]

(continued)

Table 2 (continued)

Name, main info	Goal and technology description
Precast Concrete. 2014, North Carolina (USA). Cherry Precast and Concrete Pipe & Precast. HUF RFId	To aid the state North Carolina Department of Transportation's inspections, some American companies' suppliers have integrated an RFId tag embedded in each precast concrete panel to keep track of manufacture data. The RFId led to a fast evolution of the control process. An online database (HiCAMS), accessible to all suppliers by password, allows them to view project data, orders, and delivered products [31]
HardTrack. 2014, San Francisco (USA). Shimmick Construction Co., Wake, Inc. HUF RFId	HardTrack technology consists in active RFId (UHF) tags with integrated sensors to monitor the temperature and humidity of the concrete. In San Francisco, it was used by Shimmick Construction for the concrete foundation of building project. 16 concrete slabs integrated RFId tag and sensors. Each slab must be fully cured before the next one can be poured, to prevent it from cracking due to any stresses from the next slab. HardTrack provides real-time data on the concrete temperature and a software determines the curing date of the poured concrete. This approach has proven direct benefits on construction site timing [32]
Redpoint. 2015, Boston (USA). Redpoint Positioning Corp., Skanska USA	The Redpoint technology helps the company to know when staff members go into an area of a construction site. The system also provides historical data so that management can identify workers who repeatedly enter an unauthorized area and provide them with additional training to prevent such mistakes from happening again. The development of such devices can find rapid application in buildings with restricted access areas [33]
Cluster based RFId. 2015, Montreal (CA). Concordia University	In this research, a localization method based on RFId systems which does not need infrastructure is proposed. The developed of an active RFID technology for the localization of movable objects is proposed. Building components, and equipment with an integrated RFId tag using handheld readers. By extending a Cluster-based Movable Tag Localization technique, a k-Nearest Neighbor algorithm is used [34]
Smart Construction Object. 2016, Hong Kong (CN). The University of Hong Kong	An integrated smart building component was tested and demonstrated in a real-life case in a prefabricated construction in Hong Kong. In this case, an RFId-enabled BIM system was required to track the status of prefabricated façades from off-shore manufacturing, cross-border logistics, through to on-site assembly. The tags for supply chain control included data regarding prefabrication factory, transportation routes, and a construction site [35]

(continued)

Table 2 (continued)

Name, main info	Goal and technology description
IFC-RFId. 2016, Montreal (CA). Concordia University	The mechanical room of the Genomics Centre at Concordia University is chosen for the case study. The building is modeled in BIM and the mechanical elements are added to the model. RFId tags are attached to a selected set of elements to host their related BIM information. Active and passive RFId tags are modeled in Revit under the electrical equipment category and added to the BIM model of the building [36]
The Spot-r worker safety system. 2017, East Harlem, New York (USA). Lettire Construction, Triax tech	The system developed aims to ensure the safety of workers on the construction site. When workers arrive on site, building project manager can use the real-time data to view the total number of workers per floor and zone and organize the work. Furthermore, such solution allows managers to verify if potential incident occurs. For example, the RFId system can detect sudden falls. The software's algorithms can also determine if the data is indicative of a fall or if the worker may simply have dropped the device [37]
Elbphilarmonie façade. 2017, Hamburg (DE). Permasteelisa Group	Permasteelisa, world leader in building façade technology, used an RFId tag to optimize production and construction phase in complex project. To manage a large number of different façade elements, each façade panel was tagged with an RFId that, thanks to a specific ID number, could remotely identify each individual element. In Elbphilarmonie project, such system facilitated and sped up the panel delivery to construction site that was particularly challenging, given its unique urban location and space constraints. This data will also be used for maintenance purpose [38]
MULTIfid project. 2018–2021, Università degli Studi dell'Aquila. (DICEAA), 2bite S.r.l, Pack System S.r.l	The main objective of the MULTIFId project is to create an innovative product consisting of an intelligent, low-cost, and low-emission panel, made from waste from the industrial processing of paper and cardboard. An RFId system is integrated into the panel to monitor the position of workers in risk areas, thermal performance, and monitoring humidity conditions. The academia project tested and verified RFId signal transmission through different campaign monitoring [39]
RFIBricks. 2018, National Taiwan University, Taipei. HUF RFId	RFIBricks is an academia project carried out in Taiwan University. Hsieh et al. [40] present an interactive brick system based on ultra-high frequency RFID sensing. The researchers present a system that enables geometry resolution and geolocation of the asset in a space. Although the state of research is in prototype form, the development of a dynamic user interface opens up interesting scenarios in the field of tracking and tracing components in a space

(continued)

Table 2 (continued)

Name, main info	Goal and technology description
Checked OK. 2018, Cork (IE). Anderco Liftging, CoreRFId. HF RFId	Anderco Lifting, one of Ireland's largest lift companies, is employing an HF RFId solution to improve the efficiency of inspections of the lift equipment that its customers use at construction sites. The system was developed in 2018, and the data collected is being accessed by utilities and several other customers to which Anderco provides six-month cycle inspections [41]
Flexible thermal monitoring. 2018, Turin, (IT). Polytechnic of Turin, DAUIN Department	Giusto et al. [42] investigate RFId technology for indoor climate control. Benefits of a dense deployment of pervasive temperature sensors are presented. The analysis takes into account many features, such as technology simplicity and time of development, flexibility, wired/wireless range, battery life, reliability and cost. A case study with field test shows that the RFId network is nowadays suitable for thermal monitoring
The SensX Extreme. 2019, Cupertino (USA). Smartrac and SensThys. HUF RFId	The SensX Extreme is primarily focused on the smart building and construction market. The aim is to develop technology for leak detection and concrete curing. RFId tags, which can be embedded in the roof section, can detect the presence of water, and the drying phase of concrete during paving, thus enabling higher quality and speed on site. The tags can be used in common building materials, such as gypsum board, insulation, roofing, flooring, and concrete [43]
IFC-RFID system. 2020, Theran (IR). Islamic Azad Univ., Shahid Beheshti Univ., East Carolina University	This research presents a computerized system that integrates the BIM objects in IFC and radio-frequency identification to improve building maintenance performance. The computerized system is successfully applied to the building of a soccer stadium in Theran via the proposed research methodology using a qualitative and practical approach. The research indicates how a slight effort on the implementation of the proposed system could allow a significant improvement of overall maintenance performance [44]
WoodSense. 2021, Gävle (SE). Woodsense, ByggDialog Dalarna	WoodSense provides moisture measurement using passive sensors in the form of tags that can be attached to the wood building façade, wood slab panel, and or others building components. These sensors measure the moisture on the surface and have to be scanned on site with an RFId reader. The application of such solutions is particularly effective in facilities for wooden building that require more attention to environmental phenomena [45]

service life such as building components. The service life of a tag can be as long as 15–20 year but is still shorter than that of building components. On the other hand, technological integration and the need for a battery may represent a limitation for large-scale application. Furthermore, it is evident that one of the main gap is the time limited responsibility of stakeholders in the management of an asset during its useful life. Transferring responsibility for an asset to the user once it has been sold (or the

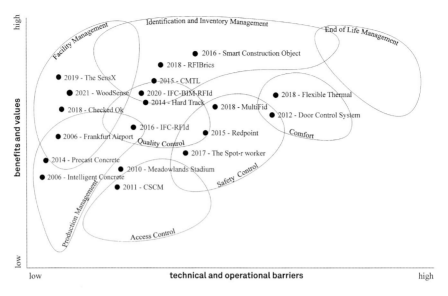

Fig. 3 Thematic areas and case studies for RFId technology in construction sector

warranty has expired) severely limits the interest of manufacturers in such technologies. However, the transition to circular approaches aimed at enhancing the value of the material and extending services (e.g. leasing, Product As a Service, Pay per Use, etc.) over time could change the current paradigm. In that case, circular solutions for inventory, management and end-of-life of an asset could require a much larger amount of information. A separate issue is the use of technology for comfort and thermo-hygrometric regulation of indoor environments. The increased attention to this topic could thus favor the application of RFId solutions for the built environment and offer new services and advantages to the user. The graph below clusters the experiences reported according to macro-topics emerged. A qualitatively identifying areas that have already been widely tested (e.g. access control) and potential areas (e.g. deconstruction) for future development in RFId is proposed. The deconstruction phase and end of life management could represent an emerging areas for RFId application. Although the technical and operational barriers makes the issue more complex, benefits and values in CE perspective could be relevant.

6 Conclusion and Future Development

More than 15 years later the first experiences of RFId in construction sector some considerations can be made. Advantages in a more rational use of resources, automation of processes, and the introduction of new dematerialized services promise new scenarios for construction sector. Industry 4.0 offers new technological tools and approaches to redesign production processes and create new opportunities for economic and environmental value generation. The digitization of the construction sector requires therefore a radical upgrade of the industrial infrastructure boosting product and process innovation (SDG 9). This objective is necessary to the achievement of SDG 12, which aims to promote more resource-efficient models. However, the impact that such technologies and the creation of big data from an environmental point of view must be considered. The most recent estimates indicate that the amount of digital data will grow from 33 zettabytes in 2018 to 175 by 2025. In this perspective, RFId and IoT technologies will contribute considerably to data creation generating an impact on the environment. Future developments in technology and data processing should thus aim to reduce their environmental impact in terms of materials consumed (e.g. to build tags and sensors) and climate-changing gas emissions (e.g. big data management). Although the challenge is complex and extremely interdisciplinary, the circular transition in the construction sector can be strongly enabled by new Industry 4.0 technologies.

References

1. European Union: Closing the loop—An EU action plan for the Circular Economy A (2015). https://eur-lex.europa.eu/resource.html?uri=cellar:8a8ef5e8-99a0-11e5-b3b7-01aa75ed71a1.0012.02/DOC_1&format=PDF. Last Accessed 23 April 2022
2. Vittoria, E.: Progetto, cultura, tecnica, in Controspazio. Gangemi Editore, Roma (1983)
3. Masino, G.: Industria 4.0 tra passato e future. In: Salento, A. (eds.) Industria 4.0: Oltre il determinismo tecnologico, pp. 23–40. TAO Digital Library, Bologna (2018). https://doi.org/10.6092/unibo/amsacta/604
4. Sachs, J.D., Schmidt-Traub, G., Mazzuccato, M., Messner, D., Nakicenovic, N., Rockstrom, J.: Six transformations to achieve the sustainable developments goals. Nat. Sustain. (2019). https://doi.org/10.1038/s41893-019-0352-9
5. Sustainable Development Goals. https://sdgs.un.org/goals. Last Accessed 23 April 2022
6. Nobre, G.C., Tavares, E.: Assessing the role of big data and the internet of things on the transition to circular economy: Part I. Johnson Matthey Technol. Rev. **64**(1), 19–31 (2020). https://doi.org/10.1595/205651319x15643932870488
7. Ingemarsdotter, E., Jamsin, E., Balkenende, R.: Opportunities and challenges in IoT-enabled circular business model implementation—A case study. Resour. Conserv. Recycl. **162**, 105047 (2020). https://doi.org/10.1016/j.resconrec.2020.105047
8. United Nations Environment Program. 2020 Global status report or buildings and construction (2020). https://wedocs.unep.org/bitstream/handle/20.500.11822/34572/GSR_ES.pdf. Last Accessed 21 March 2022
9. European Commission. https://ec.europa.eu/growth/sectors/construction_en. Last Accessed 21 March 2022
10. Ranta, V., Aarikka-Stenroos, L., Väisänen, J.M.: Digital technologies catalyzing business model innovation for circular economy—Multiple case study. Resour. Conserv. Recycl. **164** (2021). https://doi.org/10.1016/j.resconrec.2020.105155
11. Erabuild. Review of the current stat of Radio Frequency Identification (RFId) Technology, its use and potential future use in Construction, Final Report (2006). https://www.teknologisk.dk/_/media/23536_Report%20RFID%20in%20Construction%20-%20Final%20report%20EBST%20(3).pdf. Last Accessed 21 April 2022

12. Costa, F., Genovesi, S., Borgese, M., Michel, A., Dicandia, F.A., Manara, G.: A review of RFID sensors, the new frontier of Internet of Things. Sensors **21**, 3138 (2021). https://doi.org/10.3390/s21093138
13. Lu, W., Ting, L., Johan, S., Gang, W.: Design of chipless RFID tag by using miniaturized open-loop resonators. IEEE Trans. Antennas Propag. **66**(2), 618–626 (2018). https://doi.org/10.1109/TAP.2017.2782262
14. Google Trends. https://trends.google.com/trends/
15. Klaus, F.: RFID Handbook: Fundamentals and Applications in Contactless Smart Cards, Radio Frequency Identification and near-Field Communication. Wiley, United Kingdom (2010). https://doi.org/10.1002/9780470665121
16. Want, R.: An introduction to RFID technology. IEEE Pervasive Comput. **5**(1), 25–33 (2006). https://doi.org/10.1109/MPRV.2006.2
17. Marrocco, G.: The Art of UHF RFID antenna design: impedance matching and size-reduction techniques. IEEE Antennas Propag. Mag. **50**(1), 66–79 (2008)
18. Ron, W.: RFID: a technical overview and its application to the enterprise. IT Prof. **7**(3), 27–33 (2005)
19. Consolis. https://www.consolis.com/sustainability/#projects
20. Musa, A., Dabo, A.A.: A review of RFID in supply chain management: 2000–2015. Glob. J. Flex. Syst. Manag. **17**(2), 189–2281 (2016). https://doi.org/10.1007/s40171-016-0136-2
21. Zhang, J., Tian, G.Y., Zhao, A.B.: Passive RFID sensor systems for crack detection & characterization. NDT E Int. **86**, 89–991 (2017)
22. Vizinex RFId. https://www.vizinexrfid.com. Last Accessed 15 April 2022
23. Meyers, J.J., Hernandez, E.S.: Implementation of Radio Frequency Identification (RFID) Sensors for Monitoring of Bridge Deck Corrosion in Missouri. Center for transportation infrastructure and safety. Missouri University of Science & Technology Department of Civil, Architectural & Environmental Engineering (2014). https://doi.org/10.13140/RG.2.1.2528.5843
24. Swedberg, C.: McDonald's, other companies test TAG sensors' RFID temperature loggers. RFId J. (2016). https://www.rfidjournal.com/mcdonalds-other-companies-test-tag-sensors-rfid-temperature-loggers. Last Accessed 12 April 2022
25. SenseAnywhere. https://www.senseanywhere.com. Last Accessed 15 April 2022
26. Legner, C., Thiesse, F.: RFID-based facility maintenance at Frankfurt airport. IEEE Pervasive Comput. **5**(1), 34–39 (2016). https://doi.org/10.1109/MPRV.2006.14
27. Daly, D., Melia, T., Baldwin, G.: Concrete embedded RFID for way-point positioning. In: International Conference on Indoor Positioning and Indoor Navigation, pp. 15–17 (2010)
28. SKANSKA: New home for Jets and Giants. Worldwide. https://group.skanska.com/4a0579/siteassets/media/worldwide-magazine/2010/worldwide-no-2-2010.pdf. Last Accessed 21 April 2022
29. Shin, T.H., Chin, S., Yoon, S.W., Kwon, S.W.: A service-oriented integrated information framework for RFID/WSN-based intelligent construction supply chain management. Autom. Constr. **20**(6), 706–715 (2011). https://doi.org/10.1016/j.autcon.2010.12.002
30. Schüco. https://sepam-serramenti.it/pdf/progettisti/porte/door_control_system_it_en.pdf. Last Accessed 21 April 2022
31. Swedberg, C.: North Carolina transportation Dept. tracks precast concrete and samples. RFID J. (2014). https://www.rfidjournal.com/north-carolina-transportation-dept-tracks-precast-concrete-and-samples. Last Accessed 21 April 2022
32. Swedberg, C.: RFID Is the cure for San Francisco construction project. RFID J. (204). https://www.wakeinc.com/wp-content/uploads/2017/12/Reprint-RFID-Is-the-Cure-for-San-Francisco-Construction-Project-RFID-Journal.pdf. Last Accessed 21 April 2022
33. Redpoint. https://www.redpointpositioning.com. Last Accessed 30 April 2022
34. Soltani, M.M., Motamedi, A., Hammad, A.: Enhancing cluster-based RFID tag localization using artificial neural networks and virtual reference tags. Autom. Constr. **54**, 93–105 (2015). https://doi.org/10.1016/j.autcon.2015.03.009

35. Xue, F., Chen, K., Lu, W., Niu, Y., Huang, G.Q.: Linking radio-frequency identification to building information modeling: status quo, development trajectory, and guidelines for practitioners. Autom. Constr. **93**, 241–251 (2018). https://doi.org/10.1016/j.autcon.2018.05.023
36. Motamedi, A., Soltani, M.M., Setayeshgar, S., Hammad, A.: Extending IFC to incorporate information of RFID tags attached to building elements. Adv. Eng. Inform. **30**, 39–53 (2016). https://doi.org/10.1016/j.aei.2015.11.004
37. Triaxtec. https://www.triaxtec.com/wp-content/uploads/2020/04/LettireConstruction-CaseStudy.pdf. Last Accessed 12 April 2022
38. Permasteelisa Group. https://www.permasteelisagroup.com/project-detail?project=1841. Last Accessed 21 April 2022
39. Pantoli,L., Gabriele,T., Donati, F.F., Mastrodicasa, L., Berardinis, P.D., Rotilio, M., Cucchiella, F., Leoni, A., Stornelli, V.: Sensorial multifunctional panels for smart factory applications. Electronics **10**, 1495 (2021). https://doi.org/10.3390/electronics10121495
40. Hsieh, M.J., Liang, R.H., Huang, D.Y., Ke, J.Y., Chen, B.Y.: RFIBricks: interactive building blocks based on RFID. In: CHI '18: Proceedings of the 2018 CHI Conference on Human Factors in Computing Systems (2018). https://doi.org/10.1145/3173574.3173763
41. CoreRFId. https://www.corerfid.com/casestudies/anderco-lifting/. Last Accessed 21 April 2022
42. Giusto, E., Gandino, F., Greco, M.L., Rebaudengo, M., Montrucchio, B.: A dense RFID network for flexible thermal monitoring. In: 6th International EURASIP Workshop on RFID Technology (2018). https://doi.org/10.1109/EURFID.2018.8611649
43. SensThys. https://www.sensthys.com/sensx-extreme/. Last Accessed 30 April 2022
44. Kameli, M., Hosseinalipour, M., Sordoud, J.M., Ahmed, S.M., Behruyan, M.: Improving maintenance performance by developing an IFC BIM/RFID-based computer system. J. Ambient. Intell. Humaniz. Comput. **12**, 3055–3074 (2021). https://doi.org/10.1007/s12652-020-02464-3
45. WoodSense. https://en.woodsense.dk. Last Accessed 30 April 2022

Digital Tools for Building with Challenging Resources

Christopher Robeller

Abstract We present an assembly- and fabrication aware reciprocal frame construction system that exploits new possibilities of the latest generation of automatic joinery machines. Sweet chestnut wood (Castanea sativa), is a species that is currently not used for building construction in Germany. The wood of castanea sativa is highly durable and ideal for exterior conditions, but it will corrode metal connectors unless they are stainless steel. Therefore, our system uses only digitally fabricated wood-wood dovetail joints. It was inspired by Friedrich Zollingers "Zollbauweise", in its geometry as well as its philosophy—while adding a second curvature to increase stability and considering assembly constraints of the dovetail joints.

Keywords Digital timber construction · Building with less used timber species · Digital fabrication

United Nations' Sustainable Development Goals 9. Build resilient infrastructure, promote inclusive and sustainable industrialization and foster innovation

1 Introduction

Following the UN Environment Global Status Report 2017, the global need for buildings will drastically increase in the next few decades. According to the report the "global floor area" of currently approximately 2.5 trillion square feet will double to 5 trillion square feet until the year 2055 [14]. Facing such an enormous need for new buildings raises the question for new building methods with less carbon dioxide emissions, renewable resources and generally less material consumption. In

---

C. Robeller (✉)
Architektur und Bauwesen, Hochschule Augsburg, An der Hochschule 1, 86161 Augsburg, Germany
e-mail: christopher.robeller@hs-augsburg.de

© The Author(s), under exclusive license to Springer Nature Switzerland AG 2024
M. Barberio et al. (eds.), *Architecture and Design for Industry 4.0*, Lecture Notes in Mechanical Engineering, https://doi.org/10.1007/978-3-031-36922-3_19

this context, wood is a highly promising building material. While it has been used for buildings since ancient times, applications such as its use in larger, multi-story structures have seen considerable progress in recent years. These advances have been greatly supported by two developments. For once, building with sustainable materials has been incentivized by many countries, for example through funding or timber construction quotas, as well as companies and individuals choosing sustainable materials for their buildings taking personal responsibility. Also, energy prices have greatly increased and the processing of wood requires less energy than other building materials such as steel or concrete. At the same time, in many regions such as the European Union, the building industry is facing shortages of skilled workers, since fewer young people decide for physically demanding and dangerous work on building construction sites. Therefore, prefabrication of buildings in factories has increased in popularity, allowing for a more automated, safe and precise way of building compared to on-site construction. In this context, wood is particularly valuable due to its outstanding weight-to-strength ratio. Combined with its low-energy processing, this makes it an ideal material for prefabrication and transportation. "Integral attachment", the joining of components through features in their form plays a particularly import role for such prefabricated structures, since it allows not only for the transfer of stresses between parts, but also for "embedding" alignment features in the components, which greatly improve the ease, precision and safety of the final on-site assembly [13].

A major disadvantage of wood however is, that homogenous, isotropic materials such as concrete and steel can be calculated more easily than materials such as timber, which is anisotropic and hygroscopic. Depending on the type of tree species, wood may also have many defects, which make calculations even more difficult. Similarly, European forestry is focused on growing spruce trees, as straight and as free of knots as possible. However, many of the monocultures that were created this way are currently challenged by climate change. This raises the general question, can we build efficiently with less optimal, more challenging materials? In the early twentieth century, Friedrich Zollinger, city building director of the German town of Merseburg was challenged with a similar question. After World War I, resources such as high-grade materials, construction equipment and skilled workers were scarce. His approach to this problem was engineering a smart building system, using only short wood components and generally saving 40% of the material compared to a standard roof construction. The system was even designed in a way that would allow the participation of citizens in the construction process safely. In our history, technological advances have often lead to an increased demand for energy and resources, however scarcity can also be a driver for innovation.

2 Castanea Sativa

According to climate change prognoses, the forests in the warmer regions of Europe, such as the German state of Rhineland-Palatinate, will be facing great changes and challenges in the next decades. For example, the 2021 report of the department of forestry in Rhineland-Palatinate shows that only 20% of the trees in the entire state are healthy without any damage. In the year 1980, this was the case for 60% of the trees being completely healthy. Especially the spruce, currently the most important wood for building construction is greatly affected by climate change, with a prognosis of complete disappearance in this state within the century. While many of the spruce trees in this region were planted rather than growing there naturally, the most common tree species in the European forests and especially in Rhineland-Palatinate, the Beech (Fagus Sylvatica), also shows a worrying rate of damage with only 10% of the trees being fully healthy, compared to 55% in the year 1980. Therefore, it has been the subject of recent research to determine tree species which can fill the gaps and stabilize the forests during the observed and predicted climate changes during the next decades [10, 12].

The sweet chestnut (Castanea Sativa) (Fig. 2) is native and well known in the Mediterranean countries in Europe, such as Spain, France and Italy. Compared to the other European leafed trees, such as the Oak (Quercus Petraea) and the beech (Fagus sylvatica), the sweet chestnut is considerably more resistant to dry and warm climate. Due to the climate change in recent years, the occurrence of Castanea Sativa has therefore increased in many regions, including regions north of the alps where it was very rare previously [11] (Fig. 1).

3 Construction System

The construction system for our research demonstrator utilizes the *reciprocal frame method* to allow for building large floor or wall components using relatively small wooden elements. Reciprocal frames have been known for a long time, including bridges in ancient China, medieval ceiling constructions, famous sketches in Leonardo da Vinci's Atlantic Code, and the "Zollbauweise" construction methods developed by Friedrich Zollinger. However, only relatively few structures have been built using this method, most likely due to the widely available of relatively inexpensive alternatives such as steel or concrete structures, or glue-laminated wood products, which all allow for the relatively simple construction of large span structures. Many of the previously listed historical reciprocal frame structure were built due to a lack of alternatives (e.g. medieval ceilings) or scarce resources, such as the Zollbauweise, which was developed after the first world war. Facing our future challenges and modern technological possibilities, such material-saving, lightweight construction methods should be reconsidered, especially since robotic fabrication technology is very well capable of producing complex building components. Recent

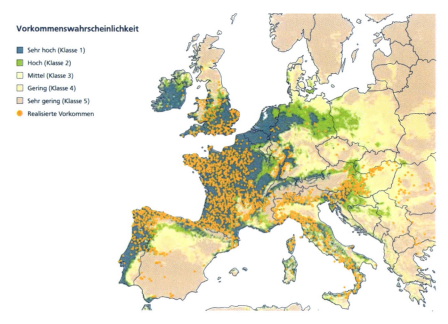

Fig. 1 European species distribution model of the sweet chestnut, illustrated using Worldclim 2.0 data. The orange dots represent the occurrence data included in the occurrence model based on the national inventory data of the respective countries. *Source* Thurm et al. [10] (use with friendly permission)

Fig. 2 Sweet chestnut hardwood, locally sourced near Annweiler, Germany

research has presented new interpretations of reciprocal frames using computational methods such as dynamic relaxation [1, 2, 5], and computer-aided structural analysis [3, 6]. Prototypes were presented using softwood and butt joints [3].

4 Assembly Constraints

Due to the high acidity of sweet chestnut wood, typical steel connectors will corrode. As an alternative, stainless steel connectors can be used, however those are considerably more costly. Reciprocal frames require a large number of connectors. Generally, achieving wood constructions using fewer chemical adhesives, more regional resources and less transportation to centralized facilities, will require a considerably number of other types of joints, such as form fitting connectors. Integral "wood-wood" connectors are a sustainable joining solution for timber structures, especially due to their previously mentioned other benefits such as a precise, simple, fast and precise on-site assembly of digitally prefabricated components. An important concept of integral "form-fitting" joints is to constrain relative motions between parts through the form of the connectors. Wood-wood connections such as mortise-and-tenon joints and dovetail tenons are so-called single-degree-of-freedom connectors (1DOF joint), where the form of the joint constrains the relative motions between the joints to only one translation vector, which allows for the assembly of the parts (Fig. 3).

In the case of dovetail tenons, the insertion of the dovetail tenon will always be from above, following the direction of gravity. Therefore, the majority of the stresses between parts is transferred through the wood-wood joint and only a minor additional joint, such as a single metal screw is needed to secure the joint in the opposite direction of its insertion. Another important feature of such integral joints

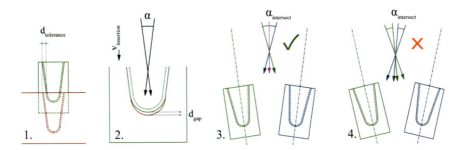

Fig. 3 a 1. Dovetail joint in front view, during assembly. Due to the V-shape of the tenon, there is a tolerance for the initial alignment. 2. Shows the fully engaged joint, where gap at the bottom allows for the complete insertion, even if the tenon is slightly too small. 3. Shows how two joints are assembled simultaneously. Due to the cone angle α., parts can be slightly rotated up to a max angle of α. 4. If parts are rotated beyond their joints cone angle α, they cannot be joined. **b** Reciprocal frame construction system, drawing shows main parameters alpha (dovetail joint cone angle) as well as beta (angle between neighboring components)—on target surfaces which are not flat, the dovetail joint angle must be equal or larger than the angle between neighboring components

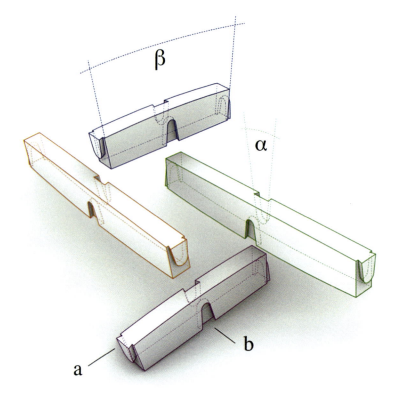

Fig. 3 (continued)

is to allow for a simple assembly. Therefore, the joint should ideally allow for some imprecision during the joining procedure, which is also called mating. Due to its V-shape, the dovetail connector perfectly combines these features, at the top of the V-shape, it allows for a simple, initial positioning of the elements because the dovetail mortise is wider than the dovetail tenon. This difference in width is gradually reduced during the mating of the parts, until a perfect fit is achieved. If a gap is left at the bottom of the V-mortise intentionally, it can be assured that even in the case of imprecise fabrication, a perfect fit of the V-joint is always achieved. A particular challenge with 1DOF joints, which have only one possible assembly direction, is to consider the overall assembly sequence of the construction. This challenge can be split into joint geometry constraints and component assembly sequence constraints.

Digital Tools for Building with Challenging Resources 337

5 Joint Geometry Constraints

Very often, parts will simultaneously connect to two or more neighboring parts. If these connections are 1DOF joints, their assembly vectors must be parallel. In the case of V-shaped dovetail tenons, the 1DOF connection is only established once the joint is fully engaged. Due to so called cone angle α, the joint can be inserted not only along one insertion vector, but along any vector within the cone angle. This means that parts can be slightly rotated and still connected with dovetail tenons, if the rotation between parts β does not exceed the cone angle α. This possibility was exploited for our research demonstrator "Castanea Sativa Pavilion", which uses a doubly-curved overall shape to increase the stability of the lightweight reciprocal frame gridshell. In this structure, the individual angles β of each component result from the target surface curvature and the length of the components. A higher target surface curvature will increase the component angles β, while a shorter length of the components will decrease β. The cone angle α of dovetail tenon joints is typically 15°. In our demonstrator, we have slightly increased this angle α to 18°, since the maximum curvature of the target surface and the maximum length of the individual components result from this angle.

6 Assembly Sequence Constraints

When multiple joints on a part can be joined simultaneously, they might still be blocked by other components in the assembly. Such mutual blocking of components in an assembly can be described through so called blocking graphs. A sequence must be found, so that every part in the assembly can be inserted. In the case of our demonstrator, due to the curvature of the gridshell, parts can only be inserted from the convex side of the shell. There is a start component in the bottom left front corner of the structure, from where all other components are inserted in a predetermined sequence.

7 CAD Plugin/Generator App

Knowing the geometrical constraints of the system, we developed a "generator app" to further explore the possibilities of this reciprocal frame gridshell system. This app was developed using the RhinoCommon Software Development Kit for the CAD Software Rhino3D. As inputs, the app requires 1. A target surface described by two sets of curves for the gridshell. This surface may be flat, singly-curved or double-curved. Also, the curvature or radii do not have to be constant, as in the original Zollinger Bauweise. We consider this an improvement, since structurally feasible shapes have a catenary cross section, with changes in curvature. The target surface

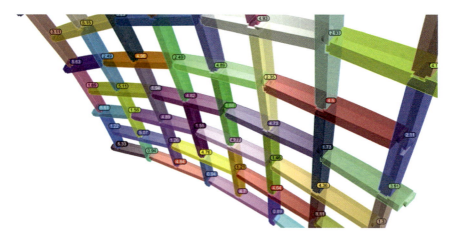

Fig. 4 Detailed 3D geometry file with joints, processed by the generator app in Rhino 3D

may also have positive or negative curvature, or both within the same surface 2. Two integer values for the gridshell subdivision of the surface in both directions. 3. The width of the beam components. 4. A minimum height of the beam components (the actual height will depend on the curvature of the target surface). 5. A selection what type of model should be generated, the geometry of the components, a detailed model including 3D joints, or an industry standard fabrication data file. All of the models can be generated simultaneously, however this will require slightly more time to process. The fabrication data file will automatically consider "raw part" sizes, which are rationalized to only a few different raw beam heights in 20 mm increments. This simplifies the processing in the factory (Fig. 4).

8 Fabrication-Aware Design

The construction system and the generator app were developed specifically for the industry standard wood joinery machines for linear wood elements, such as the most common Hundegger K2i. Other than normal, generic CNC machines for various materials and purposes, these wood joinery machines are highly specialized for linear timber beams. In particular, they have an automatic loading system with two independent gripper arms, feeding the raw parts through various stations such as sawing, milling, or drilling, each with individual motors and tools. In comparison to a typical CNC machine with only one motor, the wood machines offer much higher efficiency, while still offering very complex machining options. A very specific limitation however is presented by the gripper arms, shown in red color in Fig. 6. In order to safely hold the components during high-speed sawing, milling and drilling, the grippers need a relatively large surface area to clamp the workpieces. However, our chestnut-wood based construction system is utilizing short pieces of solid wood, following the materials naturally provided by the chestnut trees which do not grow very straight. Therefore, we had to find a compromise between the required gripper area and the locations for the 4 joints on every piece, 2 dovetail tenons and 2 dovetail mortises (Fig. 5).

Furthermore, since our generator app allows for the design of structures based on "freeform" surfaces, the individual components and their joints will all have individual shapes, lengths, components angles β and cone angles α. In a typical CNC fabrication process, this would be very challenging because only geometry is transferred from CAD to the Computer-Aided Manufacturing Software (CAM). The machining parameters are then set manually or semi-automatic, which is fine for the typical mass-production, mould- or prototype making processes. However, in the building industry each building is unique and CNC programs are created and run only for a single building component. Therefore, complete automation is critical. Our generator App was designed to directly export final production data, where no further CAM programming is needed. The files for our demonstrator, consisting of

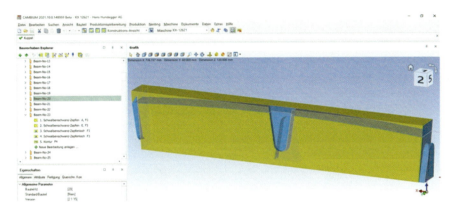

Fig. 5 Fabrication data file, processed by the generator app in Rhino 3D

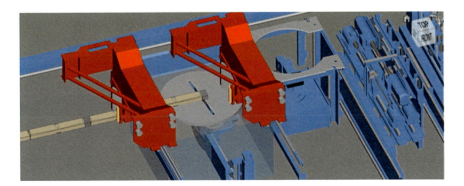

Fig. 6 Machine operation Simulation with a Hundegger K2 industry machine

147 individually shaped chestnut wood components were directly loaded into the software on a K2 industry machine.

9 Demonstrator Realization

After a series of fabrication and assembly tests, our goal was to test the production and construction of a small building structure. Due to our close collaboration with the state and regional forestry departments, we were offered the possibility to replace an old, existing forest hut along the "Chestnut hiking trail" near the town of Annweiler. The exact location is called "Wegspinne am Zollstock", which was a historical customs station, at the crossing of two old trails. This location allowed us to design a permanent structure, which was not demolished after our tests, but will serve the public for decades to come. The chestnut wood was left completely untreated, it is ideal for such exposed applications, without needing any harmful chemicals for the wood protection against decay and fungi. This strategy of combining the demonstrator with a permanent building project fit the overall sustainability spirit of the research project. First, the old forest hut was removed, since it was already in a bad condition. The foundation however, 1 3.5 m × 2.5 m concrete slab was left and re-used as a foundation for our demonstrator project. The ground surface of our demonstrator was slightly larger with 4.1 m × 2.8 m. The target surface was a doubly-curved surface forming an arch, where the cross-section was a catenary with a radius of 20 m at the bottom and 0.9 m at the top. In the other cross-section, along the surface the radius was 2.8 m up to 3.8 m, rising from an interior height of 2.7 m at the back to 3.4 m at the front. This surface was then divided into a gridshell with 6 segments in the short and 20 segments in the long direction. This resulted in 147 individually shaped beam components, with a total mass of 0.91 m^3 of wood, and a maximum component length of 0.92 m. the beam width was 60 mm with a minimum beam height of 100 mm and resulting raw part heights of 120, 140, 160, and 180 mm (Fig. 7).

The demonstrator was produced on a 2019 model Hundegger K2 industry machine with the Cambium Software system. The parts were produced in two days, the assembly was also carried out quickly and simply within the factory, since the structure was small enough to transport it to the site in one piece (Fig. 8, left). However, the system would also allow for the prefabrication and pre-assembly of multiple larger segments, which could then be combined on site, for example with a small amount of stainless-steel fasteners. In our demonstrator structure, each dovetail tenon joint was secured in place using a single self-tapping wood screw. These screws simplified the assembly, but the carry no loads in the structure and therefore costly stainless steel load bearing connectors were not necessary. Also note that other than typical Zollinger structures, our demonstrator does not have additional cross-bracing elements. Due to the double-curvature of the gridshell, the rigidity of the joints and the small size of the structure this was not necessary, however in a larger structure it would have to be added in the form of diagonal slats, just like the original Zollinger

Digital Tools for Building with Challenging Resources 341

Fig. 7 Dovetail joined parts, produced using a Hundegger K2i industry machine

System. Finally, on site, a shingled floor (Fig. 8, right) and a wooden roof were added to our demonstrator, where all of the wood was sweet Chestnut (Castanea Sativa) without any exemptions. The structure was inaugurated and opened to the public on the 30th of October 2021 (Figs. 9 and 10).

Fig. 8 Left: Assembly of the parts, Right: Sweet chestnut shingles roof cover installation

Fig. 9 Completed structure with sweet chestnut flooring and roof cover

Fig. 10 Pre-assembly of the structure in the factory

10 Conclusion

Facing the future challenges of architecture in times of climate change, a high demand for buildings structures and a lack of skilled workers in many countries, this research project wants to combine material saving lightweight constructions with renewable local materials and high-tech automated production technology. Our construction system is based on an assembly-aware algorithm, which considers the dovetail joint geometry and overall assembly sequence constraints. It is the first reciprocal frame structure made from Castanea Sativa wood and the first sweet chestnut wood structure produced on an automatic joinery machine. The vast majority of new buildings using renewable materials is currently made from softwood such as spruce. As described in this article, many areas where we currently source these softwood trees have been hit hard by the warm and dry summers of the last decade in particular [12]. Following the prognosis, we will not be able to grow these trees any more in many of these areas within the next 100 years. While many of these trees where not naturally growing in these areas, the natural trees such as beech and oak are also greatly affected by climate change, therefore research suggests to stabilize these forests with tree species like Castanea Sativa, which is highly resistant to heat and drought. The local region of our demonstrator has the largest amount of Castanea Sativa trees in Germany; however, our demonstrator was the first permanent structure using sweet chestnut wood as a structural material. We believe that further research in the architectural potential of lesser used and future proof wood species in close collaboration with the forestry experts is urgently needed.

Acknowledgements We would like to thank our generous project partners CLTech GmbH Kaiserslautern, Landesforsten Rheinland-Pfalz, Forstamt Annweiler, Haus der Nachhaltigkeit Johanniskreuz, Holzbaucluster Rheinland-Pfalz and Hundegger AG. The demonstrator project was funded by the Ministry of the Environment Rheinland-Pfalz (MUEEF). A special thanks to Jürgen Gottschall, Hannsjörg Pohlmeyer, Gregor Seitz and Michael Leschnig.

References

1. Douthe, C., Baverel, O.: Design of nexorades or reciprocal frame systems with the dynamic relaxation method. Comput. Struct. **87**(21–22) (2009)
2. Tamke, M., Riiber, J., Jungjohann, H., Thomsen, M.R.: Lamella flock. In: Advances in Architectural Geometry, pp. 37–48 (2010)
3. Franke, L., Stahr, A., Dijoux, C., Heidenreich, C.: How does the Zollinger Node really work? In: Proceedings of IASS Annual Symposia, vol. 2017, no. 11, pp. 1–10. International Association for Shell and Spatial Structures (IASS) (2017)
4. Dijoux, C., Stahr, A., Franke, L., Heidenreich, C.: Parametric engineering of a historic timber-gridshell-system. In: Proceedings of IASS Annual Symposia, vol. 2017, no. 17, pp. 1–9. International Association for Shell and Spatial Structures (IASS) (2017)
5. Song, P., Fu, C.-W., Goswami, P., Zheng, J., Mitra, N.J., Cohen-Or, D.: Reciprocal frame structures made easy. ACM Trans. Graph. (TOG) **32**(4), 1–13 (2013)

6. Kohlhammer, T., Kotnik, T.: Systemic behaviour of plane reciprocal frame structures. Struct. Eng. Int. **21**(1), 80–86 (2011)
7. Conedera, M., Manetti, M., Giudici, F., Amorini, E.: Distribution and economic potential of the Sweet chestnut (Castanea sativa Mill.) in Europe. Ecologia Mediterranea **30**, 179–193 (2004). https://doi.org/10.3406/ecmed.2004.1458
8. Conedera, M., Tinner, W., Krebs, P., de Rigo, D., Caudullo, G.: Castanea sativa in Europe: distribution, habitat, usage and threats (2016)
9. Anders, J.: Wuchsleistung der Edelkastanie (Castanea sativa Mill.) als klimaplastische Baumart in ausgewählten Beständen Ostdeutschlands. Diplomarbeit. Technische Universität Dresden, 115 S (2010)
10. Thurm, E.A., Hernández, L., Baltensweiler, A., Ayan, S., Raszto- vits, E., Bielak, K., et al.: Alternative tree species under climate warming in managed European forests (2018)
11. https://www.lwf.bayern.de/mam/cms04/service/dateien/w81_beitraege_edelkastanie.pdf
12. https://fawf.wald.rlp.de/de/veroeffentlichungen/waldzustandsbericht/
13. Robeller, C.: Integral mechanical attachment for timber folded plate structures. No. PhD THESIS. EPFL (2015)
14. UN Environment Global Status Report (2017)

Digital Deconstruction and Data-Driven Design from Post-Demolition Sites to Increase the Reliability of Reclaimed Materials

Matthew Gordon and Roberto Vargas Calvo

Abstract The research develops tools and strategies for urban mining and digital deconstruction to diminish the building sector's dependency on new natural resources. It facilitates the data capture, analysis, and characterization of secondary raw materials and defines a database system for recovered post-demolition components, promoting high-quality upcycled materials for new construction projects. A "form follows availability" digital design strategy is explored from a sparse quantity of reclaimed material. It develops a relational database from a semi-automated post-demolition item assessment, and the consequent extracted material (wood battens) is cataloged and stored before being matched and used for a new demonstrator using robotic fabrication. Each recovered element is imaged, scanned, and weighed to create a unique material health indicator. This information is presented in a user interface to help the designer filter for relevant materials. The final step of the system matches designed components with relevant stored materials by their generative design requirements. The system's flexibility is demonstrated using a construction system realizing curved surfaces from linear elements. By extracting multi-dimensional data on each wood batten and presenting their relevant indicators in a user-friendly interface, it is possible to create a dialogue between the designer and irregular shapes, augmenting the widespread use of reclaimed materials in structurally predictable assemblies. 85% of design components were well matched with the presented methods' database materials. The predictability of the system after fabrication is verified by a 10 mm maximum deviation between the as-designed and the as-built structure.

Keywords Reclaimed materials · Computer vision · Robotic fabrication · Circular economy · Relational database

United Nations' Sustainable Development Goals 9. Industry, innovation, and infrastructure · 11. Sustainable cities and communities · 12. Responsible consumption and production

---

M. Gordon · R. V. Calvo (✉)
Robotics and Advanced Construction, IAAC, 08005 Barcelona, Spain
e-mail: robert.en.var@gmail.com

1 Introduction

In order to transition to a high level of circularity in the construction industry, it is imperative to understand the number of resources we are consuming against the amount of waste we are producing. The sector is the number one consumer of global raw materials [1] while generating an alarming 25–30% of waste [2]. Resource efficiency can be increased by implementing technical innovations at multiple stages that keep our built environment circulating and contributing to local economies. For example, in the United States, 50% of all solid waste [1] created every year by construction processes could be accounted for in new buildings if we change the perspective on recyclable, reusable materials.

One solution to start using our resources smarter is to keep track of and connect the lifespans of our buildings. By better matching the supply and demand of building stocks, the more sources of high-value assets and materials the city can take for new projects, giving those resources useful second lives and market availability. Shifting the material flow back into urban areas will reduce the overall construction and demolition waste while simultaneously increasing the material recovery rate by one-third worldwide [3].

Current analysis of a demolition site is often carried out by visual inspection, along with judgments about recovery and reuse viability. The variability and clutter of worksites make it challenging to digitize this process efficiently, while post-demolition digitization suffers from increased disorder among the relevant materials themselves. Lastly, further information about material location, storage, and transport is necessary for approaching these processes cost-effectively. Creating a sufficient scope and depth of database is critical to connecting each material with a consumer while ensuring these connections are highly local and efficiently found.

Within this field, the research aims to increase the reliability of reclaimed materials by automating post demolition capture and assessment of material-specific information on a building and component level, thereby contributing to the capabilities and scale of material reuse effectiveness by reducing the data collection effort. Using a system of reality capture and analysis methods to digitize and qualify critical materials from post deconstruction sites makes it possible to better inform and plan future uses of discarded construction material off-site.

2 Methodology

The implemented methodology inverts the standard design strategy by starting the creative process from the available resources. A set of available parts informs and generates a dynamic communication between the intended design and material type, properties, place, and user. The irregularity and sparsity of the database are a catapult to unexpected architectural form and complexity.

The research explored other similar methodologies for reusing standardless and reclaimed materials. Projects like "Form Follows Availability" [4] and "Minimal-waste design of timber layouts from non-standard reclaimed Elements" [5] use the length and size data of linear elements to match the available database with the intended structure within certain design parameter flexibility. "Cyclopean Cannibalism" [6] explores the re-adaptation of material debris and creates a precedent for using a digital inventory of reclaimed material obtained through point cloud representations of each item. It also pairs the dataset with performance goals like waste reduction and fitting optimization.

"Digital fabrication of standardless materials" [7] deals with the inherent properties of standardless materials; being the natural variation, uncertainty, and unexpected geometrical and dimensional properties found in bamboo. The project advances towards utilizing non-standard components that work structurally as predicted by a parametric model. As with the other examples, digital tools help optimize and automate laborious manual evaluation and consider deeper material properties.

Based on the learnings from the previously mentioned research, the process started with selecting post-demolition material; in this case, reclaimed wood battens that were retrieved from a previous pavilion in the Institute for Advanced Architecture of Catalonia. The selection for this material was based on proximity, availability, and variability found on the site. 140 out of 2263 pieces were selected by visual judgment and stored without initial sorting. From the selection, 94 elements were fully scanned by a robotic procedure and tagged for future traceability.

The obtained point clouds and textural imagery underwent low-cost physical testing and digital evaluation to pre-select elements more appropriate to structural vs aesthetic uses. After a robotic procedure has scanned the material, the research proposes a qualitative score on each element. The synthesized qualitative data becomes a "Material Health" indicator and is communicated via web-based and design-software-integrated viewers.

A parametric design strategy was chosen considering the amount of material available. Based on a form of doubly-curved surfaces, the system exposed the scale, width, height, and local curvature parameters for adjustment. A SQL database with the information gathered in the previous step served as a digital inventory representing the physical objects and their reuse potential within the design. A multi-objective optimization algorithm ran several iterations until the chosen criteria were met.

Due to design flexibility and a division of geometric complexity, the most energy-intensive robot fabrication techniques were only necessary for producing the smaller connective components in the structure. After producing a buildable digital model, the construction began by retrieving the indicated items from the storage location and marking them by projecting rationalized information into each piece. Finally, with all the battens marked, the assembly took place in situ with the help of the digital twin.

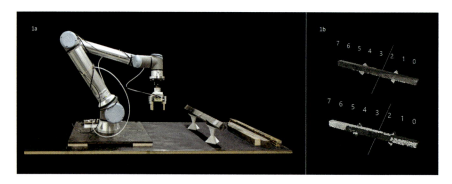

Fig. 1 **a** The UR10 in combination with the gripper-scanner tool and the scanning holder. **b** For the scanning process, each side requires seven camera captures to be registered afterward

3 Post-Demolition Item Assessment

3.1 Robotic Data Capture

The setup consisted of one UR10 robot, operating a customized pneumatic gripper mounted with a D435i Intel RGBD camera (Fig. 1a). The workspace featured a scanning stand, printed with TPU and mounted to the table magnetically, to balance between creating a steady platform and preventing damage in case of dropped materials or materials with unexpected dimensions. Intended explicitly for components with a rectangular cross-section, the stand holds components at 45° to expose multiple sides and allow for scanning in only two passes. The end tool's depth camera was mounted at an offset and at a 35° angle to prevent vision occlusion from the gripper.

Drawing from a known safe zone behind the scanning area, the robot picked up individual components for transfer to the scanning holder. While multiple angles of inclination were tested, it was found that further analysis worked sufficiently from a single row of scan captures along each half of the object. Finally, the robotically controlled camera position allowed for automatically registering captured clouds producing the final geometry (Fig. 1b).

3.2 Photogrammetry Procedure for Data Capture

While a viable automation procedure, due to the relatively low capture resolution from the Realsense camera, it was impossible to perform surface quality analysis on the reconstructed mesh or accurate point cloud measurements. In order to capture high-quality point clouds and create a reliable dataset, a photogrammetry method was also employed (Fig. 2a). An automatic subject masking was performed to increase the speed of each point cloud creation and ensure a quicker way to create a sparse cloud.

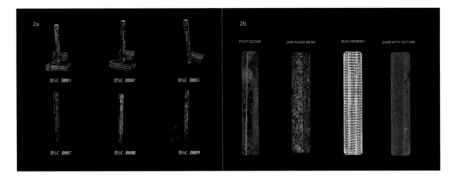

Fig. 2 a Tie points between each picture with subject segmentation. **b** Process for a usable mesh. Geometry decimation and optimization for mesh topology of each item

Each of the point clouds was produced from 25 to 35 high-resolution photographs. The resulting geometry was automatically scaled and normalized to the world axis to align the items. Each cloud was then converted into a high-resolution mesh, further decimated, and finally remeshed with a high-quality texture but low polygon count for faster processing (Fig. 2b).

3.3 Analysis

Surface curvature. Curvature analysis was based on the finalized mesh rather than the raw point data; while this resulted in a somewhat lower resolution, it led to viable efficiency within the Grasshopper environment. Consequently, an absolute curvature was extracted on every batten; the values were then averaged into a score later combined with the other analyses for a "Material Health" final score (Fig. 3a).

Curvature evaluation of mesh skeleton. The curvature evaluation of the mesh skeleton analyzed the possible wood warping in the longitudinal axis of each wood batten. The system averaged the centerline of the point cloud on which curvature analysis was performed; it could then approximate and relate them with wood warping problems depending on its location and frequency of values (Fig. 3b).

Mass comparison. Given that all methods thus described operated on surface details, an analysis of the components' mass was carried out to judge the quality of the element's interior. Given the known or estimated species of wood used (based on location, building trends, etc.), the expected mass was calculated using the measured dimensions and species average density. Simultaneously, the actual mass was measured manually or via torque sensors in the automation system. The ratio of the actual-to-predicted mass indicated how much decay may have occurred over

Fig. 3 a Surface curvature analysis. **b** Two examples of mesh curvature visualization. **c** Example mass ratios for structurally usable vs a highly decayed wooden element. **d** Example output when localizing texture defects using Mask RCNN

time. For example, the worst decayed test pieces had only 55% of the expected mass (Fig. 3c).

Textural defect detection. As not all issues were detectable via geometry, analysis was also performed on the image textures extracted in previous steps. The primary defects considered were knot holes and nail/metal connector holes, appearing visually similar and representing similar possible structural issues. Localization was performed using the Mask R-CNN algorithm [8]. This allowed each defect to be stored by its face and local position, available for deeper future structural analysis (Fig. 3d).

Element Tagging, Lifespan Tracking, and Interface. Each recovered piece received a data frame of extracted geometric, structural, and textural information applicable to its future lifetime. For aesthetics and data stability, it was decided to store the bulk of this information in an external database. This was built as a relational SQL database, with a primary table storing each element's rough dimensions and origin site, with additional associated tables storing data from each analysis method. Each table was associated by the element's UUID (universally unique identifier), thus allowing each element only to be marked by a representation of the id's 128-bit number. QR codes were chosen to store this id, given their built-in redundancy, low-cost application, and ubiquity of software to read them.

A mobile app was developed for on-site data lookup to test ways this tracking system would be accessed on-site. The app scanned the QR codes of chosen elements and retrieved a portion of the information from the MySQL database.

3.4 Database Interface

A user interface was developed to visualize the information gathered and verify the relevance of the database categories (Fig. 4a). This digital inventory viewer served as a retrieval tool and displayed indicators like mass ratio, usage, warping percentage, number of nails found, surface curvature, and in combination, a weighted synthesized score comprehending the previous indicators called "Material Health". Lastly, profile shape and dimensions are also shown (Fig. 4b).

A web-based interface (Fig. 5a) served as an approach to test the applicability of the database for a broader range of users. In this case, the information gathered could be accessed by a secondary resources supplier or remanufacturer (Fig. 5b); simultaneously, it functions as a retrieval system for designers, with filter options like minimal quality, distance, and defects to retrieve the material needed for each task. After the material has been selected, the relevant database entries can be downloaded into a design environment like grasshopper through a SQL plug-in.

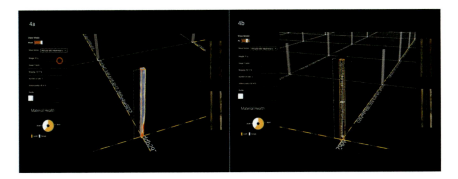

Fig. 4 **a** Database interface image on mesh view. **b** Database interface image on point cloud view

Fig. 5 **a** UI Image. The prototype viewing interface for recovered elements running on Firefox. **b** UI Image. Possible defect locations are highlighted

4 Design Strategy for Uncertain Materials

The design of the demonstrator highlighted the goals of design and material adaptability that digitization brings to the process in the form of a small-scale shelter pavilion. The design was based on approximating arbitrary input surfaces using two layers of opposingly oriented pieces. The development of this system took initial inspiration from the overlapping geometry of reciprocal structures, although this orientation does not utilize the same structural ideas. This strategy allowed for maximum flexibility to adapt to the available inventory as individual lengths can vary significantly and still fit the overall dimensions and design requirements (Fig. 6b).

Given a starting surface, it was first converted into a mesh consisting of diamond strips. These edges were then turned into initial solid 'reciprocal'-layout forms using the Grasshopper NGon plugin. These pieces were divided into two layers based on their orientation in the original surface's UV space. Each piece was then scaled to overlap with its neighbors, and offset from the surface by a calibrated amount. Thus, the two layers are connected by a series of specifically cut connector pieces (Fig. 6a).

Each connector (Fig. 6c) decomposes the vector between each main face into two simple rotations at the end faces of the connector. This simplified the digital fabrication of the connectors and allowed them to be more clearly aligned with the main pieces they attached to.

The structure considered the dimensional data to reduce the design space variants and the labor and energy required to cut the item. As with the linear elements of the structure, the connectors needed to be consequent with the usage optimization from available resources, reduce waste production, and use as much as possible the length of the item from which the connectors were going to be extracted.

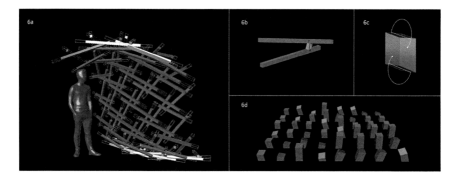

Fig. 6 **a** Digital design process from input surface to layered structure. **b** The dimensional flexibility of elements; varies in length by 0–20%, depending on inventory availability; reducing additional labor. **c** The faces of each connector only contain simple rotations to encode the overall vector being connected

Digital Deconstruction and Data-Driven Design from Post-Demolition ...

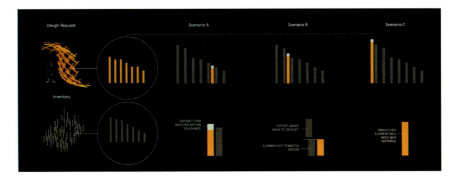

Fig. 7 Outcomes from a design piece matching itself against the database

4.1 Matching Material Stock from Availability

With this strategy for representing arbitrary surfaces with linear wooden elements, the approach needed to be realized with the available materials. With material-buyer matching already a common process for reclaimed material marketplaces, a system was developed for matching each piece to be assembled with one from the database. This was based on a top-down greedy matching system with pre-sorting by length (Fig. 7), a heuristic that generally reduces cutoff waste [9]. Each element had a degree of tolerance where it could match with elements larger than designed but never with elements shorter. As the database is arbitrarily large, the list of elements to be matched against comes from an initial reduced database selection, immediately filtering to relevant items based on dimensions, quality, or distance from storage. Particularly here, a strict quality threshold was used, given the degree of decay seen in the stock.

Each design element thus had three possible outcomes; either it was well-matched with an actual piece within tolerance, it was matched with a significantly larger piece that would need to be cut down, or no match was found and would require fulfillment with new stock. Each matching pass could then be assigned a score, with the amount of new stock and new labor required negatively impacting the viability of a particular design option given the available materials. In a larger scenario, the transport distance of each item would also serve as a negative variable.

4.2 Design Interface

The design interface contains database parameters visually integrating the database's impact within the design (Fig. 8a). Information about the number of matched items is displayed; the user is informed how much alteration or new materials may be required. An optimization algorithm runs every time the selection of the database

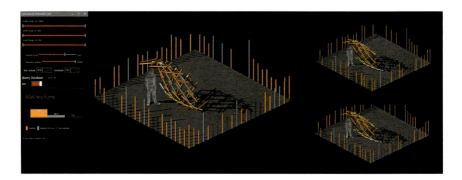

Fig. 8 a Images of the UI environment. The left panel sets parameters for an initial database filtering. During the interaction, the "Matched Items" graph will adjust accordingly, showing the percentage of design elements that were matched, matched after a cut, or not matched at all

changes [10]. The fitness for the optimization is the maximization of the matched elements while reducing the labor needed; that is, reducing the amount of cutting or any modification of the pieces.

Detail design parameters like the number of mesh subdivisions and surface curvature will vary depending on the number of elements and their properties from the database. If the database is primarily composed of long elements, the resulting design will have fewer subdivisions to accommodate the specific lengths of those elements. Surrounding the model, there is a visible representation of the database, each of the elements will turn orange if the item is found and matched for the design, gray, which means it is matched, but some labor is required, or black outline, defining a no match for that particular wood batten.

The chosen design for fabrication resulted in 85% of their components being matched with the material from the inventory, from which only 10% required to be cut to size. The second objective ensures the constructability of the system, avoiding collisions between the battens and the connectors.

5 Digital Fabrication Methods

Following this design strategy, most of the pre-assembly work was encapsulated in the connector pieces, as they contained most of the essential measurements of the design. A fabrication system was chosen such that theoretically, all necessary operations could be carried out by a human using typical woodworking tools; the inclusion of the robotic system would increase speed, safety, and accuracy. The connectors were successfully fabricated with precision using an automated method, while the rest of the construction was handled manually with the aid of a CAD interface.

Fig. 9 **a** The digital model for fabrication highlights the item to be assembled, indicating the connectors ID and the IDs of the other elements to which the piece is connected. **b** The projected user interface on the wood batten

A fabrication model was developed (Fig. 9a) to help during the assembly of the structure. The interface displays in the design space the location of the selected item, relevant connectors IDs, and the design IDs of the elements connected to that particular piece. The component meshes are color-coded as an extra layer of information depending on which oriented layer of the pavilion they belong to.

5.1 Projection Assisted Annotation

While each structural component contained a degree of variability in its total length, the intersection and connection points within its length were highly specific. This information was applied to each component using an assistive system consisting of a work table with a projector mounted overhead and displaying the digital model [11].

The projection displayed the minimum and maximum design length, connector location, id, and orientation (Fig. 9b). The connector locations were traced for all primary-layer components, and each connector was additionally pre-glued in place (Fig. 10b) for each secondary-layer component. Each layer component was assigned an incrementing id based on its layer (e.g. P2 or S5). Each connector piece could be uniquely identified by which layer components it is attached to (e.g. 2|5). Each connection point on a layer piece was likewise identified by the id of the opposite layer piece.

5.2 Demonstrator Assembly

Final assembly was performed by hand, using wood screws for all connections. Starting from the bottom diamond required multiple workers and clamping to hold

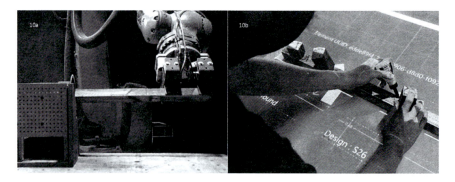

Fig. 10 a Photograph of the robotic cutting process for the connectors. **b** Pre-gluing connectors for all secondary-layer structure components

the slightly twisted quad in place (Fig. 11a). However, from there, each layer of diamonds was added progressively, with the screws at the top of each diamond added last in order to pull tension into the system if necessary. Due to the density of labeling applied during the setup phase, upcoming pieces and their orientation could be found by reading the current raw edge of the structure, reducing the need to review against the computer model often.

After the structure was completed (Fig. 11b), a point cloud was created through photogrammetry and compared to the original geometry (Fig. 12). The results from the deviation analysis showed that 85% of the area of the structure was completed within 10 mm tolerance.

Fig. 11 a The initial diamond module requires the most manual labor. **b** Image composition showcasing the material intelligence behind each fabricated item. Digital layers help improve the reliability of reclaimed materials

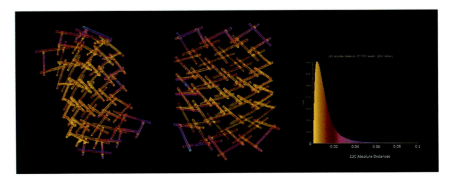

Fig. 12 Deviation analysis

6 Conclusions

This project has developed a total-lifespan set of methods and technologies for applying digitization to mass re-use construction and demolition waste. Each stage contains both an operational base for new value or greater efficiency and many opportunities for further development.

The research understands the complexity of ensuring local material banks from unused buildings and focuses intensely on one source of irregular material: wood battens from a previous pavilion. It recognizes the challenge to introduce, promote and pair the capture and serving of relevant information through databases to planners and architects. Several iterations of user interfaces create a communication channel between the profoundly technical and the creative and discovery nature of design environments, stating that a mental model shift needs to occur to create an environmentally conscious design from a material intelligence perspective.

The results obtained from the demonstrator optimized repurposed material utilization and fabrication efficiency, promoting locally sourced, readily available, and reclaimed materials in a circular approach. Ultimately, the demonstrator was a highly specific case, with a focus on realizing the form and aesthetic. The application and testing of these technologies in progressively more practical constructions will uncover further structural, spatial, and regulatory constraints to be met by the optimization and design of communication systems.

References

1. World Economic Forum: Shaping the Future of Construction A Breakthrough in Mindset and Technology. World Economic Forum, May (2016)
2. European Commission: Construction and demolition waste. European Commission. https://ec.europa.eu/environment/topics/waste-and-recycling/construction-and-demolition-waste_en. Accessed 30 Jan 2022

3. Towards the Circular Economy: Economic and business rationale for an accelerated transition. Ellen Macarthur Foundation, 1 (2013). https://www.ellenmacarthurfoundation.org/assets/downloads/publications/Ellen-MacArthur-Foundation-Towards-the-Circular-Economy-vol.1.pdf
4. Brütting, J., Senatore, G., Fivet, C.: Form follows availability—Designing structures through reuse. J. Int. Assoc. Shell Spat. Struct. **60**(4), 257–265 (2019). https://doi.org/10.20898/j.iass.2019.202.033
5. Parigi, D.: Minimal-waste design of timber layouts from non-standard reclaimed elements: a combinatorial approach based on structural reciprocity. Int. J. Space Struct. **36**(4), 270–280 (2021). https://doi.org/10.1177/09560599211064091
6. Clifford, B., McGee, W.: Cyclopean Cannibalism. A method for recycling rubble (2018). http://papers.cumincad.org/cgi-bin/works/paper/acadia18_404. Accessed 27 April 2022
7. MacDonald, K., Schumann, K., Hauptman, J.: Digital Fabrication of Standardless Materials (2019)
8. He, K., Gkioxari, G., Dollár, P., Girshick, R.: Mask R-CNN. ArXiv170306870 Cs, January (2018). http://arxiv.org/abs/1703.06870. Accessed 26 April 2022
9. Bukauskas, A., Shepherd, P., Walker, P., Sharma, B., Bregulla, J.: Form-Fitting Strategies for Diversity-Tolerant Design (2017)
10. Eversmann, P.: Robotic fabrication techniques for material of unknown geometry. In: De Rycke, K., Gengnagel, C., Baverel, O., Burry, J., Mueller, C., Nguyen, M.M., Rahm, P., Thomsen, M.R. (eds.) Humanizing Digital Reality: Design Modelling Symposium Paris 2017, pp. 311–32. Springer, Singapore (2018).https://doi.org/10.1007/978-981-10-6611-5_27
11. Huang, C.-H.: Reinforcement Learning for Architectural Design-Build—Opportunity of Machine Learning in a Material-informed Circular Design Strategy (2021)

Impact and Challenges of Design and Sustainability in the Industry 4.0 Era: Co-Designing the Next Generation of Urban Beekeeping

Marina Ricci, Annalisa Di Roma, Alessandra Scarcelli, and Michele Fiorentino

Abstract In the era of Industry 4.0, designers are expected to use new tools and approaches to innovate the design of products and services. From this perspective, the integration of design practices and technologies of the 4.0 transition can have positive implications for sustainability. Several current issues can be addressed and among them, honeybee death is relevant. Honeybees are fundamental to the ecosystem and human life. Nevertheless, their lives are extremely at risk due to exposure to several disease factors. After conducting expert interviews, the paper presents a conceptual model of intelligent beekeeping to monitor the health status of honeybees. Furthermore, after user research, the paper proposes a co-design model for urban beekeeping, scaled up to a condominium dimension, to allow condominiums and expert beekeepers to be part of an integrated design model. The critical proposition of this model is to raise awareness of the problem of honeybee death to achieve 3 out of 17 United Nations' Sustainable Development Goals: Good Health and Well-Being, Sustainable Cities and Community, and Life on Land. Early results report positive values of acceptance of the urban beekeeping practice by users and the use of IoT in managing beehives and their health status by expert beekeepers.

Keywords Urban beekeeping · Industry 4.0 · Industrial design · IoT technology · Design for sustainability · Co-design

United Nations' Sustainable Development Goals 3. Good Health and Well-Being · 11. Sustainable Cities and Community · 15. Life on Land

---

M. Ricci (✉) · M. Fiorentino
Department of Mechanics, Mathematics and Management, Polytechnic University of Bari, Bari, Italy
e-mail: marina.ricci@poliba.it

A. Di Roma · A. Scarcelli
Department of Architecture, Construction and Design, Polytechnic University of Bari, Bari, Italy

© The Author(s), under exclusive license to Springer Nature Switzerland AG 2024
M. Barberio et al. (eds.), *Architecture and Design for Industry 4.0*, Lecture Notes in Mechanical Engineering, https://doi.org/10.1007/978-3-031-36922-3_21

1 Introduction

The industry 4.0 paradigm is moving forward quickly, increasingly changing our world and the way we live and work [1]. This digital transition provides tools to address urgent issues with the help of new and emerging technologies.

The 4.0 model affects the way designers *design* as in any industrial revolution. The exponential development of digital technologies is certainly increasing the virtual component of our experience [2]. In this ever-changing scenario, design can indeed gather input, tools, and procedures that can be used to solve critical issues.

In the twenty-first century, designers must think of products and services differently from their ancestors. In Industry 4.0 enabling technologies such as the Internet of Things (IoT) provide new design opportunities to empower people and enrich their daily lives, in real-time and smart connecting people.

On the other hand, the impact of Industry 4.0 on sustainability and the way it can contribute to sustainable economic, environmental, and social development is increasingly gaining attention [1]. The 4.0 paradigm may offer opportunities for sustainability, by helping to achieve the United Nations' Sustainable Development Goals (SDGs) outlined in the 2030 Agenda. This document provides a set of guidelines for sustainable development, adopted by all United Nations member states to promote effective design for sustainability.

Thus, the paper aims to stimulate 4.0 awareness for designers who must solve real and compelling issues related to this era. Also, the paper defines the role of design in the 4.0 paradigm to achieve and disseminate innovative solutions and projects according to the 17 United Nations Sustainable Development Goals.

The honeybee can benefit from this digital 4.0 transition with the help of design, and its well-being can have positive implications for sustainability. Indeed, honeybees are the most effective pollinators of crops and are crucial in order to achieve sustainable development [3, 4]. Since the end of the twentieth century, honeybees are suffering from increasing stress factors, leading domesticated colonies to die or at least be less productive [5]. Also, several factors including parasites, bacterial and fungal infections, and pesticides [3] have been identified as drivers of honeybee death and this phenomenon triggered the need to rethink beekeeping and its tools.

Many plant species would become extinct without them, and current levels of productivity could only be maintained at great cost through artificial pollination. Domestic and wild honeybees are responsible for about 70% of the pollination of all living plant species on the planet and provide about 35% of global food production.

The future of beekeeping is to implement smart 4.0 beehive management using automated and remote tools for monitoring honeybee colonies along with beehive control mechanisms to safeguard honeybees' well-being and improve colony productivity [6].

There is also a growing awareness of beekeeping, particularly in metropolises and cities [7], called *urban beekeeping*. This practice of raising honeybees in an urban environment is blended with the co-design perspective and the creation of communities of experts and non-experts for the safeguarding of honeybees. We use co-design in a broader sense to refer to the creativity of designers and people not trained in design working together in the design development process [8].

Not all the reasons for honeybee death are fully known, and as a result, it is essential to obtain all possible information on the environmental conditions surrounding the beehives [9].

The digitalization of beekeeping first involves systems from the field of the IoT, with the development of sensors to collect and transfer honeybee-related data [5]. Then, data analysis comes into play, providing models that connect the data with the biological states of beehives. The speed and scale of IoT provide new design opportunities to empower people and enrich their everyday lives [10].

Thus, in this paper, we want to address the following Research Questions (RQs):

RQ1. What is the urban beekeeping' acceptance level of users?

RQ2. How IoT can be an appropriate tool to help beekeepers in managing and control beekeeping systems?

Through semi-structured questionnaires, we surveyed a sample of 463 participants to figure out their knowledge and awareness about beekeeping and honeybees' safeguard and to survey their acceptance of the urban beekeeping model. Also, we conducted expert interviews with expert beekeepers to survey their needs, knowledge, and acceptance of 4.0 tools for beekeeping.

The remaining paper is structured in four sections. The first describes the state-of-the-art related to smart monitoring systems for beekeeping. The second describes the methods adopted to answer the RQs. The third relates to the discussion about the role of design within the 4.0 era and its potential to reach three sustainable development goals through the conceptual model. Lastly, we report our conclusions and future works.

2 Related Work

Beekeeping has a huge impact on all agricultural field, as honeybees are the main insect pollinators and plays the important role in whole crop production and the survival of plants [11].

Recently, urban beekeeping is a rapidly developing model, involving beekeeping *in*, *of*, and *for* the city [12]. Beekeeping *in* the city concerns the importation of traditionally rural beekeeping practices into urban spaces on behalf of the beekeeper. Beekeeping *of* the city describes beekeeping consciously adapted to the urban context, often accompanied by (semi)professionalization of beekeepers and the formation of local expert communities (i. e., beekeeping associations, and communities). Beekeeping *for* the city describes a shift in mindset that addresses beekeeping for civic purposes beyond the beekeeping community itself.

Beehive monitoring is fundamental to monitoring different parameters, such as the temperature and humidity levels inside the beehives and the weight, sounds, and gases produced, which can generate important information. For example, these data can inform whether the beehives are swarming based on the temperature, whether any action is required from the beekeeper, whether the honeybees are affected by any disease, or even whether the beehives are affected moving. This last application is very useful in areas where beehives can be stolen.

Different technologies can be applied to monitor the beehives [9]. A smart beehive is a connected beehive with some intelligence, for example, a beehive capable of diagnosing health issues [5, 13].

Phillips et al. [14] encouraged the "Bee Lab project" that blends citizen science and open design with beekeeping. Their objective was to enable participants to construct monitoring devices gathering reciprocal data, motivating them and third parties. They used design workshops to provide insight into the design of kits, and user motivations, promoting reciprocal interests and addressing community problems.

Murphy et al. [15, 16] developed a fully autonomous beehive monitoring system. Their objective was to use Wireless Sensor Network (WSN) technology to monitor a colony within the beehive by collecting image and audio data and developing a multi-data source beehive monitoring system. WSN technology includes sensors, low-power processing, mobile networking, and energy harvesting. In this way, the beekeeper will obtain recorded information about in-hive conditions (e. g., during the night, winter months, etc.). The contributions of this work also include the unobtrusive monitoring of the beehive during times when the beekeeper is unable to open it.

Zacepins et al. [6] introduced and analyzed different bee colony monitoring and control systems and their combinations within the ERA-NET ICT-Agri project 'ITAPIC'. Also, they presented their vision for the implementation of Precision Beekeeping together with the smart apiary concept system based on temperature, sound, and video monitoring.

Gil-Lebrero et al. [9] designed a remote monitoring system for honeybee colonies (i. e., WBee) based on a hierarchical three-level model formed by the wireless node, a local data server, and a cloud data server. WBee is a low-cost, fully scalable, easily deployable system for the number and types of sensors and the number of hives, and their geographical distribution.

Lyu et al. [17] designed an intelligent beehive system with a real-time monitoring function of the status information of the inside and outside of the beehive (e. g., weight, attitude, etc.). The beekeeper can check the status of the beehive in real time in the monitoring center. The system reduces beekeepers' labor and improves the quality of honey.

Kontogiannis [18] developed a holistic management and control system for the apiculture industry called the Integrated Beekeeping System of holistic Management and Control (IBSMC). This system allows honeybee living conditions regulation, aiming at minimizing honeybee swarm mortality, and maximizing productivity. Within the proposed system architecture, additional security functionalities are implemented for honeybee monitoring, low energy consumption, and incident response.

Existing beehive control and management systems allow for critical thinking and rethinking of smart beehive models thanks to design approaches and 4.0 technological tools.

3 Methods

Beekeeping can be innovated and conceived as a co-design activity that can involve not only expert beekeepers but also ordinary citizens. By broadening the knowledge about beekeeping to the urban level, a series of actions can be performed that can benefit honeybees and cause benefits, combined with IoT technology.

The two research questions were explored through user research and expert interviews to answer needs, expectations, and problems and subsequently formulate a model.

3.1 User Research

We analyze the context of use by investigating the user awareness concerning honeybee importance and their behavior. Thus, we create a questionnaire using Google Forms distributed to a sample of potential users for 12 days (n = 306; 59,3% female and 40,7% male). The questionnaire provides quantitative data about user knowledge and user behavior regarding honeybees through 13 open and closed questions. As an interesting result, the user analysis demonstrates that most of the users are aware of the phenomenon of honeybee death (82,8%). Also, many users know that honeybees are fundamental to human life and sustainability (84%), but most of them have never heard of urban beekeeping (58%). A mere 30.1% have observed a beehive up close previously and most are unaware of the possibility of being able to adopt a beehive (68%) and monitor its health (82%).

Interestingly, most of the users would like to have a condominium beehive to raise honeybees in their condominiums (M = 6; SD = 0,4 on a 7-point Likert scale).

In addition, users were asked about their agreement related to the listed requirements to have a condominium beehive using a 7-point Likert Scale. The results were compiled into a graph (see Fig. 1).

3.2 Expert Interviews

We conducted expert interviews to gain information about beekeepers' IoT adoption and explore their knowledge and expectations. Expert interviews are a widely used qualitative interview method often aiming at gaining information about or exploring a

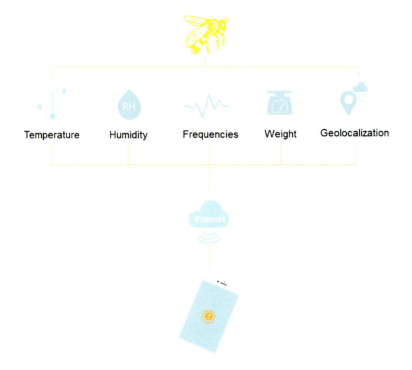

Fig. 1 Condominium beehive requirements preferences according to users on a 7-point Likert Scale

specific field of action [19] (i. e., beekeeping). Those were semi-structured interviews based on the use of a questionnaire covering the following topics:

- Generic knowledge:
  - Threats to traditional beekeeping: climate change, immoderate use of pesticides, lack of food due to excessive plowing and extensive monocultures, and spread of beehive diseases (e. g., Varroa).
  - Actual problems of urban beekeeping: swarming. For the urban environment, swarms can become a nuisance and a fright for the citizens, both for citizens' lack of knowledge about them, but also the inconvenience concerning the places where the swarming families decide to settle (e.g., inhabited places, schools, etc.).

- IoT beehive adoption: according to the expert beekeepers, there is not a wide knowledge of commercially available smart beehives and their possibilities for the environment. In their opinion, if technology could help beekeepers to manage their beehives, intelligent beehives equipped with sensors are useful. Helping beekeepers manage their beehives more conveniently, quickly, and effectively is an advantage. For example, those beekeepers with lots of beehives could optimize their intervention time and be able to intervene while helping their honeybees grow and thrive. Expert beekeepers will be willing to use intelligent beehives, based on two dependent variables: costs and ease to use.

Also, beekeepers expect IoT-intelligent beehives to:

- Remotely control the beehive's temperature, humidity, and weight to understand how the health status of the family from these parameters.
- Manage to predict swarming, and consequently avoid or control it. An abrupt decrease in weight indicates that swarming has occurred, but at that point, it is too late, at most, if possible, the beekeeper could intervene earlier to recover the swarm before it moves too far from the beehive.

The possibility of monitoring the health status of honeybees remotely, by a device (e.g., smartphone, tablet, laptop), would facilitate and improve the work of beekeepers.

- Urban beekeeping with a co-design approach: the idea of implementing smart condominium beehives as a co-design for beekeeping is very interesting according to beekeepers. However, there are some limitations and aspects to consider in the design phase:

  - Fear of condominiums.
  - Allergies of the condominiums.
  - Swarming: In the city, they can become annoying and even dangerous.
  - Distance From a legislative point of view beehives must be placed at least ten meters from public roads and at least five meters from the borders of the neighborhood, unless there is a natural barrier (e.g., hedges) or artificial (e.g., walls) at least 2 m high. In this case, the distance is nullified because the honeybees would be forced to fly above human height.
  - The non-training of condos.

Also, the smart beehive will need to have several features:

- Anti-theft system.
- Honeybee health monitoring.
- Weight of beehives.
- Swarming prediction.

3.3 The Conceptual Model

User research and expert interviews allow us to collect useful data to develop a conceptual model of a smart beehive (see Fig. 2) and to predict from a co-design perspective a useful application for the communication between condominiums and expert beekeepers.

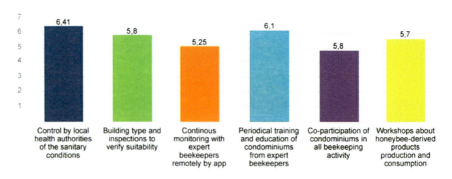

Fig. 2 IoT technology management. From honeybees to sensors, from the Internet to devices

Different parameters of the honeybee colony can be monitored: temperature, humidity, gas content, sound, vibration, etc. Continuous monitoring of some honeybee colony parameters is very challenging and not user-friendly, using it only for research purposes, not for practical implementation by the beekeepers [6].

3.3.1 Types of Sensors'

In the IoT field, the choice of appropriate sensors is essential to produce a solid base before data analysis since sensors can capture beehive-related data [5].

The design of a wireless sensor network and IoT sensors involved in this are:

- Temperature and humidity sensors: honeybees need temperatures (90–95 F) and humidity levels (50–60%) to raise their brood optimally and promote overall colony wellbeing. Below 50% humidity, eggs won't even hatch because they dry out. Additionally, higher humidity levels significantly decrease Varroa mite reproduction levels and are desirable during the brood-rearing season.
- Accelerometer: registered data about frequencies and their variations in the beehive allow for to prediction of swarming phenomena [20].
- Weight: indicates the activity level of a colony and is a good indicator of honey flow evolution and flowering.
- Geo-localization: to localize the beehive to be reached by expert beekeepers and send data via the Internet to be visualized on the application.
- Anti-theft: a tracking device consisting of a GPS tracker, a battery, a SIM card, and a motion detector. It can be placed inside the beehive (e. g., at the lid, the bottom, the body, or even inside a honeycomb). The device will stay off, without draining its battery or disturbing the honeybees via cell phone radiation. With the slightest wiggling of the beehive, the device turns on and informs users with an SMS sent to their cell phones.

Fig. 3 Mockup of the mobile app from an expert beekeeper and condominiums point of view

All data about honeybee health could be transmitted to an application (see Fig. 3) through which users, both condominiums and expert beekeepers, can be informed about their health status and any urgent problems.

3.3.2 Co-Design Approach

Co-design is an approach to designing with, not for, people. Especially in areas where technologies mature, designers have been moving increasingly closer to the future users of what they design [8]. Co-design typically works best when people, communities, and professionals work together to improve something they all care about.

From this perspective, the co-design approach could potentially be extended to urban beekeeping. The process includes several parts and starts with "Build the conditions" (see Fig. 4) for the genuine and safe involvement of people (i. e. condominium requirements).

The reference community includes ordinary citizens grouped in condominiums, supported by expert beekeepers to carry out co-creation activities. The tools employed include IoT for controlled and shared co-design for the beehive management.

4 Discussion

Design plays a key role in achieving sustainable goals because, by leveraging design culture approaches (i. e., co-design approach) and Industry 4.0 transition tools (i. e., IoT technology), it can solve current critical problems.

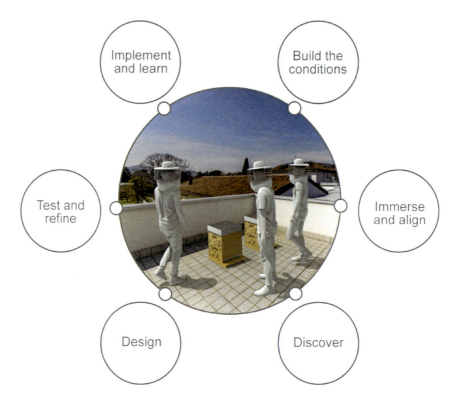

Fig. 4 The co-design process for the urban beekeeping

In this specific case, (e.g., the death of honeybees), the design allows for achieving three United Nations SDGs.

SD 3. Good Health and Well-Being

One of the sustainable goals relates to ensuring healthy lives and promoting well-being. The human being's health depends on the honeybee's health. This insect, thanks to pollination, enables new plants to be born and humans to breathe. Honeybees increase the possibility of a plant making more fruits and food for human beings.

The conceptual model developed allows for monitoring the health of honeybees in real-time, h24, to ensure their well-being, with positive implications for humans and the city. Intelligent beehives allow the recording of many variables such as:

1. Honeybee ecological variables

   - Colony life history: health monitoring, collapse events, requeening events (e. g., swarming), beekeeping tasks.
   - Colony dynamics: colony size, brood area, drone brood area, food reserve mass.

- Resource use: harvested pollen species composition, honey-embedded pollen species composition.

2. Environmental variables

   - Floral resource phenology monitoring.
   - Land use monitoring.
   - Climatic data.

SD 11. Sustainable Cities and Community

The sustainable challenges cities face can be overcome to allow them to thrive and grow by improving resource use and reducing pollution.

Urban beekeeping would allow environmental monitoring, particularly related to air quality in cities. It is necessary to constantly monitor the presence of air pollutants to allow honeybees to live and reproduce in city environments. By observing the behavior of honeybees, it is possible to evaluate the presence of heavy metals, microplastics, and fine dust in the atmosphere. All these substances are as dangerous as they are unfortunately present in the air we breathe in large and small cities.

SD 15. Life on Land

From an ecological perspective, honeybees provide a source of both direct and indirect food to a whole host of organisms, including humans [15].

Protecting the population of honeybees worldwide and enabling them to maximize their productivity is an important concern [16] since they are fundamental not only for their well-being but also for humans and vegetation.

5 Conclusions

Design is constantly evolving and using new tools and approaches to design useful products and services to solve current problems. This paper shows how a relevant problem such as honeybee death could be addressed through the integration of the 4.0 paradigm and design.

User research shows that acceptance of urban beekeeping is high among many users (RQ1). This finding is important and encouraging for the development of conceptual models scaled to the urban dimension and, more specifically, to the condominium dimension. In addition, co-design could be a potential approach to innovate the practice of beekeeping, making it accessible not only to the expert beekeeper but also to the ordinary citizen.

Interviews with experts show that IoT is an appropriate tool to help beekeepers in the management and control of beekeeping systems (RQ2). Although there is no established knowledge of smart systems for beekeeping, there is a great interest in adopting technological systems to facilitate several tasks for beekeepers.

The first step in the future would be to set up the model and design in a real beehive to see how the system responds in a real-world deployment. Future tests would also provide valuable real-world data from the beehive, which can be used for further beekeeping research.

Furthermore, it is important to test the dimension of the co-design approach within one or more apartment buildings to collect data on their involvement in activities, and use of the mobile app and the IoT management model.

Acknowledgements We would like to thank the beekeepers Nicola Ignomeriello and Cristian Scalise involved in the project DontBEEScared—stai sen'APEnsieri which concerns urban educational beekeeping in Bari, Italy. They helped us recruit participants for the expert interviews and allowed us to discover the world of urban beekeeping deeply.

Also, we would like to thank the industrial designers Federica Gentile and Adriana Romeo for their valuable effort, who developed the concept together with their colleague, Marina Ricci, during the Master's degree course in Industrial Design at the Polytechnic University of Bari (2019/2020).

References

1. Ghobakhloo, M.: Industry 4.0, digitization, and opportunities for sustainability. J Clean Prod. **252**, 119869 (2020). https://doi.org/10.1016/J.JCLEPRO.2019.119869
2. Trabucco, F.: Design. Bollati Boringhieri. (2015)
3. Howard, D., Hunter, G., Duran, O., Venetsanos, D.: Progress towards an intelligent beehive: Building an intelligent environment to promote the well-being of honeybees. In: Proceedings—12th International conference on intelligent environments, IE 2016, pp. 262–265 (2016). https://doi.org/10.1109/IE.2016.60
4. Patel, V., Pauli, N., Biggs, E., Barbour, L., Boruff, B.: Why bees are critical for achieving sustainable development. Ambio **50**, 49–59 (2020). https://doi.org/10.1007/S13280-020-01333-9
5. Hadjur, H., Ammar, D., Lefèvre, L.: Toward an intelligent and efficient beehive: A survey of precision beekeeping systems and services. Comput Electron Agric. **192**, 106604 (2022). https://doi.org/10.1016/J.COMPAG.2021.106604
6. Zacepins, A., Kviesis, A., Ahrendt, P., Richter, U., Tekin, S., Durgun, M.: Beekeeping in the future—Smart apiary management. In: Proceedings of the 17th International carpathian control conference, ICCC 2016, pp. 808–812. (2016). https://doi.org/10.1109/CARPATHIANCC.2016.7501207
7. Lorenz, S., Stark, K.: Saving the honeybees in Berlin? A case study of the urban beekeeping boom. Environ Sociol. **1**, 116–126 (2015). https://doi.org/10.1080/23251042.2015.1008383
8. Sanders, E.B.N., Stappers, P.J.: Co-creation and the new landscapes of design. CoDesign **4**, 5–18 (2008)
9. Gil-Lebrero, S., Quiles-Latorre, F.J., Ortiz-López, M., Sánchez-Ruiz, V., Gámiz-López, V., Luna-Rodríguez, J.J.: Honey bee colonies remote monitoring system. Sensors. **17**, 55 (2017). https://doi.org/10.3390/S17010055
10. Cila, N., Smit, I., Giaccardi, E., Kröse, B.: Products as agents: Metaphors for designing the products of the IoT age. In: Conference on human factors in computing systems—Proceedings, pp. 448–459. (2017). https://doi.org/10.1145/3025453.3025797
11. Klein, A.M., Vaissière, B.E., Cane, J.H., Steffan-Dewenter, I., Cunningham, S.A., Kremen, C., Tscharntke, T.: Importance of pollinators in changing landscapes for world crops. Proc. R. Soc. B: Biol. Sciences. **274**, 303–313 (2006). https://doi.org/10.1098/RSPB.2006.3721

12. Sponsler, D.B., Bratman, E.Z.: Beekeeping in, of or for the city? A socioecological perspective on urban apiculture. People Nat. **3**, 550–559 (2021). https://doi.org/10.1002/PAN3.10206
13. Meikle, W.G., Holst, N.: Application of continuous monitoring of honeybee colonies. Apidologie **46**, 10–22 (2015). https://doi.org/10.1007/S13592-014-0298-X
14. Phillips, R.D., Brown, M.A., Blum, J.M., Baurley, S.L.: Testing a grassroots citizen science venture using open design, the Bee Lab Project. In: Conference on human factors in computing systems—Proceedings, pp. 1951–1956. (2014). https://doi.org/10.1145/2559206.2581134
15. Edwards-Murphy, F., Magno, M., O'Leary, L., Troy, K., Whelan, P., Popovici, E.M.: Big brother for bees (3B) - Energy neutral platform for remote monitoring of beehive imagery and sound. In: Proceedings of the 6th IEEE International workshop on advances in sensors and interfaces, IWASI 2015, pp 106–111. (2015). https://doi.org/10.1109/IWASI.2015.7184943
16. Edwards-Murphy, F., Magno, M., Whelan, P., Vici, E.P.: B+WSN: Smart beehive for agriculture, environmental, and honey bee health monitoring—Preliminary results and analysis. In: SAS 2015—2015 IEEE sensors applications symposium, proceedings. (2015). https://doi.org/10.1109/SAS.2015.7133587
17. Lyu, X., Zhang, S., Wang, Q.: Design of intelligent beehive system based on internet of things technology. In: 3rd International conference on computer engineering, information science & application technology (ICCIA 2019). (2019)
18. Kontogiannis, S.: An internet of things-based low-power integrated beekeeping safety and conditions monitoring system. Inventions. **4**, 52 (2019)
19. Döringer, S.: The problem-centred expert interview. Combining qualitative interviewing approaches for investigating implicit expert knowledge. Int J Soc Res Methodol. **24**, 265–278 (2021). https://doi.org/10.1080/13645579.2020.1766777
20. Ramsey, M.T., Bencsik, M., Newton, M.I., Reyes, M., Pioz, M., Crauser, D., Delso, N.S., le Conte, Y.: The prediction of swarming in honeybee colonies using vibrational spectra. Sci. Rep.. **10**, 1–17 (2020). https://doi.org/10.1038/s41598-020-66115-5

Resolve Once—Output Many (ROOM): Digital Design and Fabrication at the Service of Social Equity

Blair Gardiner and Sofia Colabella

Abstract Digital technologies have focused much attention on promoting industrial efficiencies, interoperability, and decentralisation, facilitating design possibilities via complex geometries whilst eliciting precision, and embedding supply chain sustainability practice and connectivity in the ubiquity of *Smart* agency. The *Industry 4.0* proposition identifies the beneficial outcomes of technological advancement. Its basis and, by extension, the subset of digital design and fabrication, in part, is premised by an application in a market-driven competitive model approach. However, such underlying precepts may be equally applicable to invert the context to suit internal socio-economic conditions other than the economic activity to which they are directed but to which they may provide beneficial service. Less often referred to is how digital design and fabrication may be harnessed to benefit disadvantaged communities or those sectors where current or projected economic and industrial structures limit access to social equity. Leveraging the utilisation of digital technologies at play or developing within the construction industry, one may bypass some of the broader challenges of *Industry 4.0* by including an opportunity focus application to addressing social disadvantage and social equity. This chapter makes a case for performance-based and circular architecture in bringing together digital design and digital fabrication tools to enhance social equity and self-agency. It demonstrates this potential in a connected response mechanism to the housing affordability debate, which does not always include young people, particularly those at risk of homelessness. The investigation lens is the parameters for temporary independent accommodation for at-risk youth to remain within an existing support unit or family. The process mode extends by integrating a self-capacity facility. In the Australian housing sector, the historical aspects of prefabrication, land subdivision, dependant units and self-build, still supported by regulatory regimes, resonate with the potential of digital design and fabrication tools. Opportunities arise in scalability and decentralisation, upstream embedment for performance enhancement and sustainable practice, portative flexible modularity, and transformation permutation. Self-capacity lies with personalisation via participatory design customisation and engagement with fabrication and

---

B. Gardiner (✉) · S. Colabella
Faculty of Architecture, Building and Planning, The University of Melbourne, Melbourne 3050, Australia
e-mail: b.gardiner@unimleb.edu.au

assembly, serving as a training program to support employability and empowerment. By taking advantage of supply chain features of the Australian construction sector and technological advancement, one may look at the coeval opportunities that lie in applications that support social equity and inclusive responses to addressing real-world issues of social disadvantage.

Keywords Computational and Parametric design · Didactics · Circular architecture · Performance-based architecture · Digital fabrication and Social equity · Creativity · Design thinking and human–computer interaction

United Nations' Sustainable Development Goals 9. Build resilient infrastructure, promote inclusive and sustainable industrialization and foster innovation · 1. End poverty in all its forms everywhere

1 Introduction

Amongst the foundations of *Industry 4.0*, with its avidity for digital technology uptake in design and production, one may trace some of its precepts to the resonance articulated in 1776 by Adam Smith in *The Wealth of Nations* [1]. These include the notions of production efficiencies garnered by the division of labour, cross-boundary exchange and the positive but unintended outcomes of 'the invisible hand' for a collective economic and industrial benefit. So was raised 'das Adam Smith problem', for Smith had contributed an earlier seemingly contrary position in his 1759 publication in *The Theory of Moral Sentiments* [2, 3]. Whereas *The Wealth of Nations* is heralded as a seminal work in economic theory, *The Theory of Moral Sentiments* is less acclaimed. The former is seen as a validation of economic efficiencies derived from self-interest. However, the latter argued for the ethical and social dimensions of such actions.

Herein lies a potential merging, elucidation and leveraging of the Adam Smith problem for technological advancement broadened to social-economic conditions other than the economic activity to which they are directed. Innovation geared towards industrial efficiencies rarely filters beyond the self-referential domains from which they are targeted. The underlying precepts of *Industry 4.0* are less often directed to how digital design and fabrication may be harnessed to benefit disadvantaged communities or those sectors where current or projected economic and industrial structures limit access to social equity. Leveraging digital technologies presently used or developing within the construction industry, one may bypass some of the broader challenges of *Industry 4.0* by including an opportunity to address social disadvantage and social equity.

An increasingly ineluctable condition of the bespoke and customisation is to foreground output at the expense of upstream or end-use factors. The construction industry tends to adopt a reductive approach to this condition. It focuses on output processes seeking to minimise time and costs, reduce complexity and adopt conventional production modes. A consequence to design is its potential approbation to prêt-à-porter relegation. The adoption of standardisation and deemed-to-comply prescription is prefaced as desirable industrial management pursuits for time, cost and quality whilst simultaneously animadvert the construction industry on innovation and productivity. In recognition of this approach's limits, regulatory jurisdictions such as those in Australia retained deemed-to-comply provisions but also included a performance-based compliance approach. The shift from on-site to off-site manufacturing is an additional construction management response mechanism to resource and temporal management. Some regulatory jurisdictions, such as Singapore, encourage this approach in their design for manufacturing and assembly regimes, such as those promoted by its Code of Practice on Buildability, with high building design scores for standardisation, repetition, and prefabrication [4].

Amongst the consequences of these approaches is that as formal expression becomes unbounded, the upstream effort of response strategising and planning increases. However, design conception's intellectual and strategic capital is undervalued with a synonymist appropriation of consequential processes predicated on construction management, as exemplified by the uptake of *Building Information Modelling*.

The outcome's bespoke nature, the construction industry's fragmentation, with an *anything can be built* mindset, demonstrates a highly flexible sector. There are limitations to a reductivist premise of the *order from stock* myth. Is there a way to internalise capitalised costs associated with upstream design conception whilst responding to downstream production geared towards output? May we reconnect firmitas and utilitas to venustas? Is there a way to embed performance responsiveness whilst freeing design capacity beyond the design professional?

Industry processes seek to direct various permutations of commercial advantage and potential. Similar cross-utilisation to other sectors that do not offer a commercial return is rarely picked up by the funding authorities with whom the eventual cost is transferred—the public purse. In undertaking the task of a stream of *Industry 4* geared towards social equity, one must also trace the cultural roots of emergent technologies connected to geographies of place to recognise and suggest that antecedents may be discovered to certain design and technological strategies.

The project proposal outlined here lies in applying digital design to fabrication tools for an accommodation unit as a possible response to preventing youth homelessness in Australia. It identifies the questions and objectives of the project, suggests that cultural precedents backgrounds, and supports the approach adopted.

2 Youth Homelessness

An area of contention is the lack of a consistent international definition of homelessness; however, commonality occurs in aspects of stable and secure housing [5]. For statistical purposes, the *Australian Bureau of Statistics* (ABS) defines homelessness in terms of three criteria 1. that the dwelling is inadequate, 2. a lack of tenure or tenure is short and not extendable, and 3. the living arrangement does not allow control of and access to space for social relations [6]. The exaptation and coopting potential of digital-based technologies and a co-design process offer a means by which one may address the three categories of statistical definition. The locus is a customised independent accommodation unit, utilising digital design to digital fabrication methods.

Homeless persons numbers have been steadily increasing, registering a 22% increase between 2001 and 2016 [7]. As with the definition of homelessness, youth homelessness is ill-defined [6]. The ABS refers to an age group of 12−18 or 12 to 24. The findings of the 2016 Australian Census found that the numbers of homeless youth (aged 12−24 years) were consistent across Australia, with Victoria being 26% of the total homeless population, with the range being from 21 to 26% in other Australian states and territories. Youth made up 32% of total homeless persons in severely crowded dwellings, 23% of those in supported homeless accommodation, and 15% in boarding houses. One of the features of the upward trend of the homeless population suggestive of the potential of youth homeless to become structural is that "nearly 60% of homeless people in 2016 were under 35 years and 43% of the increase in homelessness was in the 25−34 age group" [7]. The result is increased demand for support services and coincides with the lack of affordable housing.

2.1 Affordable Affordability

The project does not try to offer a solution to affordability as one of Australia's most significant housing challenges [7, 8]. Instead, it focuses on how digital design and prefabrication can maximise construction process efficiency, reducing the cost of small temporary accommodation units that users can customise. The focus is on people at risk of homelessness who cannot afford affordable housing. Young persons and those experiencing short-term financial stress cannot engage with the question of housing affordability. The proposal attempts to reconcile gaps and opportunities in the formal regulations with informal building practices providing independent temporary accommodation to remain within an existing family or support network. The application aligns with government policy shifts to keep youth within support networks and avoid structural homelessness. An example is the Australian government initiative *Reconnect Program*, a community-based early intervention to prevent youth homelessness by targeted support of families and young persons who are homeless or at risk [9, 10].

2.2 Tiny House Models, Open-Source Customisation and Homelessness

The project *Resolve Once, Output Many (ROOM)* adopts the *Tiny House* or *Micro-House* model. Adherents of this form of housing argue for its responsible environmental awareness, economical housing entrée and facility as a community housing model for social interaction and support [9, 10]. It has many variations, such as those profiled in 2008 in the New York Museum of Modern Arts exhibition *Home Delivery: Fabricating the Modern Dwelling* to the open-source customisation of the *WikiHouse*. The application of independent housing units in response to homelessness is not new. The Tiny House model, either as moveable or static units, have been used as a response mechanism to access affordable home ownership or rental option. The Tiny House provides a stable, supported housing option in the *Housing First* approach to addressing homelessness in the United States and is witnessed by the advent of tiny house villages [11]. In Australia, not-for-profit organisations such as *Kids Under Cover* have been operating its *Studio Program*, which provides relocatable "one or two-bedroom studios with bathroom, built in a family or carer's home as secure and stable accommodation for young people at risk of homelessness" [12].

These approaches follow conventional systems of building and prefabrication. *ROOM* takes a similar (still very different) approach by adapting conventional building practices to digital production. It also proposes collaborative design input by including the end-user to participate in a customised output at the initial design stage.

In addition, through flat-packed delivery of simply constructed components, it facilitates self-built assembly or in conjunction with other young persons who have undertaken assembly training. This aspect of the system is seen as significant in providing the end-user with some buy-in to the unit they will occupy. It empowers the young person and provides a training mechanism which may be used for future employment or to demonstrate an individual's capacity to a prospective employer.

3 Resolve Once, Output Many

The proposition of *ROOM* (Resolve Once Output Many) is for independent secondary accommodation for a dependant young person on a property. Its integrated parametric design and CNC fabrication facilitate a customisable, cost-effective, and time-efficient housing option, lending itself to prefabrication and self-build assembly.

ROOM is in the formative stages of design and application modelling. It proposes to investigate customisation through participatory design and fabrication using digital-based technologies, aiming to produce a small housing unit incorporating ease of assembly and disassembly that is not over-reliant on technical construction skills. Its approach has commenced with the most common and familiar Australian domestic construction method of timber-framed building. It then reappraises this system to meet project objectives and, in so doing, reconfigures it as an unconventional consolidated, but conventional timber frame. The method is to recreate the frame structure using small plywood sheet sections spliced to form nodes and internodes. Design testing and verification are done by prototyping through digital modelling and structural analysis and review testing by physical scale models.

Its application is directed to vulnerable community sectors such as youth at risk of homelessness. Youth at risk of homelessness have very few housing options and reliant on private support housing providers who are under severe strain in a system that cannot meet demand. The permutations of use are extensive in addressing community disadvantage and broader housing options, such as community or cooperative housing. Key factors for the project are a minimum cost option, responsiveness to user needs and short lead times for use availability.

3.1 Adapting and Learning from Conventional Building Practice in Digital Production

Operating in the refined non-bounded space of the digital environment can only take one so far. The first stage is, therefore, to learn from conventional building practice and adapt it to digital production. The approach is not to propose retooling the construction industry, though this may eventuate. It does not offer a radical shift in industry practice but recognises that there are advantages to an industry sector considered to be overly fragmented or lacking industrial or innovation capacity in being made up of small to medium enterprises (SME). Fragmentations in the supply chain may not require each link to be connected to the initial design formulation and communication nor to a final bespoke product. A variety of dispersed SME constituents may have flexibility and rapid delivery advantages at various levels of scale and location, promoting innovation at the points where intellectual capital is best held. One construction industry sector has fully embraced digital-based design for digital fabrication capacity. It has permeated joinery manufacturing and is ubiquitous as a production method. The shift to the use of sheet materials and a fully integrated supply chain with its use of CNC machines, its move to factory production and site installation and a focus on ease of installation is not an unreasonable first port of call for prototyping. Digital craftsmanship and CNC production appear to be allied to the logic of self-sufficiency and DIY and could potentially form a new iteration on the theoretical reflections on tacit and explicit knowledge.

3.2 Enabling and Empowerment

The enabling capacity of digital technologies potentially facilitates a 'bottom-up' approach to design and education. Emergent technologies offer a means by which the end-user obtains a support mechanism to substantiate and realise their design vision to suit individual circumstances through customisation. The designer's role is adaptable to include professional design and co-design engagement through the architects being the users themselves [13].

Empowerment is a concept garnering currency in addressing homelessness. At the individual level, it is put forward as a method to rebuild self-confidence, promote social engagement, develop personal capacity and awareness of resource mobilisation, and accept consequential responsibility [33]. Adopting an empowerment process is a major factor in breaking the cycle of homelessness, as it could potentially support building resilience and providing a pathway of positive change by helping youth to help themselves. The alternative is the risk of structural homelessness and institutionalisation, negatively impacting youth self-confidence and autonomy. Additional benefits may include an informal design and construction education. Formulating an education pathway through involvement in the design and assembly process makes education a vehicle for shared community development. Self-assembly will embed training and work opportunities for youth through participation in the construction process. While it is not argued that applying digital design and fabrication is a solution to youth homelessness, it is possible that this system may provide beneficial outcomes in (1) preventing youth homelessness; (2) making the homelessness services less expensive; (3) shortening the time people will require support.

3.3 Conditional Objectives

The initial design process has adopted several challenges and conditional objectives. Although aspirational, they also aim to focus the project on the issues raised in the testing process of the design stage, construction methods, systems, fabrication, and assembly. These include:

1.1. How may readily available technologies in the construction industry be used to reduce cost, retooling and retraining, promote scalability to support housing alternatives, and reduce the impact of disadvantage vulnerability? An option is to use CNC systems available in the joinery sector of the construction industry.

    Therefore, the starting point is sheet-based material components common to contemporary commercial joinery production, using plywood for the project.

    Incorporating an existing construction sector operative method has benefits in not retooling or retraining the industry and extending market reach. It also provides a scalability method by using a current but underutilised production capacity.

    The restriction, at present, to plywood limits accounting for the anisotropic properties of timber, providing some known attributes in structural behaviour.

The project, therefore, is limited at this stage to the structural frame and its analysis. However, even given this limitation, by necessity, it has had to encompass broader project implications.

2.2. What issues are raised in adopting an integrated approach to design iteration? How may one bring the end-user upstream into the design process, and how well may one engage in front-end design with downstream supply chain issues? What degree of customisation capacity is attainable through aggregating and embedding response conditions upstream into the design stage. These include user preference, site conditions, structural behaviour, regulatory compliance, transportation, building performance and whole-of-life considerations. Digital design and fabrication were chosen to display that a given domain of variation and personalisation is available to a layperson.

It takes advantage of a fragmented construction industry. It does not require those in a construction supply chain to be knowledgeable about input application into a bespoke product and uses simple methods of assembly.

In addition, it acknowledges aspects of off-site production and integrated assembly. Such an approach has been adopted in the commercial construction sector, the most obvious example being fully finished service pods such as bathrooms or kitchens.

3.3. How far may the bourne of optimisation be pursued? It commences on the assumption that customisation is requisite on a participatory design and assembly methodology. In addition, as the outcome is seen as transition support, it facilitates a feedback loop for in-use testing and further refinement.

The digital design optioning allows wide-ranging customisation of material systems and overall formal design, including external and internal appearance. It permits user input into personal amenities, such as exterior covered space and furniture. In addition, it may include the aggregation of units for other uses or the extension of the facility where planning regulations or exemptions may apply. Optioning extends to single, aggregated, or multiple uses, such as servicing provisioning kitchens, bathrooms, and shared spaces.

Other project permutations of optimisation include the components, as the modular unit derived from standard sheet materials reduces material waste or promotes labour efficiencies and ease of assembly.

4.4. To what degree may ease of construction, material accessibility and cost and labour reduction be pursued using environmentally responsible off-the-shelf or processing sourcing geared towards circularity? The approach maximises the use of building systems and components that are commercially available off the shelf. The structural system's use of small, interdigitated components configured to form the primary framing allows for the use of material sections that may be commercially discarded. Its approach is in 'not wasting, the waste.'

In addition, the project seeks the inclusion of local recycled and repurposed materials and components. Given the current developments in material processing technology, the capacity exists to incorporate the utilisation of processed waste into 'green materials' that may be used in the construction or fitout.

5.5. To what extent can ease of assembly and disassembly not reliant on specialist construction expertise but geared towards the youth who occupy the unit be a viable self-build option? Design for Disassembly (DfD) objectives stem from acknowledging that such units are not put forward as a long-term housing solution. They are a temporary proposition as an attainable and immediate relief to permit other factors to support transit out of an at-risk situation.

DfD is geared to the components being reassembled or repurposed, thereby promoting circularity. Its endeavour to promote ease of construction minimises mechanical fastening systems. The system uses jointing precision afforded by CNC cutting, assembly, and disassembly, reducing reliance on specialist trade skills. Such a system permits the components to be used multiple times, subject to verification for structural integrity, durability, site conditions and alternate uses to that directed initially.

4 Cultural Precedents

It is premature to describe the detailed technical aspects of project conception and delivery of the *ROOM* project. Let us set aside the arguments for digital design and fabrication as a new emergent form of technology but look at it through a local historical lens. It is essential to understand the historical context of places such as Australia that may lend themselves to such technologies and the conceptual cultural framing of the approach adopted. In the Australian housing sector, the historical aspects of prefabrication, residential land tenure and an engagement with self-design and self-build, still supported by regulatory regimes, resonate with the potential of digital design and fabrication tools to be used as a temporary accommodation model.

4.1 Prefabrication

The European occupation of Australia in its housing commenced by leveraging technology and construction innovation with the importation of prefabricated buildings. Early uptake of prefabrication during the 1830s, saw British and Australian fabricators producing prefabricated timber houses for private occupation [14]. The globalisation of timber prefabrication is a feature of colonisation [16] and supply shortage arising from economic pressures and is witnessed in its transposition from the Californian Gold Rush of 1849 to the Australian context [17]. Australia's colonial history, in its urban development and response to building pressures generated in the Australian Gold Rush booms of the 1850s, is deeply connected to 'portable' buildings, often shipped from overseas fabricators based in Germany, India, New Zealand and Singapore, as well as those locally produced [15].

In Australia, an example of the recurrence of prefabrication as a housing solution may be found following the Second World War. To address the consequential privations and post-war housing shortage, the largest public housing provider, the Housing Commission of Victoria, turned to import prefabricated timber housing for one of its regional estates [18]. Local architects took up the gauntlet in design-driven output, utilising the perceived material and cost benefits derived from various forms of prefabrication. Notable during this period was the architect Best Overends' proposal for timber prefabrication in his 1946 prototype design of the *Indus House*. The Australian Navy in 1951 resorted to prefabricated housing through the importation from the Swedish manufacturer Åmåls Sågverks Aktiebolag of flat-packed houses [19]. However, the 1946 *Romcke* plywood house by the Australian plywood and timber veneer merchants Romcke Pty Ltd, and the contracting firm A V Jennings is considered the first wholly prefabricated house in Australia. Such efforts reflect an interest in precision fabrication and resource efficiency to alternate forms of housing supply [20].

This background suggests that awareness, adoption and acceptance of remote design and fabrication methods in construction have a cultural basis as a housing solution tracing back to the colonial development of European occupation in Australia. Innovation born as a response to circumstances, including housing shortage, resource access, labour and material optimisation, and the derived economic advantages has a genealogy that connects to the recent interest in the potential of digital fabrication.

4.2 Land Subdivision

The early planning history of colonial Australia through its urban demarcation of land subdivision from its European roots has left an imprint on the Australian residential suburban form. The 1788 founding of European Australia in New South Wales and its early town plan proposal decreed that "land will be granted with a clause that will prevent more than one house being built on the allotment, which will be sixty feet in front and one hundred and fifty feet in depth" [21]. Linking the foundation of white Australia to the Australian attachment for a housing form connected to a model of low-density single houses on a substantial landholding commonly identified as the 'quarter-acre' block may be tenuous [10]. However, such an attachment to this basic model of residential land tenure remains a feature of the Australian suburb. In Melbourne, Australia, although increasingly rare, residential allotments of 1,000 sqm may still be encountered, with some outer suburbs featuring larger block sizes. A 2016 study identifies the median block size for the inner city to be under 300 sqm with pockets at 300 to 500 sqm, similar to the median of the middle suburbs' [22]. However, land size and the low density that arises from the feature of single-dwelling development are further supported by the historical self-sufficiency view of residential land in its production capacity [23]. Although this model is transitioning in two ways, firstly via Australia's domestic housing market shifting to larger houses, and secondly, through its inverse by increased housing yield by land subdivision. Residential densities for Melbourne remain low, with the Victorian Planning Authority in 2018 identifying it to be 16.5 dwellings per hectare [24]. This low density suggests that capacity exists within developed residential land to provide a different form of housing accommodation supply. The approach is not to look to the further subdivision of land, with its restrictive concomitant costs and time delay, as the solution but to provide increased capacity flexibility and immediacy for existing occupiers and thereby respond to social disadvantage and accommodation stress. Therefore, in seeking response strategies to reduce potential youth homelessness, a solution may already be available in utilising available land to retain youth within existing housing environments connected to support networks.

4.3 *The* **Granny Flat**

Australian planning regulations are not unique in having a model for existing developed land to provide limited housing supply flexibility. The United States model of elder cottage (Echo) housing follows a similar model, having obtained a measure of impetus through the Australian Planner Barry Cooper's presentation in 1981 of the Australian programs at an American Association of Retired Persons forum held in Washington, D.C., [25]. The Australian concept is said to have been derived, in 1963, from a practitioner Dr Hubert Bauer working at the state mental health authority, the Victorian Mental Hygiene Department, as his response to enhancing support and housing choices for aged persons. Bauer proposed a moveable, self-contained unit that may be temporarily placed on an existing property. A key aspect of Bauer's approach was that the units be independent and autonomous, not part of an existing house but located adjacent to it, to facilitate an environment of independence and familial support. The other key feature of Bauers proposition was that the units be architect-designed. The Victorian Council for the Aged and the Rotary Club of Melbourne strongly advocated with the public housing authority, the Housing Commission of Victoria, to develop the idea. The first portable unit was built in 1975 by the Commission and was provided on favourable rental terms to a couple on their daughters' property in suburban Melbourne [26]. Bauer's concerns regarding senior person's mental health resonate today in the mental health and wellbeing of at-risk of homelessness and youth homelessness. Studies suggest that over half of young persons who have experienced homelessness have indicated some form of psychological distress, more than double that of those who have never experienced homelessness. Coping with stress is seen as a significant issue of personal concern, as are mental health, financial security, and suicide, all being reported in far higher proportions than those young persons who have never experienced homelessness [27].

The basic planning approach of dependent person's unit (*Granny Flat*) remains in local Council Planning Regulations. The property must be larger than 450 sqm, and the unit must be moveable but not be a caravan and be occupied by a dependent person. Between August 2020 and March 2021, a local government pilot program was instigated, trialling loosening restrictions to permit the construction of a secondary dwelling subject to certain conditions. These included a height limit of 5 m, a maximum floor area of 60 m, garden provisioning and no capacity to subdivide the land. As of April 2022, the pilot remains in review [28].

Therefore, a solution presents itself in providing accommodation supply to a vulnerable community sector and potentially avoiding the risk for a young homeless person's trajectory of structural homelessness. The ability to deliver the design where structural integrity and regulatory compliance are embedded into its parametric conditions facilitates a resolved customisable output. Customisable site-responsive units may be developed using digital technologies with the young person's participation as the end-user. The nature of the construction system geared towards non-technical assembly methods and the minimisation of specialist trades permits a self-build approach of assembling parts by an end-user. This approach offers a labour cost reduction and a potential training avenue for young persons.

4.4 Self-build and Self-assembly

The notion of a right to build one's own home for owner occupancy is entrenched in the Australian construction regulations [29]. Its genesis lies in the post-World War 2 period in Australia. House construction fell dramatically during the war as resources were deployed to support the war effort, resulting in a chronic shortage of housing availability. Presently, the deficiency lies in affordable housing. The proportion of total houses built annually in Australia by owner-builders steadily increased from 20% in 1948, culminating in a peak in 1954 of over 40% [30]. The Australian Bureau of Statistics estimates that in 2002–2003, the proportion of owner building as a proportion of total work done on private sector houses was 13% [31]. Although originating from a lack of housing supply, the notion of self-built housing is still perceived as advantageous through the cost savings in sweat equity and builders margin, greater project flexibility and project control, particularly at the design and design implementation stages [32].

ROOM accommodates a self-built option available under existing building regulations. Customisation through participatory design and fabrication, progressing to self-assembly, provides a potential cost–benefit and a feedback loop for the review of the process. It enables rapid and cost-effective construction, which is easy to assemble, dismantle, and reusable.

5 Australian Housing Antecedents, Digital Design and Fabrication and Social Equity

The use of emergent technologies such as those that are derived from digital design and fabrication may be a development which may lend itself to contemporary application to ongoing issues in the Australian housing sector. Such technologies may provide a quick, customisable, and accessible temporary housing option for community sectors experiencing disadvantage or as a housing adjunct for at-risk youth of homelessness. Australian historical precedents in housing include the utilisation of prefabrication as a response to construction material resources and housing supply shortages. By leveraging historical residential land plot size patterns, it facilitates the provision of temporary dependant accommodation to provide housing options for youth. Capacity lies in digital design methodologies to engage the end-user upstream in the design process, whilst parametric modelling may support customisation and optimisation. The ability to embed performance and output using readily available materials without high-level specialist construction skills lends itself to labour cost reductions via a self-build assembly method. These benefits contribute to providing affordable and quick housing options to disadvantaged sectors of the community. One tributary of the *Industry 4.0* proposition may be to use the advantages that emergent technologies provide or may already exist but are underutilised in the construction industry to address social equity in housing.

Acknowledgements This work was supported by a 2019 research development grant issued by the Faculty of Architecture, Building and Planning at the University of Melbourne, Australia.

References

1. Smith, A., Cannan, E.: An inquiry into the nature and causes of the wealth of nations. Methuen, London (1950)
2. Montes, L.: Das Adam Smith Problem: its origins, the stages of the current debate, and one implication for our understanding of sympathy. J. Hist. Econ. Thought. **25**, 63–90 (2003)
3. Smith, A.: The theory of moral sentiments. Cambridge University, Cambridge (2002)
4. Singapore building and construction authority: Code of practice on buildability. (2017)
5. Australian Bureau of Statistics 2050.0.55.002—Position Paper—ABS review of counting the homeless methodology. (2011). https://www.abs.gov.au/ausstats/abs@.nsf/Latestproducts/02613B17495C4DC4CA2578DF00228C84?opendocument
6. Australian Bureau of Statistics,: 4922.0—Information Paper—A statistical definition of homelessness. 2012. https://www.abs.gov.au/ausstats/abs@.nsf/Latestproducts/4922.0Main%20Features22012?opendocument&tabn
7. National Housing Finance and Investment Corporation: State of the Nation's Housing 2021–22. Australian Government: National Housing Fiance and Investment Corporation
8. Burke, T., Nygaard, C., Ralston, L.: Australian home ownership: past reflections, future directions. AHURI Final Report. (2020). https://doi.org/10.18408/ahuri-5119801
9. Shearer, H., Burton, P.: Towards a typology of tiny houses. Hous. Theory Soc. **36**, 298–318 (2019). https://doi.org/10.1080/14036096.2018.1487879

10. Alexander, L.T.: Tiny homes: A big solution to American Housing Insecurity. Social Science Research Network, Rochester, NY (2022)
11. Evans, K.: Tackling homelessness with tiny houses: An inventory of tiny house villages in the United States. The Professional Geographer. 1–11 (2020). https://doi.org/10.1080/00330124.2020.1744170
12. Preventing youth homelessness. https://www.kuc.org.au/what-we-do/how-we-help/
13. Arboleda, G.: Beyond participation. J. Arch. Educ. **74**, 15–25 (2020). https://doi.org/10.1080/10464883.2020.1693817
14. Lewis, M.: National trust of Australia (Vic.): La Trobe's cottage: A conservation analysis. National Trust of Australia (Victoria), Melbourne (1994)
15. Lewis, M.: Prefabrication in the Gold-Rush Era: California, Australia, and the Pacific. APT Bulletin: The Journal of Preservation Technology. **37**, 7–16 (2006)
16. Smith, R.E.: History of prefabrication: A cultural survey. Undefined. (2009)
17. Peterson, C.E.: Prefabs in the California gold rush, 1849. J. Soc. Archit. Hist. **24**, 318–324 (1965). https://doi.org/10.2307/988318
18. Survey of Post-War Built Heritage in Victoria: Stage One. https://www.heritage.vic.gov.au/__data/assets/pdf_file/0026/512288/Survey-of-post-war-built-heritage-in-Victoria-Stage-1-Heritage-Alliance-2008_Part2.pdf
19. Alshabib, A., Ridgway, S.: ASA 302 @ Georges Heights: Swedish timber prefabs in Australia. Fabrications. **30**, 323–345 (2020). https://doi.org/10.1080/10331867.2020.1826687
20. Thousands of houses may be built of plywood after war. (1944). http://nla.gov.au/nla.news-article26030997
21. Davison, G.: Australia: The first suburban nation? J. Urban Hist. **22**, 40–74 (1995). https://doi.org/10.1177/009614429502200103
22. Are Melbourne's suburbs full of quarter acre blocks?, (2016). https://chartingtransport.com/2016/05/22/are-melbourne-suburbs-full-of-quarter-acre-blocks/
23. Mullens, P., Kynaston, C.: The household production of subsistence goods. In: Troy, P. (ed.) A history of European housing in Australia. pp. 142–163. Cambridge University Press (2000)
24. What is the housing density? https://vpa.vic.gov.au/faq/pakenham-east-what-is-the-housing-density/
25. Hare, P.H.: The echo housing/Granny flat experience in the US. J. Hous. Elder. **7**, 57–70 (1991). https://doi.org/10.1300/J081V07N02_05
26. The first of the Granny Flats. (1975). http://nla.gov.au/nla.news-article55188300
27. Hall, S., Fildes, J., Liyanarachchi, D., Hicking, V., Plummer, J., Tilkler, E.: Staying home: A youth survey on young people's experience of homelessness. Mission Australia, Sydney, NSW (2020)
28. Victorian State Government, Department of Environment, Land, Water and Planning: Secondary dwelling code. https://www.planning.vic.gov.au/policy-and-strategy/smart-planning-program/rules/secondary-dwellings-code
29. State of Victoria, Victorian building authority: Owner-Builder information and study guide, (2017)
30. Dingle, T.: Self-help housing and Co-operation in Post-war Australia. Hous. Stud. **14**, 341–354 (1999). https://doi.org/10.1080/02673039982830
31. Statistics, c=AU; o=Commonwealth of A. ou=Australian B. of: Technical Note—Owner builders study (Technical Note). https://www.abs.gov.au/AUSSTATS/abs@.nsf/39433889d406eeb9ca2570610019e9a5/ab8a51a0e32e9e50ca257b960014ff1a!OpenDocument
32. Consumer Affairs Victoria, V.G.: Owner builders. https://www.consumer.vic.gov.au:443/housing/building-and-renovating/owner-builders
33. European Federation of National Organisations Working with the Homeless: Empowering ways of working: Empowerment for people using homeless services in Europe. Brussels, Belgium (2009)

From Analogue to Digital: Evolution of Building Machines Towards Reforming Production and Customization of Housing

Carlo Carbone and Basem Eid Mohamed

Abstract The construction of edifices is all about lifting, moving and setting components according to predefined patterns. The magnitude of nineteenth century industrialization produced all manner of machines, lifts and earthmovers to facilitate construction, in addition to easing the pressures on manual labor. Along the same tactical interests, the Bessemer converter and gantry cranes were invented for advancing manufacturing and facilitating standardization of building parts. Robert Le Tourneaux's Tournalayer, perhaps the most unique building machine, made it possible to mold buildings like a mega-cookie cutter by casting reinforced concrete in moveable steel formwork. The outcome of such experiments cultivated transformations in the building process, even if they were not widely utilized. Recent advancements in digital fabrication machines in the form of Computer Numerically Controlled (CNC) cutting and milling tools, bricklaying drones, and large-scale 3D printing robots, coupled with computational design processes, are driving new possibilities in design and construction. Multiple levels of design variation are feasible, reforming standardized industrial models into user-centric, and contextually driven singular designs. The chapter aims to critically examine how contemporary digitally controlled building machines are part of a spectrum of devices linked to mechanization and how they present potentials for the democratization of housing provision. Accompanied by an analysis of how the fourth industrial revolution is impacting construction, we present a detailed overview of the evolution of building machines, with a specific focus on concrete casting machines used to produce dwellings. Then, we critically analyze the parallels between traditional casting equipment invented for mass production and today's robotic fabrication to deliver inhabitable prototypes. As a conclusion to the chapter and an opening to further research, a generative framework that stems from linking digital design with production machines is proposed

C. Carbone (✉)
Faculté Des Arts, UQÀM, École de Design, Montreal, QC, Canada
e-mail: carbone.carlo@uqam.ca

B. E. Mohamed
College of Arts and Creative Enterprises, Zayed University, Abu Dhabi, UAE
e-mail: basem.mohamed@zu.ac.ae

© The Author(s), under exclusive license to Springer Nature Switzerland AG 2024
M. Barberio et al. (eds.), *Architecture and Design for Industry 4.0*, Lecture Notes in Mechanical Engineering, https://doi.org/10.1007/978-3-031-36922-3_23

for implementing customization in the industrialized housing sector, one that has long been connoted by the lack of design personalization.

Keywords Architecture · Building machines · Customization · Large-scale 3D printing · Additive manufacturing · Construction 4.0

United Nations' Sustainable Development Goals 9. Industry, Innovation, and Infrastructure · 11. Sustainable cities and communities

The submitted contribution must highlight the relevance of the research/project presented in respect of one or more key technologies of Industry 4.0.

1 Introduction: Industrialization from Machines to Construction 4.0

From Mechanization (1765) to Mass production (1870) to Automation (1969) and to the evolving concept of Digital Objects (2016), all four stages of industrialization radically changed how things were made previously. Industry 4.0 relates to the digitization of every economic sector including building design and construction. Kagermann and Wahlster's [31] definition of «data added value creation» includes technologies and practices circumscribed by information used as parameters and criteria to trace and interact with all stages, methods and tools in the building process. The use of modelling and data managing technologies in every phase of a project's development dramatically increases the rapidity and the extensive way performance criteria and intelligence can be shared, manipulated and deployed to streamline everything from design to operation.

While professionals have been quicker than builders in adopting concepts like Building Information Modeling, Virtual Design and Construction or even Design for Manufacturing principles, the uptake of 4.0 principles within the construction industry globally is asymmetrical at best. Nonetheless, architecture and construction are greatly influenced by these changes in fabrication and production methods. In design, generative data manipulation creates potential for customizing and multiplying iterations. Within fabrication, robots and cobots can be programmed to quickly output changes and design iterations to prototype shapes unthinkable until recently. Further, the link between design and fabrication increases the appetite and potential for Offsite/Smart Factory building methodologies. Within the construction phase of projects, information sharing, Ai and digital twin modelling facilitate systemic coordination and the potential to monitor every aspect of an edifice: its building, its operation and even its end of service life.

Digitalization of the building industry encompasses and streamlines the process making it possible to envision a type of file to factory to site and to operation model applied to any scope and scale of building task. A segment of industry 4.0 that is sure to further develop through these tools and methods is the complete data-informed autonomation of construction machines. It is this segment of Construction 4.0 that will be further analysed.

In every civilization, the construction of edifices, large and small, great and ordinary, has implied the development of machines to assemble buildings from heavy and sizeable components whose lifting or moving would otherwise be impossible. For instance, the treadwheel crane became an important accoutrement in the construction of gothic cathedrals for setting materials in their final positions. Machines are part of building culture [10] and in some respects even define building cultures. The steel skyscraper or the timber balloon frame would not have been possible if not for the development of steel refinement through the Bessemer converter [5] or the mechanized timber mill. Both inventions democratized technology and completely transformed cities [42, 44].

Pulleys, ropes, rollers, lifts, earth movers, chariots or wagons are just a few implements which have alleviated human effort and have forever contributed to building. Instruments rigged for lifting were first propelled by human or animal power and this defined construction capacity [25]. Industrialization converted these machines to steam, fuel or electricity and from traditional building materials to iron and steel. Self-propelled mechanization, the first industrial revolution, improved capacity and transformed building sites into veritable open-air factories.

Mechanization, then mass production and automation, influence of all these trends on building culture in general and manufactured building culture in particular was fundamental as machines enhanced division of labor and overtook manufacturing. Large components, complete subassemblies or completed building sections could be manufactured offsite and then carried to and integrated into edifices. Further, mechanization allowed building sites to be managed as ready to assemble kits. Mass production's influence on architecture and construction still impacts our current building culture through the classification and cataloguing of every type of component required in the building process. A third industrial revolution, automation, first introduced in Japan after the second world-war was as equally disruptive to production processes as Henry Ford's assembly line had been. All three eras of industrialization had machines as their central figures in advancing construction: the gantry crane, the conveyor and automatic robotic arms demonstrated how objects of every scope and size could benefit from industrial processes.

During industrialization's first three phases, an intense time of technological advancements, the building itself was also theorized increasingly as a type of integrated machine. The "*Mobile housing unit*" patented by John Hays Hammond Jr in 1947 [27] explored suspending monolithic units from two central gallow beams as shown in Fig. 1. The supporting structure served as a crane and scaffolding during construction making it possible to anchor each concrete box onto the tower. The crane also made it plausible to remove and relocate the dwelling units over time. The building as a support structure and crane essential in the building's erection is perhaps one of the most experimental visions of building to come out of the machine age.

Machines in the factory also determined dimensional standardization, repetition or normalization as each area unit of the factory participated in a serial making process. Contrary to this, on site machines made singularity possible. The tension in architecture between standardization and singularity is longstanding [9]. The connection from machines to customized or standardized or even mass customized architecture is the topic of this chapter, which looks at the links between production and customization through the lens of machines capable of generating unique architecture.

Currently machine capacity and potentials are being determined by a complete link to information technology. Artifical Intelligence (AI) in particular is being explored to define a particular breed of machines that can generate, react, engineer, design in relation to their connectivity and integrated logistical data sets. This fourth industrial

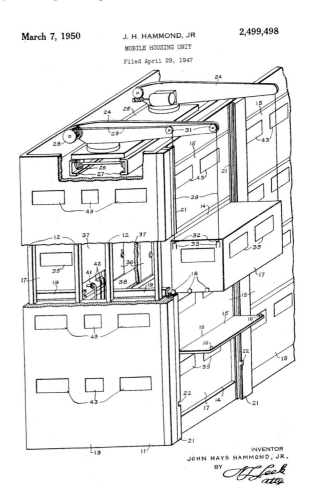

Fig. 1 Mobile housing unit (1947) John Hays Hammond. US Patent no. 2499498 (*source* Public domain)

revolution is being promoted as a new way forward for improving construction in general and driving a new era of industrialized construction experiments.

1.1 Construction 4.0

From planning to design, to procurement, to management and to off- or on-site production of edifices, every part of construction is being reformed by information technology. The literature on construction's digitization is expanding almost as quickly as information technology. Both theory and practice are understanding and accepting how its potential can federate all stakeholders to increase construction's lagging productivity and transform construction's traditional ways of making buildings. Perrier et al. [47] have made the argument that 4.0 is not only about technology but also about reviewing age old management processes that are wasteful, time consuming, fragmented and littered with too much conflict.

Many have referred to Industry 4.0 as a new digital age. The concept of 4.0 has been defined as "a new digital age for manufacturing that uses cyber-physical systems, IoT, data and services to connect production technologies with smart production processes" [19]. Although it implies newness, the novelty of 4.0 in the architecture and construction industries is certainly disrupting previous methodologies but also harmonizing and nuancing the three previous phases of industrialization though a comprehensive exchange of big data.

In the systematic literature review proposed by [4], the authors discuss drivers of construction 4.0 and indicate three principle elements that are linked to information sharing, through connectivity of devices and management processes. BIM, Big data, IoT are the three topics that together have the potential to completely reform construction's organization from a intensely disjointed process to using digital twins as the hub for all information about edifices and their service life. Processes like DfMA, Lean construction, and PLA that previously were absent from architecture and construction are being proposed though the extensive use of digital information sharing platforms. Equally as disruptive as Ford's assembly line or Taylor's task division had been to early twentieth century car production, the digital construction of the built environment is making the building process not only more efficient but increasingly metric and performance based though information toning and monitoring.

This exchange of information between people, process and tools is driving 4.0 and is enhanced by our totally connected lifestyle and culture. It is the «tool» portion of the equation that lead us to study the relationship between machines and building processes and understand if a new generation of connected tools could also reform the relationship between design, production and customization.

Paolini et al. [46] and Sartipi [49] proposed a comprehensive review of Additive Manufacturing in relation to construction 4.0 and outlined its potential to create more complex forms and buildings while simplifying the construction process. Bos et al. [7] identifies certain challenges that are all linked to machine's output. Based on previous literature and the objective of unifying design with production the present chapter looks more specifically at the issue of customization and how it related to mechanization or industrialization 1.0 and how it is being reformed toward a digital mechanization through the digitally controlled layering of material leveraged toward the potential democratization of the 3D printed building.

Developments in concrete depositing machines or 3D printers are being employed to produce completely customized homes and buildings which to date has not been possible. The links between twentieth century machinery and 3d printing are discussed as concepts in a similar spectrum of production: Using mechanization to produce architecture quickly, uniformly, and with possible complete customization. From the entire field of possibilities the link between reinforced concrete construction and numerically controlled machines that deposit it in layers to produce any form or shape truly exemplifies how two distinct eras with access to different industrial typologies examined the same subject of generating dwellings through concrete depositing machines. The recognized strength, flexibility and malleability of concrete made it a truly modern and globalized material, however the complexity associated with its casting has always fascinated and driven inventors to propose new ways of facilitating its use. The links between mechanization and concrete casting serve as an interesting case study in how material constraints influence manufacturing, design and construction.

1.2 Reinforced Concrete, Moveable Formwork and Machinery for Casting

Through three previous eras of industrialization, reinforced concrete, has commanded machines to transport, deliver, mold and set it. Mechanization 4.0 is no different, as digital tools are outlining casting machines that continue to address concrete's potentials and challenges.

Reinforced concrete is a cast artificial stone that can be conveyed and poured at variable consistencies and in any shape. This versatility inspired industrialists, architects and engineers to identify ways to integrate concrete in modern construction. From Lambot [34] and Monier [43] to Hennebique [28], much literature is dedicated to the establishment of skeletal concrete structures, esteemed as fireproof and monolithic. Hennebique's patent, an evolution from the more densely reinforced flowerpots of Monier became the most commercially available. Hennebique established a veritable concrete empire licensing his structures all over Europe and North America [51]. From these preliminary experiments, Robert Maillart's bridges and Pier Luigi Nervi's airplane hangars invoked concrete's use to reform construction and build any shape, scope and scale imaginable both by on site casting or by prefabrication. Nervi's work on prefabricated structures can be read as a manifesto on mass customization before its time. Using a process that is repeated and perfected but that could be nuanced to form multiple shapes [45]. As shown in Fig. 2, his shapes and compositions for a roof in Torre in Pietra represents a prototype of his more

Fig. 2 Agricultural Shed Roof (1946) Torre in Pietra, Pier-Luigi Nervi (*source* Authors)

well-known cupolas, using a type of geometric and modular repetition to create a process for all manner of roof structures. Further his explorations of sliding formwork allowed him to build some of the most recognized large-scale concrete slabs of the twentieth century.

Others including Felix Candela and eventually Santiago Calatrava showcased the unique potential of a magnificent material capable of any shape. Reinforced concrete and novel casting processes inspired increased productivity and affordability with the advantage of the fire protection and the durability of stone. All advantages of reinforced concrete construction, are however, mitigated by the complex formwork and temporary scaffolding that has to be edified and often removed and hauled away as waste after casting. In reaction to the time consuming nature and often wastefulness of disposable formwork, inventors such as Pier Luigi Nervi and Wallace Neff reimagined formwork as a structural and constructive process. Neff's Airform houses (Wallace, Building construction 1941) employed inflatable and pressurized plastic domes as geometric formwork onto which concrete was cast. Once cured the inflatable material could be removed and reused on another house eliminating both waste and rigid geometry. This type of inflatable formwork is still used today. Neff's airform process employed shotcrete, a process by which concrete is sprayed through a hose at a specific consistency to stick to irregular and vertical shapes.

1.3 Weight as a Constraint Outlining a Need for Machinery

At 2400 kg/m^3 conventional reinforced concrete is not only prohibitive because of wasteful formwork but also since it usually requires some form of complex mechanical process to bring it to site and deliver it into forms. Imagining a process that would reuse forms and simplify the building process was the object of two processes that in some respects can be analyzed as the predecessors to today's 3D printing machines. Thomas Edison's one pour house [17] as shown in Fig. 3, explored the potential to convey concrete vertically onto the top of a ready-made formwork and let gravity force the fluid material into the various parts of the structure. Ornamentation

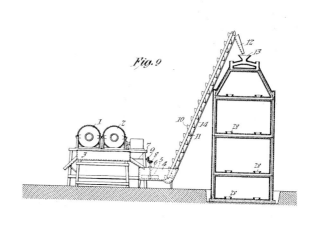

Fig. 3 Process of constructing concrete buildings (1917) Thomas A. Edison. US Patent 1,219,272 (*source* Public domain)

was shaped in the casting mold and openings distributed throughout enabled quality control and correct compacting and vibration eliminating any air pockets. A marginal number of houses were built using this system and the published advantages of a one-cast-pour would not eliminate any complexities associated with disassembling the formwork and then reassembling it on another site.

1.4 The Tournalayer

Investigating further into the idea of reusable, moveable and integrated formwork to build a type of instant structure, The Tournalayer [38] is a massive machine invented to cast houses onsite as shown in Fig. 4. A two-part house-sized form was fabricated out of steel components. Either one- story or two- story forms included networks for

Fig. 4 Tournalayer—Outer Form for house form assembly (1955) R.G. Le Tourneau (*source* Carlo Carbone private collection)

mechanical distribution and openings for windows and doors. The steel formwork was moved on-site by a tractor like machine that would then deposit the forms over a site-cast slab with infrastructure connections. Once laid over the slab, concrete was poured. The form was then lifted and removed as a cake mold once the concrete developed its initial cure. The Tournalayer was used to deliver houses in the United States, Brazil, Israel and Morocco [54]. This mega-machine was inspired by production methods used in parallel industries that were simply scaled for building production. This type of invention linked to industrialists' fascination with machinery that proposed alternatives for an optimized site casting of concrete have not changed much since the early twentieth century; Slip forms, tunnel forms [54], continue to be applied where sufficient repetition justifies the machinery's use and cost. What has changed considerably is our ability to numerically control the concrete's casting in precise ways helping to develop another generation of similar tools inspired by additive manufacturing to introduce customization to mechanization.

2 Numerically Controlled Machines

During the last two decades, advancements in digital design software and fabrication devices have outlined powerful options for implementing numerous iterations in design and predetermined variations in production towards the customization of building components. For instance, parametric design is being used to tweak and adapt parameters to simplify made-to-measure or engineered-to-order supply chain management. As early explorations reveal, precast concrete panel producers have been the first to explore how information technology could be used to numerically produce personalized details, profiles and thicknesses according to specific needs. For instance, a wide variation of possibilities could be developed using CNC milling machines for subtractive fabrication of molds for casting concrete. This technique was employed to produce the façade of the 290 Mulberry Street, a residential building in New York City designed by SHoP (2007). The architects materialized a reinterpretation of traditional brick detailing through the application of parametric design processes, coupled with advanced fabrication [50]. This process considered innovative at the time would be thought conventional compared to emerging applications of Additive Manufacturing. The project details could be viewed on architect's website.

Additive Manufacturing, and specifically 3D printing, is based on gradually layering material following a specific path to the level of building up a complete structure. The process is controlled digitally, and its quality directly related to composition and quantity of placed material.

As small-scale filament 3D printers have become common, the building industry has sought through academic and industrial partnerships to explore the potential application of AM machines in construction, with specific interest in scaling processes to envisage large-scale 3D printing. Further, based on the same principles percolating form parallel manufacturing industries including but not limited to DfMA (Design for Manufacturing and Assembly) or integrated product development, collaborative teams including architects, engineers and fabrication specialists have been experimenting with 3D printing of large building components like façade elements, structural components, and even full buildings as completely integrated processes from design to fabrication and construction. Both in academia and practice 3D printing machines are being used to conceive and make singular components, a radical change from what the manufactured architecture space has produced during more than a century of exploration. The result has been a fascinating amount of work investigating multiple processes and material possibilities in a relatively short period of time.

2.1 Concrete 3D Printing Machines

Development of concrete 3D printing could be traced back to mid-1990s, specifically in California, USA. The technique addresses several shortcomings in conventional concrete construction: complex and costly formwork, material procurement, simplified delivery, safety, and environmental impact. It is also argued that combining digital design process with 3D printing could foster the development of highly customized building components, and complex geometry [53]. While offering an opportunity to produce customized building with little labor and man power, 3D printing of concrete remains a complex undertaking, involving coordination of various parameters from machine size, information transfer to material consistency, resource availability and building system integration,most current systems have simply extruded walls without imagining the coordination of other systems.

Khoshnevis and Dutton (Khoshnevis and Dutton, Innovative Rapid Prototyping Process Makes Large Sized, Smooth Surfaced Complex Shapes in a Wide Variety of Materials 1998), Khoshnevis (Khoshnevis, Automated Construction by Contour Crafting—Related Robotics and Information Technologies 2004) introduced, then elaborated on the technique defined as Contour Crafting (CC), basically a concrete pump attached to a gantry-based system that involves the sedimentation of a fluid concrete filament. One of the unique features is the use of trowels attached to the deposition nozzles for production of accurate and smooth surfaces. This research inspired many subsequent attempts.

In 2009, Dini developed and presented D-Shape, a 3D printing machine that uses binder jetting, a powder-based technique to strengthen certain and precise areas of a large-scale sand-bed by depositing of a binding agent over these areas. The subsequent bound, calcified or cemented areas define a hardened object layer by layer. Dini filed his first patent in 2006/published in 2008 (Dini, Method and Device for the automatic construction of conglomerate building structures 2008). The potential of such a machine and technique enables the integration of voids and overhanging features towards developing complex shapes, with relatively high-resolution or smoothness [53]. Dini continued working on a wide range of objects since then.

Following these attempts, 3D printing in the building industry expanded rapidly. Langenberg's provided a comprehensive review of the remarkable growth in exploring 3D printing in architecture in the article "mapping 20 years of 3D printing in Architecture [35] where he provided a detailed infographic. An illustrated list of market available printers is provided at the end of the chapter to showcase the amount of current marketable exploration of the topic.

Tay et al. [53] published a review on 3D printing trends in building and construction, comprising a classification of different techniques based on technology, machines, materials, and scale. The paper highlighted that 3D printing full-scale building components is still emerging and is gathering notable attention in both academia and practice. The main challenge, shared in the literature, is synchronizing material consistency with the machine speed to extrude continuous layers without any deformations. Later, El-Sayedgh et al. [20] presented a comprehensive review of relevant literature exploring various 3D printing techniques in construction, while relating the benefits and challenges of each technique. Their analysis was also focused on risks but limited in terms of other challenges we have mentioned including systemic integration required in architecture.

There is a wealth of publications exploring the topic of 3D printing in construction, and any review of existing techniques would be out−dated almost as soon as it is published. The following section proposes a series of projects that were realized using 3D printing machines. The selected projects are not cited and described in an attempt to circumscribe the amount of research but to address the spectrum of possibilities all linked by the issue of mechanization and architectural singularity; the ability to deploy a machine in any context and shape any form exempt from modular constraints that have come to describe the manufactured housing undertones.

3D Printed Office, Dubai, UAE. One of the earliest projects built in 2016 by Chinese company Winsun. The building measuring $36 \times 12 \times 6$m was assembled as a set of 3D printed modular components as seen in Fig. 8. The printing machine is composed of an automated robotic arm fitted on a cartesian type portal grantry crane, and uses a proprietary cement mixture. The 3D printed modules were printed in China then delivered and assembled onsite on a granular foundation, making it possible to relocate them if necessary. Produced from 3D printed volumetric units, the office can be customized with any mechanical system and for any particular use, as it is basically an open volume. Figure 5 represents recent photos of the buildings.

3D Printed House, the University of Nantes, France. An experimental 3D printed house by Batiprint3d at the University of Nantes, the project has caught attention for its use of polyurethane expandable foam as a customizable formwork into which reinforced concrete is poured as shown in Fig. 9. The hybrid system is relatively inexpensive and avoids costly and specifically rectilinear formwork, which is usually discarded. The 3d printed formwork is deposited in any shape and once cured with the concrete infill creates a strong bond and a superior insulated wall. Even though it is particularly suited to contexts where material procurement and delivery are difficult, this simple construction method can be deployed in any setting. The robot is mounted on an automated vehicle and precisely controlled with laser precision for making any

Fig. 5 Office of the future, Dubai, UAE, produced by company Winsun (*source* Basem Mohamed private collection)

Fig. 6 Illustration of the WASP big delta structure (2012) WASP (*source* Authors)

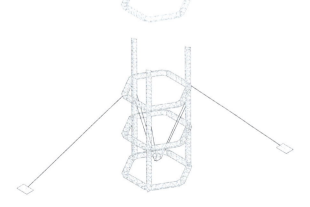

vertically extruded shape where compression is maintained as the principle acting force.

3 BigDelta, WASP

The additive manufacturing process is interpreted as a cultural and contextual endeavor through an organization known as WASP (World's advanced Saving Project). The organization has been developing and researching giant 3D printers such as their Big Delta; the 12 m, scaled Delta type tall printer uses a robotic nozzle equipped with a mechanical mixer, which combines material into a paste and extrudes it through a nozzle. The material is a mixture of cement and mud, similar to a daub clay composite but with an extrudable consistency. The nozzle deposits layer upon layer of material in a precisely digitally controlled pattern, extruding shapes as the layers increase. The Big Delta's printing format impedes linear forms that are possible with Gantry systems, however it offers more flexibility in terms of topology and forms (Fig. 6).

Municipality building, Dubai, UAE. Apis Cor, an innovative startup company is complementing the extensive list of onsite casting inventions and has garnered remarkable interest, thanks to its innovative printing machine; a telescopic robotic arm mounted onto a central rotating crane. The company has to date produced one 38m^2 dwelling structure in a small Russian town, a municipal building in Dubai that is considered the largest on-site 3D printed building as shown in Fig. 7, and a competition proposal for Mars habitat. The Apis Cor printer shown in Fig. 7 is based on a flexible machine that could be transported and positioned on any site. The machine uses a moveable column crane that has a circular printing range of 132 m^2. Analogous to a computer controlled concrete pump, the printer's nozzle is controlled to deposit a lightweight concrete mixture in horizontal strata. The company argues in favor of their lean construction process, which reduces material use and waste as material is mixed, generated and used on-site and as needed. A small mixing plant is located in the printer's range and material is transferred to the printing nozzle. A cellular truss matrix maximizes voids in the structure reducing both mass and volume. These voids could theoretically be used as a network for passing cables and mechanical systems.

Quake Column. Printing buildings as a whole can be dimensionally and logistically prohibitive. A more feasible avenue for 3D printing in construction is making personalized components that can be easily assembled and scaled toward larger structures. Architectural components can be produced with intricate geometries and made-to-order. *Emerging Objects* has been exploring 3D printing for construction as a way of giving new life to age-old construction methods. They have developed a masonry unit, which requires no mortar. The 3D printed mixture of sand, sawdust, ground-up tires, salt, pulverized bone are bound into a type of concrete piece that fits together precisely into a giant 3D puzzle. The firm was inspired by ashlar stonework in which each stone is precisely cut and dry bedded to form a robust structural system for walls or columns (Emerging objects n.d.).

Informed by the study and analysis of ancient Incan stonework, the Quake Column combines precisely defined, printed and numbered elements to facilitate assembly as shown in Fig. 8. Various geometric patterns are possible and could be modified as needed, improved through generative design parameters and shared with the click of a mouse. As complex geometry no longer requires the steady hand of the stonemason, each individual unit didactically displays its shape relating to a whole individualized pattern. Each chunk's geometry is completely self-locking.

This type of file to component production method nuances the traditional debate between on and offsite construction as a large-scale 3d printer produces project specific components reforming the ideals of normalization and standardization normally associated with architectural production.

ICON: Housing For The Homeless. ICON is a US based construction technology company focused on large format concrete 3D printing. The company has developed

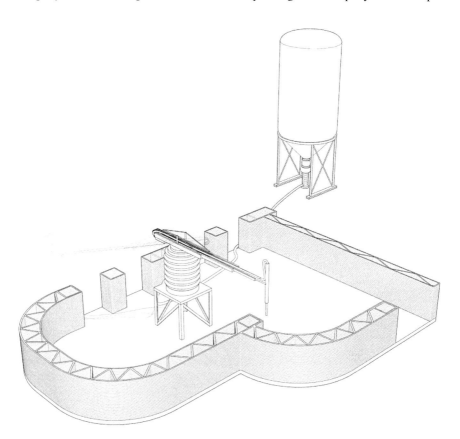

Fig. 7 An illustration of the Apis cor 3D printing machine (*source* Authors)

Fig. 8 An illustration of quake column morphology by Emerging Objects (2014) (*source* Authors)

an advanced 3D printing model that employs a gantry system designed specifically for deposition of concrete, and operated through an interface-based software system to print structures using a cement-based mix designed to be extruded without slumping, increase bonding between layers, and that hardens quickly.

In 2018, ICON delivered its first 3D printed house prototype in Austin, Texas using the Vulcan printer. The hope is that this could be leveraged towards a wider utilization for building housing for the homeless. Later in 2019, the company introduced its first 3D printed community project in Tabasco, Mexico. In collaboration with New Story and Echale, the project delivered 50 housing units, each was printed in 24 h.

4 Towards Reforming Construction: Open Source Customization, and Rationalized Manufacturing

As 3D printing technology is brought to mainstream building, Edison's dream of a one-cast concrete house seems conventional. The very idea of production in architecture has been inspired by the multiple perspectives and evolutions of parallel industries. From mass production, to lean production and then digital fabrication, the ability to make architecture a commodity would reform construction culture, thus

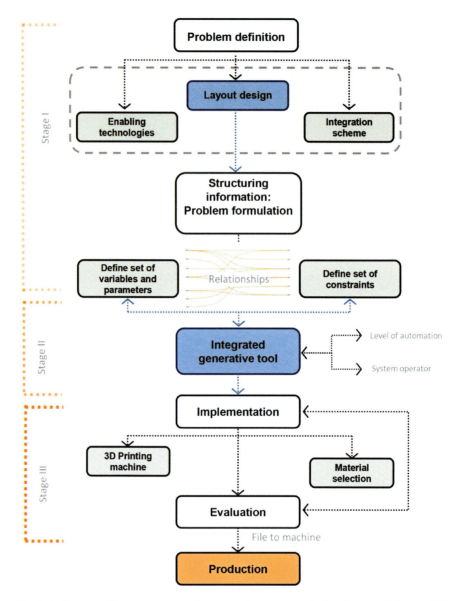

Fig. 9 A diagrammatic representation of a proposed mass customization framework (*source* Eid M., B. 2022)

allowing for alternatives to address issues of productivity and affordability. The implications for traditional industrialized construction ideas are significant. The dream of the factory-made building or house consistently rested on manufacturing models that placed an importance on modularity and repeatability to achieve some type of cost effectiveness. The conventional paradigm of mechanization and mass production only ever attained marginal competitiveness with onsite fabrication, which has always been argued as offering greater flexibility. Connected, transportable, and moveable machines such as 3D printers offer a new paradigm to reform traditional archetypes of manufactured building as the process is controlled through a digital workflow. The fabrication process is streamlined from a collaboratively detailed 3D design model to a machine code that is then translated into a physical object which can be produced multiple times or just once without substantially changing the work-flow as would be the case for producing a one-off house in a modular factory.

Moving from mass customization to a type of individualized production is at the heart of this budding business model for architecture. Age old prefabrication theories are being challenged by the possibilities of bringing a unique vision of mechanization to the construction site. The tension between design, production and fabrication is in a sense erased in favor of a digital design thread, which weaves new architectural potentials. Further this digital thread can be invested by anyone with the capacity to share or harvest crowd knowledge in favor of multiple iterations. Streamlined 3D printing has already brought the factory into the home. It now has the potential to bring the factory to the construction site.

Additive Manufacturing combined with reinforced concrete eliminates the need for costly formwork and the requisite orthogonal geometries that are easier to shape. The nozzle head could be controlled to deposit or spray a material in any direction or angle. Further using horizontal or arched or vaulted decking as part of the final structure makes it possible to envision structures that are more than just vertical extrusions but that arrange continuity of form from walls to floors and roofs, which until now remain absent for 3D printing explorations applied to building. Most explorations in this sense offer a very narrow spectrum of architectural possibilities. This opens an area for future research in matters of how machines and materials could be used to complete more comprehensive and complex forms.

To speak to the advancements of pioneers in reinforced concrete construction the reduction of weight compensated by resistant shapes finds in 3D printing new possibilities of geometrical matrices optimizing material use. Each wall or surface can be realized with an efficient infill pattern to make use of geometric patterns' and material strength. Analogous to trussed elements in a structural steel trellis framework or spaceframe or even in a tightly weaved fabric pattern, concrete can be deposited to reproduce such patterns largely increasing strength per unit of weight ratio. Furthermore contextually informed parameters can be used to generate shapes that are filled according to site conditions, seismic or climatic constraints. The same shaped building could have a 70% infill in the arctic while only a 40% infill would be required in another context.

Along with these structural possibilities, specialized mixtures, even from localized volcanic ash or industrial waste such as slag, creating a greener concrete, can be harvested locally, mixed on site and localized in every corner of the globe with minimal site disturbance.

4.1 Mass Customization of Housing: New Systems and Machines

When looking at recently completed attempts in employing 3D printing in construction it is evident that housing as a typology has gained particular focus given its appropriate scale for experimentation, in addition to the continuous call for exploring new ideas to tackle issues of quality, affordability, sustainability and customization. While most projects we studied are clear with regard to the type of machine used for production, whether a gantry system or a robotic machine, very little information is provided about the design process, and how it could be configured to deliver customized designs. Primary constraints relate to the machine or printable area, and material deposition characteristics. Accordingly, we believe that further developments in orchestrating a relationship between a novel design process and 3D printing machines could lead to new production protocols, and dimensional parameters that could inform the process from design to delivery.

To give a broader perspective, the growing role of Artificial Intelligence (AI) in architecture, coupled with the capacities of adaptive generative design tools, and supported by new modes of fabrication in the form of large-format 3D printing, are all promising to offer a new opportunity for implementing customization in architecture, and specifically housing, thus potentially reforming current models towards addressing many of the contemporary challenges. Appropriately, a design to production workflow could offer an alternative to conventional industrialized housing models that has faced obstacles with regard to perceived quality, and customization. Whether 3D printing of a whole housing unit, or components to be assembled on site, the process transforms the prefabricated housing production paradigm.

It is evident that customization in the housing industry has always been connected to advancements in design and fabrication technologies. Following every new wave of techniques and machines, the interest in mass customization arises. One of the early efforts to explore enabling technologies for mass customization of housing is the work by House_n [36]. This former digital media and housing research group at MIT's Department of Architecture defined three necessary elements for the mass customization of housing. First, a preference engine for customer profiling. Second, a design engine that employs computational set of rules encoded into a shape grammar to generate design solutions in response to the profiling process. Finally, a production system that relies on digitally controlled machines for construction. Powered by the rise of more accessible computational tools and digital fabrication machines in architecture, much of the research efforts explored mass customization of housing from three different angles [48].

As a response to rethinking a model to integrate all three levels, rather than only highlighting the potential of computational design tools or digital fabrication machines, we propose open framework for mass customization of housing that builds on previous efforts, while taking advantage of recent technologies. The framework is outlined by integrating large-format 3D printing, redefining the traditional relationship between design and production. Figure 9 represents a schematic diagram for a proposed mass customization framework, one that is enabled by the utilization of large- format 3D printing machine for production.

The framework is structured based on the following three stages:

- Stage I: Identifies the various levels and activities that are required to implement mass customization of housing. The initial objective is to define the level of customization based on a company's readiness with regard to available design and production technologies with the means by which communication with homebuyers will take place. This could happen on two levels: first, configuration of pre-defined designs with framed limitations and second, spatial layout following particular design rules and criteria.
- Stage II: Explores the process of selecting a design methodology to formulate a solution space. The framework that relates to the 3D concrete printing machines proposes exploring the integration of adaptive generative design tools towards developing design solutions. Adaptive generative design tools, such as Finch3D (Finch3D n.d.) [41], given its potential in extending current parametric CAD/BIM tools through advanced computational design logic and AI, Such tools offer to automate the design process through a well- informed decision support model. Given that the framework is structured to capitalize on large-format 3D printing capacities, the generative tool could be programmed to comply with inputs describing user profiling, while being controlled by the parameters and constraints of the 3D printing machine.
- Stage III: Proposes the implementation of the framework and validating its capability to deliver customized designs in response to the requirements defined in the preceding stages, this stage would require deciding on the production process and device, orchestrating the relationship between the design system and production machine, and testing various material options before prototyping a final design.

4.2 Challenges and Potentials

The idea of a customized inhabitable prototype certainly transforms traditional industrialized housing. Further, linking design parameters with generative design solutions and programming an automated output is symbolic of construction 4.0's potential. The digital process integrates design with production relating fields that are normally discordant and that in itself is transformative in a fragmented industry. Architecture and construction remain however, deeply material-based and any advances in technology are often mitigated by entrenched material constraints.

As is the case for reinforced concrete in general and 3D printed concrete, systems that have been explored require layering of material creating structures that are based on massive construction as opposed to filigree construction. As this becomes an advantage with respect to thermal inertia and mass, it is a complete transference from modernity's penchant for the skeletal structures and their intrinsic flexibility and adaptability. In massive bearing structures changes are difficult to make over time; structures become obsolete more rapidly. If the structures become obsolete the question of their end of service life becomes significant. Concrete can be dismantled then crushed or pulverized to be reused in concrete over and over again, however it is a time and resource intensive process making it unlikely to be used in contexts that lack the necessary tools or resources.

Change is a permanent need in residential structures. From minor adjustments to major retrofits, adapting to the need for variation is a fundamental requirement of architecture. Looking at 3D printing from this perspective opens avenues for further research. For the moment, cable network channels, service integration or structural matrices have been studied for optimizing current construction. It is possible to reimagine 3D printing from the perspective of making structures more adaptable. Concrete deposited in interrupted layers or channeling for service replacement, can be designed in the patterns but still monolithic construction impedes disassembly of certain components at the end of their service life. Such a challenge also relates the discussion regarding machine size, and choosing a technique that offers greater efficiencies; on-site 3D printing, or 3D printing in a factory-like facility then assembly on site. While most of the recent projects were 3D printed on site, production in a controlled environment addresses the notion of quality control that has always been considered an advantage of industrialized building. In fact, the first 3D printed office in Dubai was 3D printed in parts and assembled on site. Additionally, one of the recently completed projects, 3D printed house in Sharjah, UAE, was 3D printed on-site, yet within a specially built tent to control the printing environment [6].

5   Discussion and Conclusion

Even with the high-tech nature of Additive Manufacturing, sourcing local materials to be deposited according to contextual traditions or patterns paves the way for an interesting union between information technology and traditional construction, thus potentially inducing a type of site-specific manufacturability. From mass-production to a type of customized manufacturing model, unlocks a new era of building machines.

The geneology from the Tournalayer, gantry crane and other modern experiments to large scale 3d printers is clear. Machines were central to the modernization of building culture and to the development of modern construction methods. Experiments in fabricating architecture through mechanization, to automation and robotization often are linked to normalization and repetitive production, but 3d printing offers a different path forward. The democratization of 3D printers puts fabrication in the laps of the many—as was the case for prototypes such as the Robo-hand [39] sourcing the needs and sharing knowledge on how to respond to them—with a basic understanding information technology. The issue of customization, a longstanding burden on the fabrication of architecture is being turned on its head with the help of machines that can address a type of intrinsic singularity programmed into a nozzle moved in any direction depositing concrete layers onto any surface in any context.

References

1. Hwang, D., Yao, K.-T., Yeh, Z., Khoshnevis, B.: Mega-scale fabrication by contour crafting. Int. J. Ind. Syst. Eng. **1**(3), 301–320 (2006)
2. n.d. Apis cor. Accessed Dec 2021. https://www.apis-cor.com/dubai-project
3. n.d. Batiprint3D. Accessed 2020. https://www.batiprint3d.com/en
4. Begić, H., Galić, M.: A systematic review of construction 4.0 in the context of the BIM 4.0 premise. Buildings **11**(8), 337 (2021)
5. Bessemer, H.: (1865). USA Patent US49055A
6. Boissonneault, T.: A tour of the 3D printed houses in Sharjah, UAE built with CyBe. (2019). Accessed Feb 2022. https://www.3dprintingmedia.network/cybe-construction-3d-print-houses-sharjah-uae/
7. Bos, F., Wolfs, R., Ahmed, Z., Salet, T.: Additive manufacturing of concrete in construction: potentials and challenges of 3D concrete printing. Virtual Phys. Prototyp. **11**(3), 209–225 (2016). https://doi.org/10.1080/17452759.2016.1209867
8. Bredt, R.: Catalogue of cranes. Ludwig Stuckenholz publisher, Düsseldorf (1894)
9. Davies, C.: The prefabricated home. Reaktion Books, London (2005)
10. Davis, H.: The culture of building. Oxford University Press, New York (2006)
11. Dini, E.: D shape: The 21st century revolution in building technology has a name. (2009). http://www.cadblog.pl/podcasty/luty_2012/d_shape_presentation.pdf. Accessed 2018
12. Dini, E.: Method and Device for the automatic construction of conglomerate building structures. (2008). Italy Patent ITPI20050031
13. Dini, E., Nannini, R., Chiarugi, M.: WIPO (PCT) Patent WO2006100556A2, (2005)
14. Duarte, J.P.: A discursive grammar for customizing mass housing: the case of Siza's houses at Malagueira. Autom. Constr. **14**, 265–275 (2005)
15. Customizing mass housing: a discursive grammar for Siza's Malagueira houses. Massachusetts Institute of Technology, Cambridge, MA (2001)
16. Duarte, J.P.: Towards the mass customization of housing: the grammar of Siza's houses at Malagueira. Environ. Plann. B. Plann. Des. **32**, 347–380 (2005)
17. Edison, Thomas. 1917. Process of constructing concrete buildings. United States Patent 1219272.
18. Eid Mohamed, B., Carbone, C.: Mass customization of housing: A framework for harmonizing individual needs with factory produced housing. Buildings 12, 955 (2022)
19. El Jazzar, M., Urban, H., Schranz, C. and Nassereddine, H.: Construction 4.0: A roadmap to shaping the future of construction. In: Proceedings of the 37th International Symposium on Automation and Robotics in Construction (ISARC 2020), pp. 1314–1321. Kitakyushu, Japan (2020)
20. El-Sayegh, S., Romdhane, L., Manjikian, S.: A critical review of 3D printing in construction: benefits, challenges, and risks. Arch. Civ. Mech. Eng. **20** (34). https://doi.org/10.1007/s43452-020-00038-w
21. Emerging objects. (n.d.). Accessed Dec 2021. http://emergingobjects.com/project/quake-column/

22. Everett E., Henderson Jr.: Making the tool to make the thing: The production of R.G. Letourneau's prefabricated concrete homes. Offsite: Theory and practice of Architectural Production, pp. 145–148. ACSA Fall Conference Proceedings, Philadelphia
23. Finch3D. (n.d.). Accessed March 2022. https://finch3d.com/
24. Figliola, A.: Envision the construction sector in 2050. Technological innovation and verticality. Firenze University Press, TECHNE (2019)
25. Fitchen, J.: Building construction before mechanization. MIT Press, Cambridge (1986)
26. Gramazio, F., Kohler, M., Willmann, J.: The robotic touch: how robots change architecture. Park Books, London (2014)
27. Hammond, Jr John Hays.: United States Patent US2499498A. (1947)
28. Hennebique, F.: France Patent 223546. (1892)
29. Icon. (n.d.). Accessed February 2022. https://www.iconbuild.com/
30. Jr., Henderson, E.: Making the tool to make the thing: The production of R.G. Letourneau's prefabricated concrete homes. In: Quale, J., Ng, R., Smith, R.E. (eds.) Offsite: Theory and practice of Architectural Production, ACSA Fall conference proceedings
31. Kagermann, H. and Wahlster, W.: Ten years of industrie 4.0. Sci **4**(3), 26. https://doi.org/10.3390/sci4030026
32. Khoshnevis, B.: Automated construction by contour crafting—Related robotics and information technologies. Autom. Constr. **13**, 5–19 (2004)
33. Khoshnevis, B., Dutton, R.: Innovative rapid prototyping process makes large sized, smooth surfaced complex shapes in a wide variety of materials. Mater. Technol. **13**(2), 53–56 (1998)
34. Lambot, Joseph-Louis.: France (1855)
35. Langenberg, E.: Mapping 20 years of 3D printing in architecture. (2015). Accessed 2021. https://www.elstudio.nl/?p=1904
36. Larson, K., Tapia, M.A. and PintoDuarte, J.: A new epoch: automated design tools for the mass customization of housing. A+U 366, 116–121 (2001)
37. Leatherbarrow, D.: Uncommon ground: architecture, technology, and topography. MIT Press, Cambridge (2002)
38. Letourneau, R.G.: Outer form for house form assembly. United States Patent US2717436A. (1952)
39. MakerBot.: Mechanical hands from a makerbot: The magic of robohand. 3/22. (2013). Accessed 13 April 2021
40. www.makerbot.com. https://www.makerbot.com/stories/engineering/robohand/.
41. McSweeney, E.: Automation finds home in building design. Financial Times. (2020). https://www.ft.com/content/e36ba45e-f973-11e9-a354-36acbbb0d9b6.
42. Merwood-Salisbury, J.: Chicago 1890: The Skyscraper and the modern city. University of Chicago Press, Chicago (2009)
43. Monier, J.: France Patent 77165. (1867)
44. Monteyne, D.: Framing the American Dream. J. Arch. Educ. **58**(1), 24–34 (2004)
45. Nervi, P-L.: Structural prefabrication. Italy Patent 377969. (1939)
46. Paolini, A., Kollmannsberger, S., Rank, E.:Additive manufacturing in construction: A review on processes, applications, and digital planning methods. Addit. Manuf. **30**. https://doi.org/10.1016/j.addma.2019.100894
47. Perrier, N., Bled, A., Bourgault, M., Cousin, N., Danjou, C., Pellerin, R., Roland, T.: Construction 4.0: a survey of research trends. J. Inf. Technol. Constr. **25**, 416–437 (2020)
48. Piroozfar, P.A.E., Piller, F.T.: Mass customisation and personalisation in architecture and construction. Routledge, New York (2013)
49. Sartipi, F., Sartipi, A.: Brief review on advancements in construction additive manufacturing. J. Constr. Mater. **2** (4), (2020). https://doi.org/10.36756/JCM.v1.2.4
50. Sharples.: Manufacturing material effects: Rethinking design and making in architecture. In: Kolarevic, B., Klinger, K. (eds.). Routledge, New York (2008)
51. Slaton, A.E.: Reinforced concrete and the modernization of american building, 1900–1930. Johns Hopkins University Press, Baltimore (2001)
52. Stephens, A.A.: United States Patent US3357685A. (1966)

53. Tay, Y.W., Daniel, B.P., Paul, S.C., Mohamed, A.N.N., Tan, M.J., Leong, K.F.: 3D printing trends in building and construction industry: a review. Virtual Phys. Prototyp. **12**(3), 261–276 (2017)
54. Tracoba.: Tunnel Form concrete formwork. France Patent FR1337089. (1963)
55. Neff, W.: United States Patent US2335300A. (1941)
56. Neff, W.: Building construction. United States Patent US2335300A. (1941)
57. WASP. (n.d.). Accessed Sep 2020. https://www.3dwasp.com/en/giant-3d-printer-bigdelta-wasp-12mt/#bigdelta
58. Winsun builders. (n.d.). https://www.winsun3dbuilders.com/project/3d-printed-office-in-dubai/
59. Van Wuyckhuyse, H.J.: Machine for shaking moulds filled with concrete. United States Patent 3357685. (1967)

Virtual, Augmented and Mixed Reality as Communication and Verification Tools in a Digitized Design and File-To-Factory Process for Temporary Housing in CFS

Monica Rossi-Schwarzenbeck and Giovangiuseppe Vannelli

Abstract This work presents a research project in which Cold-Formed Steel building components for temporary post-emergency housing are developed and realized with a digitalized workflow. This starts from early design ideas (peacetime), includes file-to-factory production and assembly processes (emergency relief/early recovery) and leads to the disassembly of building components and their reuse (reconstruction). The key element of the entire process is the Information Model. This is the place of the interoperability that, during the different stages, interfaces with different devices including Virtual, Augmented and Mixed Reality tools as well as file-to-factory processes for the industrial production. Aim of the paper is to show how visualization tools (like interactive Whiteboard, Tablet, Cardboard, Oculus Rift, Hololens 2, and Cave) can be used not only to realistically and immersively represent the project, but also to optimize design, production and construction processes. Indeed, these devices can also be used to improve the communication between the involved stakeholders, to enhance participatory processes, to help in decision-making, to verify a digitalized design and manufacturing process and to train workers. To achieve this goal, the innovative workflow is presented in chronological order, highlighting the purposes for which the selected tools were applied, analyzing their characteristics, potential, limits, software, interfaces, involved users and costs. The results comprise not only the application itself, but in particular the advantages and challenges evaluation of the use of the selected tools in a design project in order to improve future applications.

Keywords VR AR and MR · BIM and interoperability · File-to-factory · Cold-formed steel · Temporary housing

United Nations' Sustainable Development Goals 9. Industry Innovation and Infrastructure · 11. Sustainable Cities and Communities · 13. Climate Action

M. Rossi-Schwarzenbeck (✉)
University of Applied Science HTWK Leipzig, Leipzig, Germany
e-mail: monica.rossi@htwk-leipzig.de

G. Vannelli
Department of Architecture, University of Naples "Federico II", Napoli, Italy

© The Author(s), under exclusive license to Springer Nature Switzerland AG 2024
M. Barberio et al. (eds.), *Architecture and Design for Industry 4.0*, Lecture Notes in Mechanical Engineering, https://doi.org/10.1007/978-3-031-36922-3_24

1 Introduction and State-of-the-Art

1.1 Industry 4.0 and the Architecture, Engineering and Construction Sector

During the Hannover Fair in 2013, the German Federal Ministry for Economic Affairs and Climate Action and that one for Education and Research, together with other partners, established the Industry 4.0 platform to support industrial production with innovative information and communication techniques in order to make it faster, more efficient and more flexible [1, 2]. The platform has been a huge success and a driving force not only in Germany, but also in the rest of Europe, in the digitization also of sectors outside manufacturing. Industry 4.0 has now taken an important place in the economy. Companies see this as an opportunity to increase their competitiveness, drive technological development and, in particular, optimize production and business processes. On the contrary, in the AEC sector the digitization of building design, production, construction and more generally of processes is taking place in a slow and uneven manner [3]. This is attributable to several factors including two main ones: the high number of involved stakeholders (often characterized by different interests, applied workflows and used tools) and the fact that each building is a unique piece often made on site and with wet technologies, so that an industrial production (either serial or customized) is not always applicable. In order to reduce this gap, numerous initiatives such as Planen und Bauen 4.0, started in 2015, have been set up with the aim to introduce digital construction processes in the AEC, to consider the entire life cycle of buildings and transform new business models for real estate projects [4].

Therefore, the application of innovative paradigms and digitized processes has become more widespread in recent years. However, this is often done in a fragmented manner, only at certain stages of the process or by only some of the involved stakeholders. Industry 4.0 in AEC is a challenge, but also a big opportunity, which, in order to work efficiently, must cover with a collaborative and holistic approach the entire designing, construction and management process.

1.2 Digitization in the AEC: Challenge and Opportunity

Digitization in the AEC sector is taking shape particularly in the following applications: Building Information Modelling (BIM), digitized production (file-to-factory) and buildings project representation (Virtual, Augmented and Mixed Reality).

The use of BIM—consisting of an innovative collaborative method of design, construction and management that allows integrating all information about a building in one single three-dimensional digital model—is becoming increasingly widespread, particularly in complex and large-scale building projects for public clients [5, 6]. The creation of a digital twin of the to-be-built/built building facilitates communication between the involved stakeholders, reduces the possibility of design and construction mistakes, limits unforeseen cost increases, and can also be subsequently used for Facility Management [7, 8].

The digitized production of components using the file-to-factory method is nowadays customary in many sectors where the entire production takes place in the factory, such as the automotive industry. In the AEC, building realization mainly takes place on the construction site, and file-to-factory manufacturing is limited to components that can be produced in the factory and then assembled on the construction site. Not all building materials are suitable for this type of production, although interesting applications in concrete, steel, brick, plastic, wood and fiber materials have been realized in recent years [9, 10]. The most significant difference between this type of production and the traditional industrial one is the ease in making unique pieces. According to the 2030 Vision for Industry 4.0, digitized manufacturing is becoming increasingly more customized and less serial [11].

In recent years, the rapid spread of three-dimensional representation software and visualization tools has meant that these have reached the general public. Initially, users were almost exclusively video game players, now immersive representation tools are also routinely used in completely different sectors such as museums, education, medical, military, robotics, marketing, tourism, urban planning and civil engineering [12]. In the AEC these tools are generally applied for realistic, experimental and immersive visualization of urban, architectural or industrial design projects, although they have great potential in improving the design process by being applied in the stakeholder engagement, design support, design review, construction support, operation and management and workers training [13].

1.3 Virtual, Augmented Und Mixed Reality: Typologies and Characteristics

The above mentioned digital technologies for the representation of an architectural project are all based on a three-dimensional model of buildings and/or components. A 3D representation can take place on a two-dimensional medium (such as paper, tablet or screen) or the third dimension can be rendered more realistically with, for example, the aid of 3D glasses and the screen of a cinema or of a special television. When this third dimension comes to life, it becomes another type of representation such as Virtual, Augmented and Mixed Reality.

Virtual Reality (VR) is a technology that provides a 360-degree world in which the viewer can look in all directions and can also change his perspective by bending down or walking towards something. VR needs a tool (e.g. VR-glasses) and is immersive. This means that the user is totally immersed in the virtual world and perceives no other space outside this environment [14].

Augmented Reality (AR) is used to describe a combination of technologies that enable real-time mixing of computer-generated content with live video display. AR inserts additional objects or information into the real field of view and can be immersive (e.g. using Hololens 2) but also not (e.g. using a tablet or smart phone) [15].

Mixed Reality (MR) is an extension of the AR. The user is able to move and perform actions at the same time and in an integrated manner, in both the real and digital worlds, interacting and manipulating physical and virtual objects [16].

2 Research Purpose, Methodology and Case Study

2.1 Workflow Development: Information Model as Place of Interoperability

BIM, file-to factory production and VR, AR and MR are certainly important resources for the AEC, but at present, they are often used independently of each other, without exploiting their synergies. In order to better exploit the potential of these tools in the AEC sector, the aim of this paper is to specifically develop for a real case study an appropriate workflow able to improve the entire design, production, construction and management process using different types of representation tools as well as file-to factory production devices. A key element in digitized processes is the interoperability. In the glossary of the Industry 4.0 it is defined as the ability of different components, systems, technologies, or organizations to actively work together for a specific purpose [17]. In the proposed workflow, the place of interoperability is the Information Model. It is able to interface with different devices, it is the common element between representation tools and digitized production one (Fig. 1) and can be used by the different stakeholders, facilitating their communication and interaction. The Information Model is able to filter the information and provide the user with only the needed information for example, it provides the representation tools with information like form, materials, surfaces, colors, and the production one with information like processes, procedures, thicknesses [18].

2.2 Case Study: Design, Production, Construction and Implementation Process of Post-Disaster Temporary Housing in CFS

With the aim to develop and test a workflow that is capable of exploiting the full potential of the information model, of the various immersive and non-immersive visualization tools and of the file-to-factory production, a complex project is essential as case study. It must be characterized by a high level of complexity and a long time-span and in which interoperability and communication play an important role (because of the many stakeholders involved). Due to that the design of modular [19], incremental, flexible, dry-assembled, easily removable and reusable temporary

Virtual, Augmented and Mixed Reality as Communication ... 415

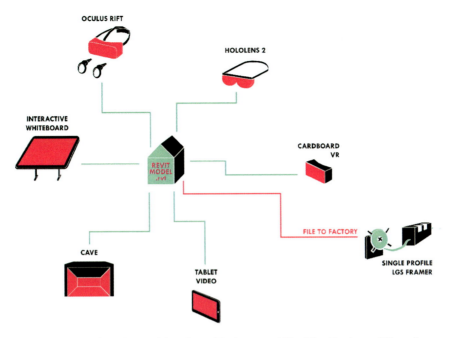

Fig. 1 Building information model as place of the interoperability. Visualization and file-to-factory tools that interface with the information model in the workflow developed

housing, using a construction system in Cold-Formed Steel (CFS) [20–22], for the construction of residential settlements to be built quickly following a disaster [23] (Fig. 2) was chosen as case study. These dwellings are conceived, according to the "building by layers" [24] logic, to be flexible and expandable over time, and at the end of their use, which is planned to be temporary, the wall-panels can be reused both to build new structures and to redevelop existing architecture by improving their seismic and energy performances [25].

The design of the housing units is part of a more complex process in which time plays an important role and is carried out in three phases: Peacetime (before the disaster, which as such is unpredictable), Emergency Relief/Early Recovery (following the disaster) and Reconstruction (Fig. 3).

The simulated process is characterized by a large number of involved stakeholders—that vary over time—and by different time phases whose duration is difficult to predict in advance as they are linked to unpredictable circumstances. In order to develop a workflow appropriate to the specific case study, it is necessary to analyze and predict the roles and relationships that stakeholders play during the process. These are the Italian Department of Civil Protection, designer (architects and engineers), university research centers, enterprises, local administrations, workers and inhabitants.

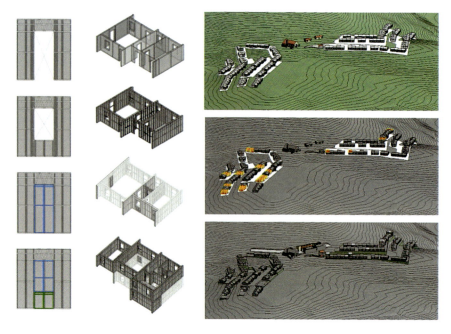

Fig. 2 Application case study. Design of the panel as building element, of the housing system and of the housing settlement in Sant'Eusanio Forconese (L'Aquila, Italy)

Fig. 3 Timeline of the design, production and implementation process: Peacetime, emergency relief/early recovery and reconstruction

3 Experimentation

3.1 Peacetime

In the first phase of the simulated process—called Peacetime, since it takes place in a timeframe prior to a possible emergency—the Italian Department of Civil Protection (DPC), supported by a university research center, is issuing a call for proposals for post emergency housing. The best proposal is chosen and optimized until it is ready to be manufactured and built immediately succeeding the emergency (see Fig. 4).

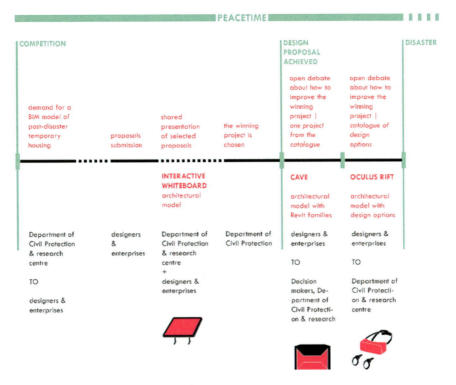

Fig. 4 Workflow and tools in the peacetime

The DPC is the national body that deals with post-disaster recovery and relief and in the simulated experimentation process constitutes the contracting and bidding body. The partner that could scientifically support the entire process should have a BIM laboratory equipped with the applied visualization tools, as it is the case for the Leipzig University of Applied Sciences (HTWK Leipzig, Germany). The call also requires that design proposals contain not only the project of modular dwellings, but also the design of dry-assembled and easily removable and reusable building components for their construction.

Then, design proposals must be submitted jointly by designers and enterprises that will be responsible for manufacturing the components. Additionally, the design must be presented in form of an Information Model able to be managed with a BIM working methodology. In this regard, the contracting authority also provides in the call an accurate Employer´s Information Requirement (EIR) [18] based on which the participants in the competition develop an appropriate Building Execution Plan (BEP).

Fig. 5 Mobile interactive whiteboard with multi-touch display applied for the presentation as well as for the improvement of selected proposals (3D visualization)

After the submission of design proposals, designers and enterprises of the selected projects present their proposals to the principals with the support of a mobile interactive whiteboard with multi-touch display. This whiteboard constitutes a useful visualization tool for presenting the project to a small number of people, facilitating communication and interaction through the model (Fig. 5). The informative architecture model can be opened in the proprietary format in the modeling software (e.g. Revit or ArchiCAD) as well as Industry Foundation Classes (IFC) file in a checking software (e.g. Solibri Model Checker) and can be verified and superimposed on specialized models (e.g. technical building equipment or constructive model).

Following the selection of the winning project, open debates can be held in which the DPC, designers, enterprises, and decision makers participate. These meetings could take place at the university research center and aim at optimizing the winning design in order to make it executable. Therefore, following a disaster, the housing units to be built can be quickly chosen and/or adapted to the specific situation and the necessary building components can quickly go into production.

Then, the proposed experimentation simulates a possible process using a project developed as part of a doctoral thesis elaborated within a collaboration between the University of Naples "Federico II" and the HTWK Leipzig while Irondom srl—a CFS component producing company—has been the manufacturer enterprise partner.

Two representation tools could be used in the open debates: a Cave Automatic Virtual Environment and the Oculus rift. The Cave is a VR Space in which a discrete number of people (up to about 30) with the support of special 3D glasses can simultaneously immerse themselves in the virtual reality of the design proposal so that they can discuss it and possibly optimize it together (Figs. 6 and 7). The Oculus Rift is an immersive VR tool in which the moving around and interaction with the building

Virtual, Augmented and Mixed Reality as Communication ...   419

elements (thanks to the two touch controllers) is greater. The model is editable and it is possible to write issues and to see model information. However, this is an experience that an individual performs alone and there is no interaction with the rest of the stakeholders, who see the model on the screen (Fig. 8).

Fig. 6 Cave automatic virtual environment and associated 3D glasses applied for the open debate between designers, enterprises and members of department of civil protection to improve the winning project (VR space)

Fig. 7 Cave. View with and without building envelope

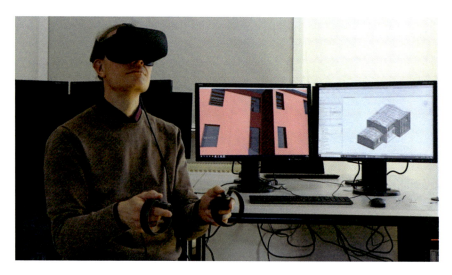

Fig. 8 Oculus rift (immersive VR). One person has a VR vision and people can see the 3D model on the screen

3.2 Emergency Relief and Early Recovery

Following the disaster—in case of the simulation an earthquake that occurred in central Italy (a seismic prone area, struck by several events in recent decades (1997, 2009 and 2016))—the process for the rapid construction of housing units starts (Fig. 9).

In the emergency relief phase, the site where the housing could be built is identified with the support of the local public administration. Within the experimentation an area located in the municipality of Sant'Eusanio Forconese (L'Aquila, Italy) was chosen. With the support of Hololens 2 it is possible to visualize the information model with different design options in the lot chosen for the settlement. DPC and the local government in order to define the exact placement of housing units in the lot can use this immersive AR tool, which can also be called MR as it allows interactions with both the real and virtual worlds (Fig. 10). At this stage, the quantity and types of housing units to be built and consequently the CFS building components to be produced are defined, based on number and needs of the future inhabitants.

A non-immersive AR visualization is also possible with the use of a tablet or iPad. This inexpensive tool, which even non-technicians often already have, can be used e.g. by workers (who may also be non-specialists and volunteers participating in the post-emergency construction) and future inhabitants to visualize the dwellings on the site. Unlike with the Hololens 2, the model is only partially editable with the tablet (Fig. 11).

In order to involve future inhabitants in the process and allow them to move virtually into their future homes, Cardboard VR glasses can be used. These are extremely inexpensive VR visualization tools that can be used with any type of smartphone,

Virtual, Augmented and Mixed Reality as Communication …   421

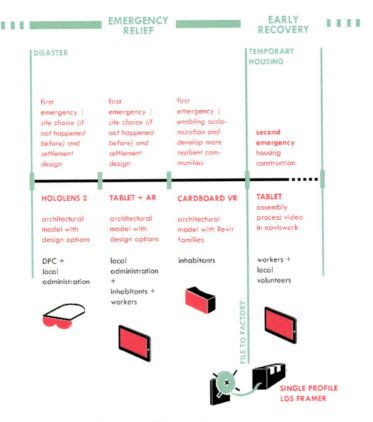

Fig. 9 Workflow and tools in the energy relief and early recovery

Fig. 10 Hololens 2 used to visualize the dwelling in the lot (immersive AR)

Fig. 11 Tablet used to visualize the dwelling in the lot (non-immersive AR)

a device currently owned by any type of user. This visualization is immersive but not interactive and does not allow you to edit the model, view information or write issues (Fig. 12).

As soon as the exact needs are defined, production based on file-to-factory processes can start. Panel-walls and other CFS elements can be produced by the enterprise—Irondom srl, in case of the simulated process—using a CNC machine (e.g. Arkitech AF I 200P Framer (Fig. 13)). This is a single profile LGS framer that is able to read the information of the building components from the information model and produce them through a digitized process based on customization and on-demand production (Fig. 14). Such a light production system opens the possibility of structuring flying factories [26] where to pre-assemble the building components. Moreover, this approach to the production process ensures—in accordance with Industry 4.0 goals—a high mass customization capability [27] and combines

Fig. 12 Cardboard VR-glasses used with the smartphone by future inhabitants

Virtual, Augmented and Mixed Reality as Communication ... 423

process flexibility, resource saving and high productivity. In post-disaster condition, also considering the amount and the variety of needs (variables over time) and of involved stakeholders, this is a meaningful workflow.

At this point, assembly and construction can begin on the construction site, carried out by skilled workers as well as volunteers. To assist the training of volunteers, videos made from the information model and viewable on a tablet or iPad can be prepared to illustrate the assembly of prefabricated elements (Fig. 15).

Fig. 13 Arkitech AF I 200P framer. Manufacturing process file to factory

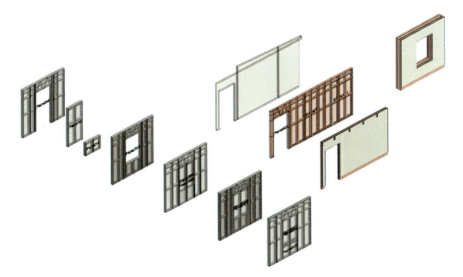

Fig. 14 Example of CFS customized panels/wall produced with a single profile lgs framer in a file-to-factory process

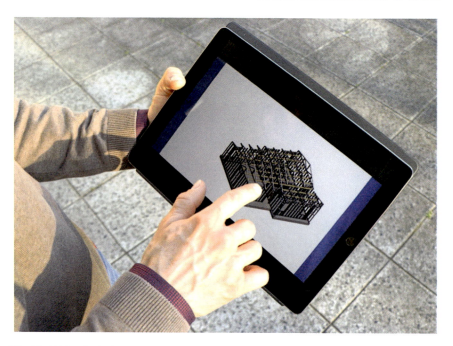

Fig. 15 Tablet displaying assembly process videos used by workers (3D visualization)

3.3 Reconstruction

While displaced communities are living in temporary housing, the reconstruction begins and includes rehabilitation and seismic improvement of lightly damaged buildings and demolition and reconstruction of severely damaged or collapsed buildings (Fig. 16).

At this stage, Virtual, Augmented and Mixed Reality tools (particularly the Cave and the Hololens 2) can be useful devices for stakeholders to jointly make decisions regarding future developments (Figs. 17 and 18).

The duration of this phase is difficult to predict as it depends widely on reconstruction policies and other external factors. Thus dwellings—which were intended to be temporary—are sometimes used for a longer period than expected and in the meantime the needs of the inhabitants may change (e.g. the number of members of a household increases). Otherwise, it may happen that two temporary housing units adjacent to each other, become vacant at different times, because inhabitants of different households may return to their permanent (redeveloped or reconstructed) home at different times. These changes in requirements over time may result in the need to expand one housing unit or to disassemble two adjacent ones at different times.

Virtual, Augmented and Mixed Reality as Communication …

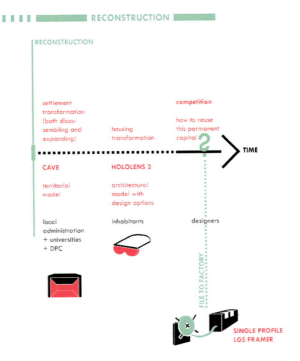

Fig. 16 Workflow and tools in the reconstruction

Fig. 17 Meeting in the cave using the urban model to identify housing units that need to be expanded, modified or disassembled/removed

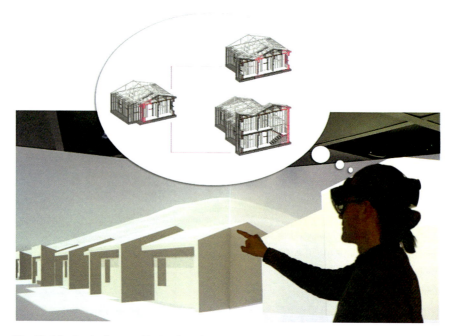

Fig. 18 Meeting in the cave. Evaluation of possible extensions of a housing unit

To respond quickly and easily to such needs, the housing units as well as the construction system in CFS components are designed with a high level of transformability and reversibility. Additionally, the reuse of CFS components, disassembled from temporary housings at the end of their use, for seismic improvement of pre-existing damaged buildings is conceivable. The reuse and recycling of materials and components makes the entire developed process part of a circular economy approach [28].

4 Results and Discussion

The outlined experimentation achieved several results on three different levels: process innovations (development of a digitized workflow that includes the use of VR, AR and MR tools and file-to-factory processes), design innovations (design of a catalogue of flexible, extendable, transformable and removable residential buildings) and product innovations (fine-tuning of different types of customized CFS panels, dry-assembled on site, removable and reusable).

In relation to the topic of this chapter, the development of the workflow is the most interesting result. Indeed, the case study demonstrated how it is possible to manage a complex, time-extensive process with numerous stakeholders based on an information model capable of interacting with different AR, VR and MR devices as well as with customized file-to-factory production tools. The application also showed that visualization devices could be used not only to realistically and immersively represent an architectural project, but also to optimize design, production and construction process.

For this process to work best, it is necessary for the client to define the Employer's Information Requirements (EIR) in detail, as well as for the designers to develop an appropriate Building Execution Plan (BEP). The Level of Definition (Level of Geometry and Level of Information), as well as the software and the exchange formats between them, are important points that must be defined in advance so that communication takes place without errors and the stakeholders receive only the information that is relevant to them.

The choice of which visualization tools to apply, at which stage, for which purpose and with which stakeholders must also be based on an understanding of their peculiarities. The tools used in the experimentation differ from each other in some fundamental characteristics such as typology, involved people (both in terms of amount and skills), possibility to check the model information, ability to modify the model, possibility to write issues, used software, formats and costs.

In particular, the number of users who can simultaneously use a device influences its application in a meeting among decision makers (even 20–30 people) or, for example, for solving a problem by an expert who is working alone. The ability to modify the model as well as the ability to write issues are key features at the stage in which the building system or design still need to be optimized, while they have no importance at the presentation stage of the project itself. Costs of the devices are also important: these range from €1.00 for Cardboard VR glasses to around €300,000 for the Cave. The Cardboard can be a gadget that can be distributed free of charge to residents; the cost of the Cave can be met by, for example, a university research center that rents it out to design groups to conduct their meetings or presentations. Other tools such as Hololens or Oculus Rift have costs that are also affordable for design firms or enterprises.

In fact, this research, in addition to testing, intends to analyze the potential and limitations of the visualization tools in order to facilitate their selection for possible future applications. To this end, their characteristics are presented in detail in Figs. 19 and 20.

In relation to design innovation and product innovation, it is possible to say that the use of a digitized workflow, in accordance with the principles of Industry 4.0, based on an information model, or rather on the overlay of specific information models (architectural model, structural model, plant engineering model, etc.) has certainly facilitated the development of a housing and a construction system capable of reacting to changing conditions and needs as well as a file-to-factory production. In fact, the use of the BIM methodology makes it possible to highlight any problems and collisions in advance. This is a fundamental requirement for a complex project such as the experiment performed, which has a high degree of flexibility and unpredictability due to a disaster event and uncertain development over time. On the production side, the information model also plays an important role by providing inputs to the CNC machine and allowing easy customization.

Fig. 19 Used visualization and production tools and their characteristics

Fig. 20 Used visualization and production tools and their characteristics

In conclusion, the developed workflow certainly yielded positive results that—after an appropriate adaptation to the specific situation as well as the needs of stakeholders and their relations—could be also applied to other complex projects. To this end, the digital information model becomes the place of interoperability and communication, while AR, VR and MR devices play an important role as facilitators for example in communication, in problem identification and resolution, in worker training, and in on-site assembly. Additionally, it is certainly desirable for future applications to extend the workflow of the digitized process to other phases such as facility management.

Acknowledgements The presented research is by its nature characterized by the interaction of numerous partners. In this context, we would like to thank the people and institutions that in various ways contributed to its success. We would therefore like to thank the team of Irondom srl with whom a fruitful two-way transfer of knowledge took place. We also thank the Department of Civil Protection and the mayor of the municipality of Sant'Eusanio Forconese for the exchange of information and materials necessary for the success of the research. We would like to thank Christian Irmscher (BIM employee at the HTWK Leipzig) for sharing his knowledge on BIM, Virtual and Augmented Reality. We would also like to thank Sergio Russo Ermolli and Angela D'Agostino, supervisors—together with Monica Rossi-Schwarzenbeck—of the doctoral thesis conducted by Giovangiuseppe Vannelli at the University of Naples "Federico II" in collaboration with HTWK Leipzig and Irondom srl. The research has been carried out the framework of a PhD with Industrial Characterization financed with P.O.R. funds Campania region FSE 2014/2020, Axis III—Specific Objective 14 Action 10.4.5.

References

1. Industry 4.0 Homepage. www.plattform-i40.de. Accessed 13 Oct 2021
2. Reinheimer, S.: Industrie 4.0, Herausforderungen, Konzepte und Praxisbeispiele, Springer Vieweg, Wiesbaden (2017)
3. Rahimian, F.P., Goulding, J.S., Abrishami, S., Seyedzadeh, S., Elghaish, F.: Industry 4.0 Solutions for Building Design and Construction. A Paradigm of New Opportunities, 1st edn. Imprint Routledge, London (2021)
4. Planen-Bauen-4.0 Homepage. https://planen-bauen40.de/. Accessed 13 Oct 2021
5. Borrmann, A., König, M., Koch, C., Beetz, J.: Building information modeling. In: Technology Foundations and Industry Practicie, 1st edn. Springer International Publishing, Cham (2015)
6. Hausknecht, K., Liebich, T.: BIM-Kompendium. Building Information Modeling als neue Planungsmethode, 2nd edn. Fraunhofer IRB Verlag, Stuttgart (2020)
7. Teicholz, P. (ed.): BIM for Facility Managers, 1st edn. John Wiley & Sons, New Jersey (2013)
8. Pinti, L., Codinhoto, R., Bonelli, S.: A review of building information modelling (BIM) for facility management (FM): implementation in public organisations. Appl. Sci. **12**(3), 1540 (2022)
9. Kaiser, A., Larsson, M., Girhammar, U.A.: From file to factory: innovative design solutions for multi-storey timber buildings applied to project Zembla in Kalmar, Sweden. Front. Archit. Res. **8**, 1–16 (2019)
10. Naboni, R., Paoletti, I.: Advanced Customization in Architectural Design and Construction. Springer International Publishing, Berlin (2015). BMWi: 2030 Vision for Industrie 4.0 Shaping Digital Ecosystems Globally Federal Ministry for Economic Affairs and Energy (BMWi), Berlin (2019)
11. Wen, J., Gheisari, M.: Using virtual reality to facilitate communication in the AEC domain: a systematic review. Constr. Innov. **20**(3), 509–542 (2020)
12. Davila, J. M., Delgado, Lukumon, O., Demianc, P., Beach, T.: A research agenda for augmented and virtual reality in architecture, engineering and construction. Adv. Eng. Informatics **45**, 101122 (2020)

13. Tea, S., Panuwatwanich, K., Ruthankoon, R., Kaewmoracharoen, M.: Multiuser immersive virtual reality application for real-time remote collaboration to enhance design review process in the social distancing era. J. Eng. Des. Technol. **20**(1), 281–298 (2022)
14. Mekni, M., Lemieux, A.: Augmented reality: applications, challenges and future trends. In: Applied computer and applied computational science Conference Proceedings, pp. 205–214 (2014)
15. Wang, X., Aurel, M.: Mixed Reality in Architecture, Design, And Construction. Springer Science + Businnes Media, Berlin (2009)
16. Standards of the Association of German Engineers (VDI), "Industrie 4.0 Begriffe / Terms and definitions", Düsseldorf (2022)
17. Baldwin, M.: The BIM Manager: A Practical Guide for BIM Project Management. Beuth Verlag, Berlin, Wien, Zürich (2019)
18. Mellenthin Filardo, M. Krischler, J.: Basiswissen zu Auftraggeber-Informationsanforderungen (AIA), bSD Verlag, Dresden (2020)
19. Smith, R.E.: Prefab Architecture: A Guide to Modular Design and Construction. John Wiley & Sons, Hoboken (2010)
20. Grubb, P.J., Gorgolewski, M.T., Lawson, R.M.: Light Steel Framing in Residential Construction. Building Design using Cold-Formed Steel Sections, The Steel Construction Institute, Ascot (2001)
21. Yu, C.: Recent Trends in Cold-Formed Steel Construction. Woodhead Publishing, Sawston (2016)
22. Landolfo, R., Russo Ermolli, S.: Acciaio e sostenibilità. Progetto, ricerca e sperimentazione per l'housing in cold-formed steel, Alinea editrice, Firenze (2012)
23. Antonini, E., Boeri, A., Giglio, F.: Emergency driven innovation. Low Tech Buildings and Circular Design, Springer, Cham (2020)
24. Brand, S.: How Buildings Learn: What Happens After They're Built. Viking, New York (1994)
25. Formisano, A., Vaiano, G.: Combined energy-seismic retrofit of existing historical masonry buildings: the novel "DUO system" coating system applied to a case study. Heritage **4**, 4629–4646 (2021)
26. Ciribini, A., Di Guida, G. M., Lollini, R., Aversani, S., Gubert, M., Miorin, R.: Moderni Metodi di Costruzione, ReBuild Italia, Milano (2019)
27. Tang, M., Zhang, M.: Impact of product modularity on mass customization capability: an exploratory study of contextual factors. Int. J. Inf. Technol. Decis. Mak. **16**(4), 939–959 (2017)
28. Antonini E., Boeri. A., Lauria M., Giglio F.: Reversibility and durability as potential indicators for circular building Technologies. Sustainability **12**(18), 7659 (2020)

Digital Processes for Wood Innovation Design

Fabio Bianconi, Marco Filippucci, and Giulia Pelliccia

Abstract The study reports the outcomes of a research activity focused on digitization techniques and in particular on the value of computational design, with the aim of implementing innovative product and process solutions resulting from an integrated design approach. The digitization paths focused on representative strategies for digital optimization of the architectural form of wooden houses as a function of context, based on research on generative modelling and evolutionary algorithms for multi-objective optimization applied to the architecture of wooden houses. With such an approach, centered on artificial intelligence or at least on augmented computational intelligence, it was possible to achieve a process of mass customization of meta-planning solutions of wooden architectures, based on the morphological and energetic selection of the best configurations, identified according to the context. These results were made accessible through a web-based configurator that provides the designer with initial configurations from which starting the real project. The studies are projected to the definition of a prototype of the "breathing house," characterized by its moisture-responsive wooden panels, with the identification of innovative solutions capable of reacting passively to changes in humidity according to the "natural intelligence" of the material, whose morphological transformation, empirically studied and digitally transcribed to identify performance solutions, generates well-being for living.

Keywords Meta-design · Energy optimization · Responsive architecture · Timber construction · Digital representation

United Nations' Sustainable Development Goals 9. Build resilient infrastructure, promote inclusive and sustainable industrialization and foster

F. Bianconi · M. Filippucci · G. Pelliccia (✉)
Department of Civil and Environmental Engineering, University of Perugia, 06073 Perugia, Italy
e-mail: giulia.pelliccia@outlook.it

F. Bianconi
e-mail: fabio.bianconi@unipg.it

M. Filippucci
e-mail: marco.filippucci@unipg.it

innovation · 11. Make cities and human settlements inclusive, safe, resilient and sustainable · 12. Ensure sustainable consumption and production patterns

1 Introduction

The digitization processes accompanying the recent industrial revolution, in the context of AEC (Architecture, Engineering and Construction), are based on representation: the design of form and enrichment through multiple information are amplified with the interactivity, modifiability and multimedia inherent in digital [1]. The goal is to make the information implicit in the form visible [2, 3]. The design is asked first and foremost for a total freedom but also for a full awareness, as validation tools for the construction phase, to the point of generating a digital clone that replicates reality to continuously derive information in the dynamic integration between hardware and software [4]. The built does not become an object in itself, but stands precisely through its digital replication in relation to its environment, through the processing of causal associations used to create predictions [5], in a cycle marked by data-information-knowledge [6]. Data, defined as the new oil [7], promotes the architecture's resilience.

Such processes result in multiubiquity [8] arising from enabling technologies [4, 9] that lead to a "Cyber-Physical Production System" [4, 10] inherent in the "Smart Factory" [11], which are bringing interesting substantial benefits in terms of digital production [12] and automation [13], but also in what concerns mass customization [14] and stimulating innovation [15]. This activates a digital thread [16] where design becomes a continuous process overlapping the built.

The field of research has thus increasingly shifted from the physical to the virtual space: the morphogenesis from which the model is originated, where the multiple information converges, changes the relationship between sign and image as well as the way of reading and interpreting reality, observing it from different points of view, analyzing it with integrated skills, studying its multiple variables and their mutual interaction. In Industry 4.0 [13], the new paradigm of digital tectonics is developing a new coincidence between virtual representations and their fabrication (Fig. 1), in what appears to be a true cultural revolution [17, 18], because "fabrication is not a modelling technique, but a new way of doing architecture" [19].

Profoundly affecting this transformation are the advances in computational design [20] and in particular parametric logics [21], all based on a different interpretation of the value of data: if the digital revolution has shifted the focus from models to visualization, then parametric design allows to reclaim the infinite potential of the model [1, 22] because data is not secondary but rather the main element that leads to form. "Through a well-designed system of rules, generative design systems have the capability of maintaining stylistic coherence and design identity while generating different designs" [23], rethinking relationships that are "intelligible because its members exhibit a common order resulting from the operation of the same generative principles" [24]. By varying parameters in defined but always open processes, the

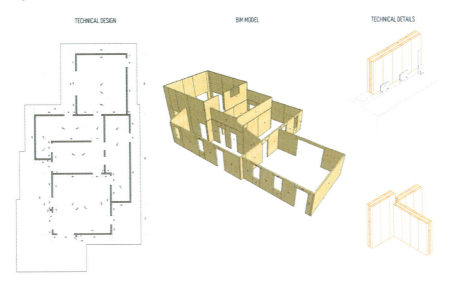

Fig. 1 Digital wood design and digital manufacturing

algorithm is a tool for configuring, rather than a static morphology, a generative pattern characterized by diversity.

Thus, a data-driven design [25] is developed where one relates to "complex systems" not by looking for simplifications or optimal solutions, rather by creating a combinatorial explosion [26], contrasting this Evolutionary Engineering marked by Multiscale Analysis [26–28] as an alternative to the "iterative and incremental" standard. Optimization strategies [29, 30] are structurally and environmentally optimized form-finding solutions [31] that enable the designer to visualize and evaluate thousands of design options and variants [32] through the materialization of the issue of architectural complexity [33], from the perspective of mass customization [34–43].

Representation is proposed in its definition of architecture and design as "*Lineamenta*" [44], understood according to Leon Battista Alberti's diction [45] as a system of signs. Artificial intelligence [46, 47] and in particular the logic of optimization inherent in evolutionary algorithms [48] acquires a central role in this relationship: computational design is enhanced thanks to the augmented intelligence of the digital [49–51], which finds perfect exemplification in wooden constructions, from design to realization.

Thus, in this context, wood emerges as a material that is absolutely congenial to digital logics, versatile, and fully adapted to contemporary needs for performance and customization (Fig. 2). Therefore, digital representation is added to the properties of matter and its Natural Intelligence (NI) [49–51]. Nature offers not forms but processes for thinking about form [52] and teaches how to create it [53, 54] in efficient structures [55, 56] explaining how the roles of design [57, 58] are truly adaptive and optimized [59]. NI defines paradigms that could be transposed not as "ignorant copying of forms"…but in the recognition "that biomimicry teaches that

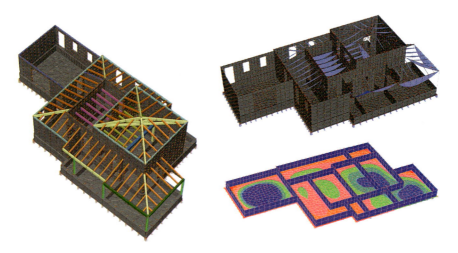

Fig. 2 Analysis and simulation in digital wood models

form is the most important parameter of all" [60], because at all levels it builds responsive and adaptive forms to preserve material and energy resources through the use of modular components combined with low-energy structural strategies [61].

2 The Research

Digitization has been the focus of the entire research project started in 2016 with the then wooden construction start-up Abitare+, in the aim of achieving innovative solutions that would characterize the quality of its offerings. The study is focused on digitization techniques and, in particular, on the value of computational design, with the aim of implementing innovative product and process solutions as the result of an integrated approach to design. The digitization paths addressed various issues, starting with the identification of representative strategies for digital optimization of the architectural form of wooden houses according to the context.

The research was developed at the Department of Civil and Environmental Engineering of the University of Perugia, a path that has found an important recognition in the "BIM&DIGITAL Awards 21" promoted by CLUST-ER BUILD and DIGITAL&BIM Italia. The award was intended to report as excellent the innovative path that through the new digital techniques of representation has led to the development of solutions for new generation wooden housing, capable of responding to the multiple performances and needs of living today. This collaboration was supported by funds from the 2014–2020 POR FESR of the Regione Umbria aimed at supporting the creation and consolidation of innovative start-ups with a high intensity of knowledge application. The sustainability of innovation processes is then strengthened by the support for research guaranteed by national regulations, which provides tax credit for a significant part of the investments made.

2.1 Meta-design and Mass Customization of Wooden Houses

One of the first approaches developed jointly with the company is based on computational design research to address the need to promote a design culture of wooden constructions. Indeed, the research is aimed at designers, who in the national context often face difficulties in designing wood constructions due to the absence of specific training: it is then intended to provide meta-planning solutions, a basis for then developing the project. Evolutionary principles have been applied aimed at informing the design and customization process in the early stage of design. The main goal is to design a comfortable house characterized by high energy performance taking maximum advantage of the passive use of sun and wind. For this purpose, a web-based interface has been developed allowing to explore the design alternatives of a specific construction model by visualizing and downloading a series of multimedia files.

The developed parametric process generates a wide variety of architectural solution, and each of them differs from the others mainly for their orientation, size, type of ceiling, roof slope and shape of the glazing elements. In this case, the definition of rule-based design emerges from a study of local codes and CLT construction systems; indeed, while defining the geometrical rules of the model, constraints have been encoded in a way that each solution follows codes dimensioning and affordable fabrications methods. As a result, each house in the series is unique in shape and size, even if it shares with the others the same building system characterized by CLT panels and a fixed number of manufacturing operations, in both the factory and the building side.

The integrated process proposed in this research has been entirely developed with Grasshopper, introducing in each phase of the project different add-ons for analysis, representation, and interoperability. The definition of the performance criteria started with the study of a construction model through the definition of detailed solutions and sizing of structural elements. In this phase, the construction cost was computed with the company through the definition of a series of parametric costs for the elements constituting the structural system and the envelope, while energy performances were evaluated through advanced energy analysis by estimating building energy consumption, comfort, and daylighting. The goal of an environmental optimization is to ensure a satisfactory comfort with the minimum use of energy, through the adaptation of the architectural organism to its context and its inhabitants. Natural lighting then becomes one of the major driving forces in this design process, which aims to reinforce circadian rhythms and to reduce the use of electric lighting by introducing daylight into space, and it results an effective reduction of energy consumption and comfortable spaces (USGBC, 2013) (Fig. 3). In this research, due to time constraints, 15 generations of 100 individuals were evaluated with the climate data of Perugia,

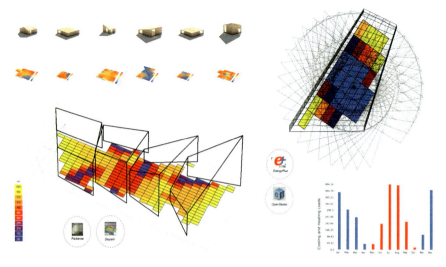

Fig. 3 Lighting, cooling and heating loads have been analyzed for the different geometric solutions through Honeybee for Grasshopper

Italy. Comparing the less and the most efficient solutions, taking into account that they have different areas, volumes, grazing rations, orientations, the results in saving on energy consumption per square meter was 380%, and saving on construction costs per square meter was 240%.

The proposed workflow highlights the centrality of the research approach, based on computational and collaborative strategies, as a link between design teams and modern construction companies. The analyzed solutions have then been represented in a web-based catalog through which the technology owned by modern construction companies can be shared with designers to achieve an integrated approach. Within a wider collaborative workflow, encompassing smart manufacturing principles and integrated design strategies, the research and development project focuses on the analysis and representation of data. As a result, design teams can take part in the process of mass-customization and start an integrated design process to optimize the product and achieve further design customizations.

2.2 Multi-objective Optimization of Architectural Elements

The logic of mass customization applied to architectural morphogenesis can be similarly applied to the design of building elements. Focusing on the envelope as an essential element of buildings, the research examines perimeter walls by analyzing them from the point of view of both winter and summer behavior and verifying the absence of the of interstitial condensation. Two building systems, Platform-Frame and X-Lam, were analyzed with the aim of optimizing, through the creation of special

Digital Processes for Wood Innovation Design

algorithms, their stratigraphy in order to provide the company with a set of diversified solutions that take into account both cost and energy performance [62, 63].

The combination between the thicknesses and types of materials that compose the wall is the basis of the analysis. Large amounts of data can thus be analyzed and combined and to get solutions that simultaneously present the best values of the parameters chosen as inputs, returning the required outputs. By varying the parameters, the outputs describe the summer and winter behavior of the wall through thermal transmittance U, periodic thermal transmittance Y_{ie}, decrement factor f and time shift φ, according to UNI EN ISO 13786:2018 [64] in addition to the verification of interstitial condensation by Glaser diagram according to UNI EN ISO 13788:2013 [65]. The total cost was then used as a benchmark for comparison with the optimized packages. The stratigraphy optimization process was conducted using Octopus, which allows evolutionary principles to be applied to parametric modelling in order to optimize specific parameters [66, 67]. In the two cases considered, Octopus calculated about 5.000 possible solutions; among them, only those belonging to the Pareto Front [68] were selected. Packages were then divided according to performance, as defined by DM 26/6/2009 "National guidelines for energy certification of buildings" [69]: excellent performance ($\varphi > 12$ h, $f > 0.15$) and good performance ($10 < \varphi < 12$ h, $0.15 < f < 0.30$). Depending on the final cost, the most suitable walls were identified, narrowing down to those with a lower or slightly higher cost than the standard reference package (Fig. 4).

Through the proposed optimization and selection method, it was possible to obtain walls with significantly better performance than the standard ones, even at a lower cost. For Platform-Frame, in particular, a stratigraphy can be obtained with excellent

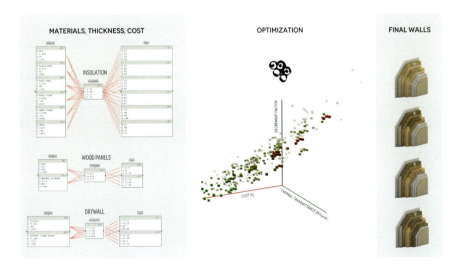

Fig. 4 The optimization process through Octopus for Grasshopper combines materials and thicknesses while maximizing time shift and minimizing thermal transmittance, decrement factor and costs

performance with a saving of 0.7% and one with good performance with a saving of 3.8%. For X-Lam, on the other hand, an excellent stratigraphy can be obtained with a saving of 1% and a good one saving 13% of the costs as compared to the standard wall. If, instead, the improvement in energy performance is considered without limiting the cost, Platform-Frame can be improved up to 20% in transmittance and 31% in time shift, and for X-Lam up to 28% in transmittance and 22% in time shift. Thus, the parametric design and muti-parametric optimization tools proved to be, once again, essential to process a large amount of data and select the best performing solutions according to specific needs.

Experiments in digital form-finding are linked to BIM modelling, aimed at defining digital manufacturing processes. The construction is linked to representation through BIM interchange models that define the characteristics of the envelope and structural elements, which in turn find information from the algorithms developed in generative modelling.

2.3 The Breathing House

Alongside digital simulations carried out to achieve the ultimate goal of improving the energy consumption of wooden buildings, solutions can also be found in exploiting the properties of the materials themselves. This is the case, for example, of technological smart materials, which have the ability to react to a particular stimulus and change some of their properties to adapt to the external conditions [70, 71]. Smart materials also include natural materials such as wood, which is a sustainable, easily available and low-cost solution [72]. In fact, thanks to its hygroscopic properties, wood expands and shrinks as ambient humidity changes. One can therefore think of exploiting these properties for the control of hygrometric well-being in indoor environments [73, 74].

A typical example of hygroscopic behavior is showed by pine cones, whose scales bend due to the different reaction to moisture of the two tissues from which they are composed (fibres and sclereids) [75–78]. It is therefore possible to replicate these properties by making an artificial composite that takes advantage of the hygroscopicity and anisotropy of wood (active layer) combined with a material that does not react to moisture or has a lower hygroscopic expansion coefficient (passive layer). A single layer of wood will always show some degree of shrinkage/expansion proportionally to changes in moisture; by coupling it with a material that does not undergo deformation, its response can be pre-programmed in order to achieve the desired configuration at a given moisture content, and from a simple dimensional change will result in bending [79, 80]. The carried tests on various specimens with different configurations and characteristics were used to create a prototype of a modular ceiling panel which is pre-programmed to bend for relative humidity other than 40% (Fig. 5). Double-layered wood panels react passively to changes in humidity and, therefore they can be considered as low-cost, low environmental impact and technological

Digital Processes for Wood Innovation Design

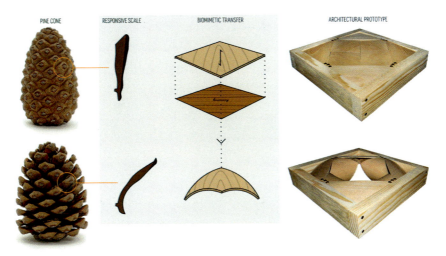

Fig. 5 The biomimetic transfer of properties from the pine cone to an artificial prototype can be used in responsive architectures to passively react to humidity variations

elements that may be able to improve indoor hygrometric comfort without additional energy consumption. In particular, it is intended to propose a model of natural ventilation, complementary to the use of air conditioning systems, especially for the regulation of humidity, which is thus regulated thanks to bioclimatic principles that exploit the convective motions of warm and humid air and the chimney effect.

These principles and goals have since been transferred to the field of 4D printing [81–83] of wood-based filaments, joining the 3 dimensions in space with a fourth dimension, time which allows the composite to adapt to environmental humidity. Additive printing allows a total customization, making it possible to design the hygroscopic deformations. Using the commercially available filament LAYWOOD, composed of 40% recycled wood powder and 60% PLA (polylactic acid) [84], the direction of expansion of the composite is drawn by the deposition of the filament itself, which can then follow the desired pattern and result in complex deformations. Various printing properties radically affect the result obtained, from layer height to infill, feed rate to z-offset with respect to the printing plane, and so on. By varying these parameters, different curvatures and curvature velocities can be obtained for the same design (Fig. 6).

The production of multiple specimens to compare the hygroscopic deformations, both in natural and printed wood, and the research aim of applying these composites to the improvement of indoor well-being then led to the creation of a test room where these panels and optimized solutions for wood walls could be tested.

Fig. 6 Different curvatures and velocity of curvature of similar 3D printed specimens with different printing properties

2.4 The Wooden Test Room

Digitally optimized solutions as well as the idea of a "breathing house" involve multiple aspects that need to be analyzed but also verified. For this need, a temporary wooden test room was built at the Engineering Pole of the University of Perugia.

This structure was developed on a single level with Platform-Frame structure, of about 20 m^2 and average height 2.4 m, characterized by a glazed opening in the south direction. On the roof, about 15 m^2 of thin-film photovoltaic panels with a storage battery are needed for the heat pump, to cool and heat, simulating the winter and summer indoor thermo-hygrometric conditions typical of a residential environment. The north-facing wall is removable and can be replaced with other walls characterized by different stratigraphies.

The peculiarity of this Platform-Frame test room is that the monitored north-facing wall is removable for testing with different stratigraphies chosen from the various solutions optimized by the algorithm in the simulation phase. Monitoring was carried out through heat flux sensors, thermocouples, humidity and temperature probes during the summer period (Fig. 7). The acquired data were used for the determination of the in situ thermal transmittance to be compared with the one simulated by the algorithm, referring to UNI ISO 9869 [85], according to which it is possible to obtain the thermal resistance from the ratio of the summation of the surface temperature difference between outside and inside and the summation of heat fluxes.

Moisture-responsive wooden panels made of beech and larch wood were installed at the ceiling of the test room. Thanks to the convective motions of moist air rising toward the false ceiling, which causes the panels to bend due to the difference in humidity, air flows inside a cavity and is carried outside by exploiting the chimney effect.

Measurements were made on an optimized wall that best combines the most popular insulation materials on the market. Thermal transmittance was then calculated from the acquisitions, and considering the 10% uncertainty rate of the direct measurement, due to multiple factors, as well as the fact that the stated values of the

Digital Processes for Wood Innovation Design

Fig. 7 The test room has been designed and built to experiment in reality what had been digitally simulated, concerning in particular the responsive false ceiling and the optimized timber walls

thermal conductivities of the insulating materials also exhibit percentages of variability, it was concluded that the actual behavior is similar to that obtained from the simulations. The information is then returned from the built to the digital in the final section of the study, in order to build a real-time monitoring system of what is happening within the test room and integrate that data into the model, creating a digital twin in the BIM environment [86].

The test room certainly represents one of the clearest paradigms of a contemporary way of doing research, a space set up to transfer innovation to the market in which Abitare+ presents itself by offering innovative and performing solutions. At the end of this first part of the research it is interesting to highlight that the data acquired confirmed the reliability of the calculations made by the algorithm, with a small percentage of error due to field measurements.

2.5 Representation and Communication of the Optimized Models

One of the key aspects of the research was the collection of data and its representation through ways that are accessible to everyone: the algorithms created have their own complexity and specificity so that even the company's engineers cannot manage the information. In fact, during optimization processes, a huge amount of data comes into play, and visualizing this data is crucial for understanding, comparing and sharing the results of the approach carried out.

In this context, the combination of data visualization became an effective way to enhance the decision-making process [87] while design space catalogs, which present a collection of different options for selection by a human designer, have become a commonplace in architecture in the perspective of the design democratisation [88]. The aim is to create an open source design [89–91] as meta-project for adaptable and mass customized housing [92]. The user interface developed in this research is based on Design Explorer, an open source project realized by CORE Studio Thornton Tomasetti, that allows to intuitively visualize and effectively navigate the design space of parametric models developed in Grasshopper, Dynamo, and Catia. These tools can support the designer in the complex problem-solving processes, through the combination of the designer's preferences with the great amount of information owned by modern construction companies, thus filling the gap between technological advances and design practice. Furthermore, their usability and effectiveness will grow along with advances in Building Information Modeling (BIM), performance simulations and parametric design and hopefully, in the next future, a similar data-driven approach will help the designer to deal with increasingly complex projects and achieve both performance and aesthetic expression.

For both the meta-planning solutions of the houses and the multi-objective optimizations of the walls, the same representative action was carried out, which allows the selection of parameter ranges for the different elements that characterize their geometry (Fig. 8). At the strategic level, the company then decided to make public the results of the mass customization of the houses while leaving for internal use the dynamic catalogs of the wall element combinations, which become a performance enhancement to the solutions of each house as a result of the executive design, in terms also of production, which the company refines at the end of the design phase.

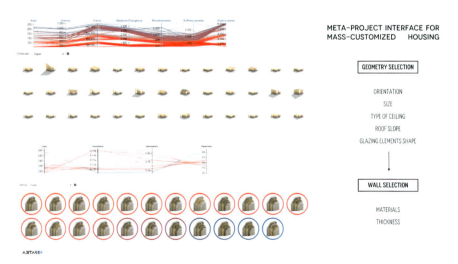

Fig. 8 The web interface allows the users to select the optimal solutions, choosing the best geometries and walls depending on their performance requirements or budget

2.6 Representation and Communication of the Environmental Simulations

The data, analysis and the whole logic of performance and efficiency that is demanded of buildings today does not meet the needs of living. The design of the home must be communicated comprehensively to both engineers and clients. Technical drawings and renderings, which are the main tools for communicating design, do not guarantee customization, which instead is inherent in digital language.

The experience conducted is projected in the logic of the serious game [93–96], aimed at gathering information on the impact of simulations. In fact, the interaction between the user and the model is monitored by an analysis of the sensations related to the different configurations: in the interactive experience, sensors applied to the fingers are able to assess the galvanic skin response (GSR) [97–100] to pick up variations in its micro-sensation, including, for example, involvement and stress. This information, after the interactive visit, is cross-referenced with data collected "in game" regarding the user's location and point of view to understand what events triggered the changes recorded by the sensors.

In order to create a process of relations and communication between the user and the company, an immersive model was then created with the Unreal Engine graphics engine of a dwelling chosen from the optimized ones that is made interactive and customizable according to the client's needs.

Starting from the catalog shapes, transformed into a design, materials were applied and the interior of the building was enriched with furniture. In addition, some geometric variations to the morphology of the house were reproduced to create different configurations of the spaces to be made available later during the immersive experience. Thus, all the functions necessary for the different types of variations were programmed, as well as those that allow the user to move through the space, others for recording all the actions, movements and points in the scene on which the user focused most, the functions necessary to allow the execution of the experience with a VR visor and the logic related to the graphical interface. Finally, some numerical data and interactive graphs were included that adapt following the user's choices and describe the impact they have on some parameters related to the house (such as energy costs, construction costs, comfort parameters), with the aim of providing awareness of the effects of decisions (Fig. 9).

3 Conclusions

Wood constructions reverse some conceptions linked to the building processes proper to our architectural culture, which is deeply tied to craftsmanship. This issue concerns the transition between architecture and fabrication: concepts such as smart manufacturing [101], robotic fabrication [102] virtual prototyping [103], automatic rule-based design [104] and virtual building design [105] are rapidly reshaping the relationship

Fig. 9 Starting from the design, the project can be 3D modelled and reproduced through virtual reality before the construction, to show the results and help in the decision-making phase

between architecture and construction, where we are increasingly seeing a direct transition from design to fabrication thanks to digitization and robotics [106].

Wood is naturally connected to the intelligence of biology [72]. It is a renewable resource [107] with an aesthetic value, workability, flexible, relatively light, versatility, low thermal conductivity, but it presents also undesirable characteristics for its sizing limits and deformations, anisotropy, hygroscopicity and degradation. Through its engineering aimed at the homogenization of its characteristics, wood represents a performing solution that integrates fabrication as a generative paradigm into the design process [108]. For these conditions, wood represents one of the most important field of application of parametric design [109–113] where "non-standard timber structures can be efficiently aggregated from a multitude of single timber members to foster highly versatile timber constructions" [114]. In hybridization and integration of digital wood design, the innovative tools involve a transformation of paradigm and form, connections and limits.

The processes of digitization and the value of digital techniques of representation are increasingly being contextualized in the innovations implemented in relation to the issues of contemporary building, many of which may find optimal solutions in wood. As a natural, sustainable, inexpensive, and extremely versatile material, wood is well suited not only for use as a material in construction, but also for experimenting with generative design and digital fabrication solutions.

The presented research ranges from multi-objective optimization of forms and construction details, to empirical research on responsive panels, and finally focus on forms of sharing and communicating the obtained results. The presented path thus describes the collaborative process implemented that has created a development of products, processes and services to meet the need for innovation that characterizes the new Industry 4.0 applied to architecture. The data and experiments are enriched with dynamic simulations in immersive reality, with the interactive variation of possible morphological and perceptual configurations, which can be used by the company to show the client the impact of the final project, reinforcing the full involvement of the end user, designer and/or owner, in the choices thus made aware of their impacts.

Representation presents itself as the field of existence of research: from design to construction, and beyond into the digital twin, drawing understood as a model is enhanced by the logics of digital. The transdisciplinary language of representation, open to different knowledges based on form, suited to bring out the underlying relationships, presents itself as the lifeblood of Industry 4.0 and the contemporary logics of doing architecture.

References

1. Bianconi, F.: Segni Digitali. Morlacchi, Perugia (2005)
2. Bianconi, F., Filippucci, M.: Il disegno degli olivi tra forma e luce. Le potenzialità analitiche della rappresentazione parametrica nell'interdisciplinarità della ricerca. Drawing form and light of olive trees. The analytic potentiality of parametric representation into the interdisc, in Territori e frontiere della Rappresentazione/Territories and frontiers of Representation, UID, Ed. Roma: Gangemi Editore, pp. 439–450 (2017)
3. Filippucci, M., Rinchi, G., Brunori, A., Nasini, L., Regni, L., Proietti, P.: Architectural modelling of an olive tree. Generative tools for the scientific visualization of morphology and radiation relationships. Ecol. Inform. **36**, 84–93 (2016). https://doi.org/10.1016/j.ecoinf.2016.09.004
4. Broy, M.: Cyber-Physical Systems Innovation durch Software-Intensive Eingebettet Systeme, Acatech Disk, pp. 1–141 (2010). http://www.acatech.de/de/publikationen/berichte-und-dokumentationen/acatech/detail/artikel/cyber-physical-systems-innovation-durch-softwareintensive-eingebetete-systeme.html%5Cn. https://doi.org/10.1007/978-3-642-14901-6
5. Bohn, R.E.: Measuring and managing technological knowledge. IEEE Eng. Manage. Rev. **25**(4), 77–88 (1994). https://doi.org/10.1016/b978-0-7506-7009-8.50022-7
6. Ackoff, R.L.: From data to wisdom. J. Appl. Syst. Anal. **16**, 3–9 (1989)
7. Forbes: Customer Engagement: Best of the Best (2015). https://www.forbes.com/forbesinsights/sap_customer_engagement/index.html
8. Iansiti, M., Lakhani, K.L.: Digital Ubiquity: How Connections, Sensors, and Data Are Revolutionizing Business,CFA Dig.,vol. 45, no. 2 (2015)
9. Ackerman, E.: Fetch robotics introduces fetch and freight: your warehouse is now automated. IEEE Spectrum (2015). https://spectrum.ieee.org/automaton/robotics/industrial-robots/fetch-robotics-introduces-fetch-and-freight-your-warehouse-is-now-automated
10. Conti, M., et al.: Looking ahead in pervasive computing: challenges and opportunities in the era of cyberphysical convergence. Pervasive Mob. Comput. **8**(1), 2–21 (2012). https://doi.org/10.1016/j.pmcj.2011.10.001
11. Wang, S., Wan, J., Li, D., Zhang, C.: Implementing smart factory of industrie 4.0: an outlook. Int. J. Distrib. Sens. Netw. **12**(1), 3159805 (2016). https://doi.org/10.1155/2016/3159805
12. Hartmann, B., Narayanan, S., King, W.P.: Digital Manufacturing: The Revolution will be Virtualized. McKinsey&Company (2015)
13. Kamarul Bahrin, M.A., Othman, M.F., Nor Azli, N.H., Talib, M.F.: Industry 4.0: a review on industrial automation and robotic. J. Teknol. **78**(6–13) (2016). https://doi.org/10.11113/jt.v78.9285

14. Bianconi, F., Filippucci, M., Buffi, A.: Automated design and modeling for mass-customized housing. A web-based design space catalog for timber structures. Autom. Constr. **103** (2019). https://doi.org/10.1016/j.autcon.2019.03.002
15. Lenka, S., Parida, V., Rönnberg Sjödin, D., Wincent, J.: Digitalization and advanced service innovation : how digitalization capabilities enable companies to co-create value with customers. Manage. Innov. Technol. **3**, 3–5 (2016)
16. Lidong, W., Guanghui, W.: Big data in cyber-physical systems, digital manufacturing and industry 4.0. Int. J. Eng. Manuf. **6**(4), 1–8 (2016). https://doi.org/10.5815/ijem.2016.04.01
17. Oxman, R.: Theory and design in the first digital age. Des. Stud. **27**(3), 229–265 (2006). https://doi.org/10.1016/J.DESTUD.2005.11.002
18. Oxman, R., Oxman, R.: The New Structuralism: Design, Engineering and Architectural Technologies. Wiley, New York (2010)
19. Oxman, R., Oxman, R.: Introduction. In: The New Structuralism: Design, Engineering and Architectural Technologies, pp. 14–24. Wiley (2010)
20. Filippucci, M., Bianconi, F., Andreani, S.: Computational design and built environments. In: Amoruso, G. (Ed.) 3D printing: breakthroughs in research and practice, pp. 361–395. IGI Global, Hershey (2018). https://doi.org/10.4018/978-1-5225-1677-4.ch019
21. Schumacher, P.: The Autopoiesis of Architecture: A New Agenda for Architecture, vol. II. John Wiley & Sons, West Sessex (2012)
22. Empler, T., Bianconi, F., Bagagli, R.: Rappresentazione del paesaggio : modelli virtuali per la progettazione ambientale e territoriale, vol. 1. DEI Tipografia del Genio Civile, Roma (2006)
23. Granadeiro, V., Duarte, J.P., Correia, J.R., Leal, V.M.S.: Building envelope shape design in early stages of the design process: integrating architectural design systems and energy simulation. Autom. Constr. **32**, 196–209 (2013). https://doi.org/10.1016/j.autcon.2012.12.003
24. Taylor, M.C.: The Moment of Complexity: Emerging Network Culture. University of Chicago Press, Chicago (2001)
25. Brown, N., Mueller, C.: Designing with data: moving beyond the design space catalog. In: Acadia 2017 Discipline + Disruption, pp. 154–163 (2017)
26. Mina, A.A., Braha, D., Bar-Yam, Y.: Complex engineered systems: a new paradigm. In: Complex Engineered Systems, pp. 1–21. Springer, Berlin, Heidelberg (2006). https://doi.org/10.1007/3-540-32834-3_1
27. Bar-Yam, Y.: Complexity Rising: From Human Beings to Human Civilization, a Complexity Profile. Cambridge (1997)
28. Bar-Yam, Y.: Multiscale variety in complex systems. Complexity **9**(4), 37–45 (2004). https://doi.org/10.1002/cplx.20014
29. Kolarevic, B., Malkawi, A.: Peformative Architecture. Routledge, London (2005)
30. Aish, R., Woodbury, R.: Multi-level interaction in parametric design. In: Smart Graphics, pp. 151–162. Springer (2005). https://doi.org/10.1007/11536482_13
31. Bergmann, E., Hildebrand, S.: Form-Finding, Form-Shaping, Designing Architecture. Mendrisio Academy Press, Mendrisio (2015)
32. Self, M., Vercruysse, E.: Infinite variations, radical strategies. In: Fabricate 2017 Conference Proceedings, pp. 30–35 (2017)
33. Scheurer, F.: Materialising complexity. Archit. Des. **80**(4), 86–93 (2010). https://doi.org/10.1002/ad.1111
34. Pine, B.J., Slessor, C.: Mass Customization: The New Frontier in Business Competition. Harvard Business School, Boston (1999)
35. Duray, R., Ward, P.T., Milligan, G.W., Berry, W.L.: Approaches to mass customization: configurations and empirical validation. J. Oper. Manage. **18**(6), 605–625 (2000). https://doi.org/10.1016/S0272-6963(00)00043-7
36. Zipkin, P.: The limits of mass customization. MIT Sloan Manage. Rev. **42**(3), 81–87 (2001). ISSN: 1532-9194
37. Anderson, D.M.: Build-to-order and mass customization: the ultimate supply chain management and lean manufacturing strategy for low-cost on-demand production without forecasts or inventory. CIM Press, Cambria (2002)

38. Dellaert, B.G.C., Stremersch, S.: Marketing mass-customized products: striking a balance between utility and complexity. J. Mark. Res. **42**(2), 219–227 (2005). https://doi.org/10.1509/jmkr.42.2.219.62293
39. Salvador, F., De Holan, P.M., Piller, F.: Cracking the code of mass customization. MIT Sloan Manage. Rev. **50**(3), 71–79 (2009)
40. Willis, D., Woodward, T.: Diminishing difficulty: mass customisation and the digital production of architecture. In: Corser, R. (Ed.) Fabricating Architecture : Selected Readings in Digital Design and Manufacturing, pp. 184–208. Princeton Architectural Press (2010)
41. Nahmens, I., Bindroo, V.: Is customization fruitful in industrialized homebuilding industry? J. Constr. Eng. Manage. **137**(12), 1027–1035 (2011). https://doi.org/10.1061/(ASCE)CO.1943-7862.0000396
42. Knaack, U., Chung-Klatte, S., Hasselbach, R.: Prefabricated Systems : Principles of Construction. Birkhäuser, Basel (2012)
43. Page, I.C., Norman, D.: Prefabrication and Standardisation Potential in Buildings (SR 312). Branz, Wellington (2014)
44. Bianconi, F., Filippucci, M., Pelliccia, G.: Lineamenta. Maggioli, Santarcangelo di Romagna (RN) (2020)
45. Alberti, L.B.: De re aedificatoria. Nicolai Laurentii Alamani, Firenze (1443)
46. Rechenberg, I.: Evolurionsstratgie. Holzmann-Froboog, Stuttgart (1973)
47. Mitchell, M.: An Introduction to Genetic Algorithms. MIT Press, Cambridge (1998)
48. Fasoulaki, E.: Architecture: a necessity or a trend? In: 10th Generative Art International Conference (2007)
49. Goel, A.K., McAdams, D.A., Stone, R.B.: Biologically Inspired Design. Springer, London (2014). https://doi.org/10.1007/978-1-4471-5248-4
50. López, M., Rubio, R., Martín, S., Croxford, B.: How plants inspire façades. From plants to architecture: Biomimetic principles for the development of adaptive architectural envelopes. Renew. Sustain. Energy Rev. **67**, 692–703 (2017). https://doi.org/10.1016/J.RSER.2016.09.018
51. Vattam, S., Helms, M.E., Goel, A.K.: Biologically-Inspired Innovation in Engineering Design: A Cognitive Study. Atlanta (2007). http://hdl.handle.net/1853/14346
52. Oxman, R.: Performative design: a performance-based model of digital architectural design. Environ. Plan. B Plan. Des. **36**(6), 1026–1037 (2009). https://doi.org/10.1068/b34149
53. Barthel, R.: Natural forms-architectural forms. In: Nerdinger, W. (Ed.) Frei Otto Complete Works, pp. 16–32. Birkhäuser Architecture, Basel-Boston-Berlin (1967)
54. Bhushan, B.: Biomimetics: lessons from nature—an overview. Philos. Trans. A. Math. Phys. Eng. Sci. **367**(1893), 1445–1486 (2009). https://doi.org/10.1098/rsta.2009.0011
55. Wester, T.: Nature teaching structures. Int. J. Sp. Struct. **17**(2–3), 135–147 (2002). https://doi.org/10.1260/026635102320321789
56. Knippers, J., Speck, T.: Design and construction principles in nature and architecture. Bioinspir. Biomim. **7**(1) (2012). https://doi.org/10.1088/1748-3182/7/1/015002
57. Mattheck, C.: Design in Nature : Learning from Trees. Springer, Berlin Heidelberg (1998)
58. Mazzoleni, I.: Architecture Follows Nature: Biomimetic Principles for Innovative Design. CRC Press, New York (2013)
59. Pawlyn, M.: Biomimicry in Architecture. RIBA Publishing, London (2011)
60. Vincent, J.: Biomimetic patterns in architectural design. Archit. Des. **79**(6), 74–81 (2009). https://doi.org/10.1002/ad.982
61. Pearce, P.: Structure in Nature is a Strategy for Design. MIT Press, Cambridge (1979)
62. Seccaroni, M., Pelliccia, G.: Customizable social wooden pavilions: a workflow for the energy, emergy and perception optimization in Perugia's parks. In: Digital Wood Design. Innovative Techniques of Representation in Architectural Design, vol. 24, pp. 1045–1062. Springer (2019). https://doi.org/10.1007/978-3-030-03676-8_42
63. Bianconi, F., Filippucci, M., Pelliccia, G., Buffi, A.: Data driven design per l'architettura in legno. Ricerche rappresentative di algoritmi evolutivi per l'ottimizzazione delle soluzioni multi-obiettivo. In: Atti del XIX Congresso Nazionale CIRIAF. Energia e sviluppo sostenibile, pp. 61–72 (2019)

64. UNI EN ISO 13786:2018 Thermal performance of building components—Dynamic thermal characteristics—Calculation methods (2018)
65. UNI EN ISO 13788:2013 Hygrothermal performance of building components and building elements—Internal surface temperature to avoid critical surface humidity and interstitial condensation—Calculation methods (2013)
66. Wang, W., Zmeureanu, R., Rivard, H.: Applying multi-objective genetic algorithms in green building design optimization. Build. Environ. **40**(11), 1512–1525 (2005). https://doi.org/10.1016/j.buildenv.2004.11.017
67. Wright, J.A., Loosemore, H.A., Farmani, R.: Optimization of building thermal design and control by multi-criterion genetic algorithm. Energy Build. **34**(9), 959–972 (2002). https://doi.org/10.1016/S0378-7788(02)00071-3
68. Censor, Y.: Pareto optimality in multiobjective problems. Appl. Math. Optim. **4**(1), 41–59 (1977). https://doi.org/10.1007/BF01442131
69. Decreto Ministeriale 26/6/2009—Ministero dello Sviluppo Economico Linee guida nazionali per la certificazione energetica degli edifici (2009)
70. Addington, M., Schodek, D.L.: Smart Materials and New Technologies: For the Architecture and Design Professions. Architectural, Oxford (2005)
71. Loonen, R.C.G.M., Trčka, M., Cóstola, D., Hensen, J.L.M.: Climate adaptive building shells: state-of-the-art and future challenges. Renew. Sustain. Energy Rev. **25**, 483–493 (2013). https://doi.org/10.1016/J.RSER.2013.04.016
72. Ugolev, B.N.: Wood as a natural smart material. Wood Sci. Technol. **48**(3), 553–568 (2014). https://doi.org/10.1007/s00226-013-0611-2
73. Holstov, A., Bridgens, B., Farmer, G.: Hygromorphic materials for sustainable responsive architecture. Constr. Build. Mater. **98**, 570–582 (2015). https://doi.org/10.1016/J.CONBUILDMAT.2015.08.136
74. Reichert, S., Menges, A., Correa, D.: Meteorosensitive architecture: biomimetic building skins based on materially embedded and hygroscopically enabled responsiveness. Comput. Des. **60**, 50–69 (2015). https://doi.org/10.1016/J.CAD.2014.02.010
75. Burgert, I., Fratzl, P.: Actuation systems in plants as prototypes for bioinspired devices. Philos. Trans. A Math. Phys. Eng. Sci. **367**(1893), 1541–1557 (2009). https://doi.org/10.1098/rsta.2009.0003
76. Reyssat, E., Mahadevan, L.: Hygromorphs: from pine cones to biomimetic bilayers. J. R. Soc. Interface **6**(39), 951–957 (2009). https://doi.org/10.1098/rsif.2009.0184
77. Song, K., et al.: Journey of water in pine cones. Sci. Rep. **5**(1), 9963 (2015). https://doi.org/10.1038/srep09963
78. Dawson, C., Vincent, J.F.V., Rocca, A.-M.: How pine cones open. Nature **390**(6661), 668 (1997). https://doi.org/10.1038/37745
79. Rüggeberg, M., Burgert, I.: Bio-inspired wooden actuators for large scale applications. PLoS ONE **10**(4), e0120718 (2015). https://doi.org/10.1371/journal.pone.0120718
80. Vailati, C., Bachtiar, E., Hass, P., Burgert, I., Rüggeberg, M.: An autonomous shading system based on coupled wood bilayer elements. Energy Build. **158**, 1013–1022 (2018). https://doi.org/10.1016/J.ENBUILD.2017.10.042
81. El-Dabaa, R., Salem, I.: 4D printing of wooden actuators: encoding FDM wooden filaments for architectural responsive skins. Open House Int. (2021). https://doi.org/10.1108/OHI-02-2021-0028
82. Correa, D., et al.: 4D pine scale: biomimetic 4D printed autonomous scale and flap structures capable of multi-phase movement. Philos. Trans. R. Soc. A Math. Phys. Eng. Sci. **378**(2167) (2020). https://doi.org/10.1098/rsta.2019.0445
83. Sydney Gladman, A., Matsumoto, E.A., Nuzzo, R.G., Mahadevan, L., Lewis, J.A.: Biomimetic 4D printing. Nat. Mater. **154**(4), 413–418 (2016). https://doi.org/10.1038/nmat4544
84. Le Duigou, A., Castro, M., Bevan, R., Martin, N.: 3D printing of wood fibre biocomposites: From mechanical to actuation functionality. Mater. Des. **96**, 106–114 (2016). https://doi.org/10.1016/j.matdes.2016.02.018

85. ISO 9869-1:2014 Thermal insulation—Building elements—In-situ measurement of thermal resistance and thermal transmittance—Part 1: Heat flow meter method (2014)
86. Bianconi, F., Filippucci, M., Pelliccia, G.: Wood and generative algorithms for the comparison between models and reality. Int. Arch. Photogramm. Remote Sens. Spat. Inf. Sci. **XLIII-B4-2**, 409–415 (2021). https://doi.org/10.5194/isprs-archives-XLIII-B4-2021-409-2021
87. Tsigkari, M., Angelos, C., Joyce, S.C., Davis, A., Feng, S., Aish, F.: Integrated design in the simulation process. Society for Computer Simulation International, San Diego (2013)
88. Kolarevic, B.: From mass customisation to design "Democratisation". Archit. Des. **85**(6), 48–53 (2015). https://doi.org/10.1002/ad.1976
89. Weber, S.: The Success of Open Source. Harvard University Press, Cambridge (2005)
90. Ratti, C., Claudel, M.: Open Source Architecture. Thames & Hudson, New York (2015)
91. Rajanen, M., Iivari, N.: Power, empowerment and open source usability. In: Proceedings of 33rd Annual ACM Conference on Human Factors Computing Systems—CHI '15, pp. 3413–3422 (2015). https://doi.org/10.1145/2702123.2702441
92. Lawrence, T.T.: Chassis+Infill: A Consumer-Driven, Open Source Building Approach for Adaptable, Mass Customized Housing. Massachusetts Institute of Technology (2003)
93. Bianconi, F., Filippucci, M., Cornacchini, F.: Play and transform the city. Sciresit.it **2**, 141–158 (2020). https://doi.org/10.2423/i22394303v10n2p141
94. Larson, K.: Serious games and gamification in the corporate training environment: a literature review. TechTrends **64**(2), 319–328 (2020). https://doi.org/10.1007/s11528-019-00446-7
95. Smith, J., Sears, N., Taylor, B., Johnson, M.: Serious games for serious crises: reflections from an infectious disease outbreak matrix game. Global Health **16**(1) (2020). https://doi.org/10.1186/s12992-020-00547-6
96. Alvaro Marcos Antonio de Araujo Pistono, R.J.V.B., Santos, A.M.P.: Serious games: review of methodologies and games engines for their development. IEEE Xplore (2021). https://ieeexplore.ieee.org/abstract/document/9140827/. Accessed 28 Jun 2021
97. Altıntop, Ç.G., Latifoğlu, F., Akın, A.K., İleri, R., Yazar, M.A.: Analysis of consciousness level using galvanic skin response during therapeutic effect. J. Med. Syst. **45**(1) (2021). https://doi.org/10.1007/s10916-020-01677-5
98. Sanchez-Comas, A., Synnes, K., Molina-Estren, D., Troncoso-Palacio, A., Comas-González, Z.: Correlation analysis of different measurement places of galvanic skin response in test groups facing pleasant and unpleasant stimuli. Sensors **21**(12) (2021). https://doi.org/10.3390/s21124210
99. Iadarola, G., Poli, A., Spinsante, S.: Analysis of galvanic skin response to acoustic stimuli by wearable devices. In: 2021 IEEE International Symposium on Medical Measurements and Applications, MeMeA 2021—Conference Proceedings (2021). https://doi.org/10.1109/MeMeA52024.2021.9478673
100. Chen, F., Marcus, N., Khawaji, A., Zhou, J.: Using galvanic skin response (GSR) to measure trust and cognitive load in the text-chat environment. In: Conference on Human Factors in Computing Systems—Proceedings, April 2015, vol. 18, pp. 1989–1994. https://doi.org/10.1145/2702613.2732766
101. Davis, J., Edgar, T., Porter, J., Bernaden, J., Sarli, M.: Smart manufacturing, manufacturing intelligence and demand-dynamic performance. Comput. Chem. Eng. **47**, 145–156 (2012). https://doi.org/10.1016/J.COMPCHEMENG.2012.06.037
102. McGee, W., Ponce de León, M. (Eds.): Robotic Fabrication in Architecture, Art and Design 2014. Springer Science & Business Media, Cham (2014)
103. Li, H., et al.: Integrating design and construction through virtual prototyping. Autom. Constr. **17**(8), 915–922 (2008). https://doi.org/10.1016/J.AUTCON.2008.02.016
104. Eastman, C., Lee, J., Jeong, Y., Lee, J.: Automatic rule-based checking of building designs. Autom. Constr. **18**(8), 1011–1033 (2009). https://doi.org/10.1016/J.AUTCON.2009.07.002
105. Popov, V., Juocevicius, V., Migilinskas, D., Ustinovichius, L., Mikalauskas, S.: The use of a virtual building design and construction model for developing an effective project concept in 5D environment. Autom. Constr. **19**(3), 357–367 (2010). https://doi.org/10.1016/J.AUTCON.2009.12.005

106. Gramazio, F., Kohler, M.: Made by Robots : Challenging Architecture at the Large Scale AD. John Wiley & Sons, London (2014)
107. Dangel, U.: Turning Point in Timber Construction : A New Economy. Birkhäuser, Basilea (2016)
108. Gramazio, F., Kohler, N., Oesterle, S.: Encoding material. In: Oxman, R., Oxman, R. (Eds.) The New Structuralism : Design, Engineering and Architectural Technologies, pp. 108–115. Wiley (2010). https://books.google.it/books?id=035HAQAAIAAJ&q=The+new+Structuralism:+Design,+Engineering+and+Architectural+Technologies+AD&dq=The+new+Structuralism:+Design,+Engineering+and+Architectural+Technologies+AD&hl=it&sa=X&ved=0ahUKEwjirozclondAhVMkiwKHQFfBbEQ6A
109. Menges, A., Schwinn, T., Krieg, O.D. (eds.): Advancing Wood Architecture. Routledge, London (2017)
110. Kaufmann, H., Nerdinger, W. (eds.): Building with Timber: Paths into the Future. Prestel Verlag, Munich (2011)
111. Weinand, Y.: Advanced Timber Structures : Architectural Designs and Digital Dimensioning. Birkhäuser, Basel (2016)
112. Chilton, J.C., Tang, G.: Timber Gridshells: Architecture, Structure and Craft. Routledge, London (2016)
113. Vierlinger, R.: Towards AI drawing agents. In: Ramsgaard Thomsen, M., Tamke, M., Gengnagel Christoph Faircloth, B.S.F. (Eds.) Modelling Behaviour, pp. 357–369. Springer International Publishing, Cham (2015). https://doi.org/10.1007/978-3-319-24208-8_30
114. Willmann, J., Gramazio, F., Kohler, M.: New paradigms of the automatic: robotic timber construction in architecture. In: Menges, A., Schwinn, T., Krieg, O.D. (Eds.) Advancing Wood Architecture. A Computational Approach, pp. 13–28. Routledge, London (2017). https://doi.org/10.4324/9781315678825-11